INTERMITTENCY IN TURBULENT FLOWS

The Isaac Newton Institute of Mathematical Sciences of the University of Cambridge exists to stimulate research in all branches of the mathematical sciences, including pure mathematics, statistics, applied mathematics, theoretical physics, theoretical computer science, mathematical biology and economics. The research programmes it runs each year bring together leading mathematical scientists from all over the world to exchange ideas through seminars, teaching and informal interaction.

INTERMITTENCY IN
TURBULENT FLOWS

edited by

J.C. Vassilicos

University of Cambridge

CAMBRIDGE
UNIVERSITY PRESS

CAMBRIDGE UNIVERSITY PRESS
Cambridge, New York, Melbourne, Madrid, Cape Town, Singapore,
São Paulo, Delhi, Dubai, Tokyo, Mexico City

Cambridge University Press
The Edinburgh Building, Cambridge CB2 8RU, UK

Published in the United States of America by Cambridge University Press, New York

www.cambridge.org
Information on this title: www.cambridge.org/9780521159425

First published 2001
First paperback edition 2010

A catalogue record for this publication is available from the British Library

ISBN 978-0-521-79221-9 Hardback
ISBN 978-0-521-15942-5 Paperback

CONTENTS

Contributors

Takahiro Adachi, OTEC Laboratory, Faculty of Science and Engineering, Saga University, 1 Honjo-machi, Saga 840–8502, Japan

L. Biferale, Dipartimento di Fisica and INFM, Università di Roma, 'Tor Vergata', Via della Ricerca Scientifica 1, I–00133, Roma, Italy

Peter N. Blossey, Department of Mechanical and Aerospace Engineering, University of California, San Diego, 9500 Gilman Drive, La Jolla, CA 92093–0411, USA

G. Boffetta, Dipartimento di Fisica Generale and INFM, Università di Torino, via Pietro Giuria 1, I–10125, Roma, Italy

O. Brausch, Institute of Physics, University of Bayreuth, D 95440, Bayreuth, Germany

F.H. Busse, Institute of Physics, University of Bayreuth, D 95440, Bayreuth, Germany

Patrizia Castiglione, Laboratoire de Physique Statistique, Ecole Normale Supérieure, 24, rue Lhomond, 75231 Paris, France

A. Celani, Dipartimento di Fisica Generale and INFM, Università di Torino, via Pietro Giuria 1, I–10125, Roma, Italy

M. Cencini, Dipartimento di Fisica and INFM, Università di Roma, 'La Sapienza', Piazzale Aldo Moro 2, I–00185, Roma, Italy

Michael Chertkov, T13 and CNLS, Los Alamos National Laboratory, NM 87545, USA

Sergio Ciliberto, Laboratoire de Physique, CNRS, Ecole Normale Supérieure de Lyon, 46, Allée d'Italie, 69364 Lyon, France

C.R. Doering, Department of Mathematics, University of Michigan, Ann Arbor, MI 48109–1109, USA

Gregory Falkovich, Department of Physics of Complex Systems, Weizmann Institute of Science, PO Box 26, Rehovot 76100, Israel

Krzysztof Gawędzki, IHES, 35, route de Chartres, F–91440, Bures-sur-Yvette, France

J.D. Gibbon, Department of Applied Mathematics, Imperial College of Science and Technology, Huxley Building, 180, Queen's Gate, London SW7 2BX, UK

Marie-Christine Guegan, Physics Department, Eindhoven University of Technology, PO Box 513, 5600 MB, Eindhoven, The Netherlands

M. Jaletzky, Institute of Physics, University of Bayreuth, D 95440, Bayreuth, Germany

Javier Jiménez, School of Aeronautics, Universidad Politécnica, 28040 Madrid, Spain *and* Center for Turbulence Research, Stanford University, Stanford, CA 94305, USA

A. Juel, Center for Non-Linear Dynamics and Department of Physics, University of Texas at Austin, TX 78712, USA

Marie-Caroline Jullien, Laboratoire de Physique Statistique, Ecole Normale Supérieure, 24, rue Lhomond, 75231 Paris, France

Michael Kholmyanksy, Department of Fluid Mechanics and Heat Transfer, The Iby and Aladar Fleischman Faculty of Engineering, Tel Aviv University, Ramat Aviv 69978, Israel

Shigeo Kida, OTEC Laboratory, Faculty of Science and Engineering, Saga University, 1 Honjo-machi, Saga 840–8502, Japan

Emmanuel Lévêque, Laboratoire de Physique, CNRS, Ecole Normale Supérieure de Lyon, 46, Allée d'Italie, 69364, Lyon, France

John L. Lumley, Sibley School of Mechanical and Aerospace Engineering, Cornell University, Ithaca NY 14853–7501, USA

Hideaki Miura, OTEC Laboratory, Faculty of Science and Engineering, Saga University, 1 Honjo-machi, Saga 840–8502, Japan

F. Moisy, Laboratoire de Physique Statistique, Ecole Normale Supérieure, 24, rue Lhomond, 75231 Paris, France

T. Mullin, Department of Physics and Astronomy, University of Manchester, Manchester M13 9PL, UK

K. Ohkitani, Research Institute for Mathematical Sciences, Kyoto University, Kyoto 606–8502, Japan

Jérôme Paret, Laboratoire de Physique Statistique, Ecole Normale Supérieure, 24, rue Lhomond, 75231 Paris, France

T. Peacock, Department of Computer Science, University of Colorado at Boulder, Boulder, CO 80309–0430, USA

W. Pesch, Institute of Physics, University of Bayreuth, D 95440, Bayreuth, Germany

Alain Pumir, INLN, CNRS, 1361, route des Lucioles, F–06560, Valbonne, France

D. Queiros-Conde, Department of Applied Mathematics and Theoretical Physics, University of Cambridge, Cambridge CB3 9EW, UK

Gerardo Ruiz Chavarria, Departamento de Física, Facultad de Ciencias, UNAM, 04510, Mexico DF, Mexico

Boris I. Shraiman, Lucent Technology, 700 Mountain Avenue, Murray Hill, NJ 07974, USA

Adrian Staicu, Physics Department, Eindhoven University of Technology, PO Box 513, 5600 MB, Eindhoven, The Netherlands

Patrick Tabeling, Laboratoire de Physique Statistique, Ecole Normale Supérieure, 24, rue Lhomond, 75231, Paris, France

Arkady Tsinober, Department of Fluid Mechanics and Heat Transfer, The Iby and Aladar Fleischman Faculty of Engineering, Tel Aviv University, Ramat Aviv 69978, Israel

J.C. Vassilicos, Department of Applied Mathematics and Theoretical Physics, University of Cambridge, Cambridge CB3 9EW, UK

D. Vergni, Dipartimento di Fisica and INFM, Università di Roma, 'La Sapienza', Piazzale Aldo Moro 2, I–00185, Roma, Italy

A. Vulpiani, Dipartimento di Fisica and INFM, Università di Roma, 'La Sapienza', Piazzale Aldo Moro 2, I–00185, Roma, Italy

Willem van de Water, Physics Department, Eindhoven University of Technology, PO Box 513, 5600 MB, Eindhoven, The Netherlands

H. Willaime, Laboratoire de Physique Statistique, Ecole Normale Supérieure, 24, rue Lhomond, 75231 Paris, France

Introduction

The last decade has seen what is perhaps an unprecedented and multifaceted flurry of activity in the search to describe and understand the intermittency of turbulent flows and other dynamical systems. The aim of the Workshop on Intermittency in Turbulent Flows and Other Dynamical Systems held at the Isaac Newton Institute for Mathematical Sciences, Cambridge, UK, from June 21 to 24, 1999, was to capture and summarize these developments, encourage cross-fertilization of ideas and lay out research directions for the future. This volume provides a record of the Workshop and an overview of our current understanding of the subject.

Different turbulence problems, for example Navier–Stokes turbulence, are encountered in different nonlinear systems described mathematically by different partial differential equations (PDE), for example the Navier-Stokes equation, or systems of ordinary differential equations (ODE). Nevertheless, the dynamics of many nonlinear systems gravitate around a few common central themes: intermittency, order/coherence and disorder. These features affect scalings and lead to deviations from Gaussian behaviour. Intermittency may be the universal outcome of a large class of nonlinear systems, however the universality properties of specific nonlinear systems, that is the dependencies of the intermittent structure on initial and boundary conditions, remain open questions. What is the appropriate kinematic description of intermittency, and how different is it for different nonlinear systems, different boundary and different initial conditions? The challenge is that a general kinematic description of intermittency should encompass both order and disorder and mixtures of both. Furthermore, what are the dynamics of intermittency and the mechanisms that create it? To what extent and in what sense is intermittency related to deviations from self-similar dynamics and to dissipative properties of the nonlinear system? What are self-similar dynamics and what are their dissipative properties?

Intermittency in some form or another is a feature of nonlinear ODEs and PDEs. ODEs are effectively the common focus of the chapters by Blossey & Lumley and by Mullin, Juel & Peacock. Blossey & Lumley discuss the intermittent production of turbulence in the turbulent boundary layer's wall region which is dominated by a few large-scale coherent structures that break down intermittently. A Proper Orthogonal Decomposition of the Navier–Stokes equation leads to a system of coupled nonlinear ODEs which can be used to predict breakdown of structures and thereby act to reduce drag on the wall. Mullin, Juel & Peacock are concerned with low-dimensional chaos and temporal intermittency found in Taylor-Couette and variant flows and they claim that it is impossible to completely reduce these dynamics to a low-order set of ODEs. Busse, Brausch, Jaletzky & Pesch focus on convection layers with and without rotation where phase/weak turbulence appears and is related to heteroclinic orbits. The rest of the Workshop was on internal intermittency in PDEs and turbulence experiments.

The main questions were: how can we relate scaling exponents of structure functions to near-singular flow structure (e.g. vortex tubes but also other straining structures)? Is the near-singular flow structure an imprint of finite-time singularities of the Euler equations assuming they exist? In his chapter Ohkitani proves the existence of finite-time singularities in the inviscid Burgers equation without use of the Hopf–Cole solution and applies his method to the Euler equation with a Taylor-Green vortex initial condition to show that if a finite-time singularity exists it cannot be too weak. A condition for the non-existence of finite-time singularities of the Navier-Stokes equation was derived by Doering & Gibbon at the Isaac Newton Institute during the turbulence research programme (January 6 to July 2, 1999) and a detailed account is given in their chapter here. Other PDEs that were discussed for their intermittency properties during the Workshop are the advection-diffusion equation for scalar fields (chapter by Gawędzki) and the advection-stretching-diffusion equation for magnetic fields (chapter by Falkovich).

Results from a variety of new experiments were presented at the workshop, and can be found in the chapters of Ciliberto, Lévêque & Ruiz Chavarria, Queiros-Conde & Vassilicos, Jullien, Castiglione, Paret & Tabeling, and Kholmyansky & Tsinober. In their investigation of velocity structure functions in a turbulent boundary layer flow near a solid plate, Ciliberto *et al.* report that mean shear significantly influences the scalings of structure functions which therefore depend on the distance from the plate. However, the relative scaling exponents remain constant in the turbulent logarithmic sublayer. Queiros-Conde & Vassilicos have measured turbulence fluctuations and velocity profiles at large distances in the wakes of 3D fractal grids where the turbulence has had the time to travel a large number of turnover times. Their main result is that velocity structure functions are more slowly varying functions of two-time separation for larger fractal grid dimensions and that this is not an effect of mean shear. Jullien *et al.* have developed an experimental setup where a 2D turbulent flow is generated in a large aspect ratio container with stable stratification and friction exerted on the fluid by the bottom wall. The flow is forced by an array of magnets placed below this wall, and according to the scale of the forcing, the turbulent flows produced can have either $-5/3$ or -3 energy spectra. Detailed measurements of both velocity and scalar statistics are presented and compared with theory. Kholmyansky & Tsinober have made atmospheric measurements of all velocity derivatives at the top of a 10-metre tall tower. This is the first time that measurements of all velocity derivatives have been made at a Reynolds number as high as $Re_\lambda = 10^4$. Their conclusion is that the properties of enstrophy and strain production, geometrical statistics, the role of concentrated vorticity and strain and depression of nonlinearity are the same, qualitatively, as they are at $Re_\lambda = 10^2$. However, Kholmyansky & Tsinober's measurements of structure functions conditional on high values of turbulent velocity fluctuations reveal a significant correlation between small and large scales, thereby pointing to non-universality of turbulent flows. The analysis of experimental longitudinal velocity data by Jiménez, Moisy, Tabeling & Willaime leads to results that are at odds with the model of a self-similar multiplicative cascade.

It may be that longitudinal structure functions, which have focussed research interest since the seminal works of Kolmogorov, are not the most appropriate tools for studying intermittency and scalings. In particular they are not very instructive as to the spatio-temporal flow structure of the turbulence. It is for this reason that an entire battery of new ideas, concepts and methods were presented and discussed at the Workshop. The oldest of these new approaches are not more than 20 years old, and the most recent date from the second half of the 1990s: wavelets, extended self-similarity, numerical vortex tube visualization techniques, transversal and inverted structure functions, multi-point correlators and geometrical statistics.

The experimental evidence presented in the chapter of van de Water, Staicu & Guegan indicates that power-law scalings are better defined for transverse than they are for longitudinal structure functions, and that transverse scaling exponents are smaller than the Kolmogorov predictions. Vulpiani, Biferale, Boffetta, Celani, Cencini and Vergni introduce the study of statistics of separations across which the velocity difference has a certain value. These statistics are called inverted structure functions and lead to improved results for Richardson's two-point turbulent diffusion law. However, moving from two-point to multi-point Lagrangian statistics may be a good way to probe more of the turbulence spatial flow structure, and Chertkov, Pumir & Shraiman study multi-point correlators and tetrad dynamics. Finally, Kida, Miura & Adachi present an automatic tracking scheme for unambiguous visualization of flow structure in 3D numerical simulations.

One conclusion from the Workshop is that there are now many new ideas and experiments and new ways to interrogate turbulent fields which may be leading to exciting developments. Another conclusion is that we must be prepared to do away with the concept of universal turbulence, or at least qualify it so as to allow for qualitative rather than quantitative universality.

I, and I believe all the speakers at the Workshop, are extremely grateful to my fellow organizers of the Isaac Newton Institute's turbulence research programme, Geoff Hewitt, Peter Monkewitz and Neil Sandham; to Keith Moffatt and the wonderful staff of the Isaac Newton Institute; to ERCOFTAC, the European Commission, the Royal Academy of Engineering and the Industrial Working Group under the chairmanship of Michael Reeks for their support; and to the Isaac Newton Institute for sponsoring and hosting the Workshop.

J.C. Vassilicos

Control of Intermittency in Near-wall Turbulent Flow

Peter N. Blossey and John L. Lumley

1 Introduction

The boundary layer exhibits intermittency at the interface between the turbulent fluid inside the boundary layer and the irrotational flow outside the boundary layer as well as internal intermittency. However, one finds in the boundary layer a third type of intermittency as well: the intermittent production of turbulence in the wall region of the turbulent boundary layer. The production of turbulence near the wall is dominated by a few, large-scale coherent structures which break down intermittently, resulting in peaks in turbulent production and the generation of small-scale turbulence — seen in experiments and simulations as bursts of Reynolds stress. Experimental work focusing on the breakdown of the coherent structures has led to the identification of a characteristic time scale, the mean inter-burst period. Lumley & Kubo (1985) collected the available data on the inter-burst period and found that the product of the bursting period with the turbulent skin friction at the wall was approximately constant. Thus, the shorter the bursting period becomes, the more active the coherent structures near the wall, and the higher the skin friction. If the bursting period is prolonged, the coherent structures are less active, and the skin friction will decrease. In this paper, we construct a control algorithm which identifies when the coherent structures are tending towards bursting and intervenes to suppress the burst, thereby increasing the inter-burst duration and decreasing the drag at the wall. Our control relies on the identification of the coherent structures in the boundary layer and on the prediction of their strength based on available information, namely the shear stress at the wall. However, some insight into the dynamics of the large scales of the turbulence in the wall layer — along with an estimate of their strength — is necessary to recognize the signs of bursting in the coherent structures.

Since the wall region is dominated energetically by just a few coherent structures, it is tempting to model the dynamics there including only these large-scale structures, while parameterizing the smaller scales of the turbulence. Many different techniques have been used to identify coherent structures in turbulent shear flows, from visualization and pattern matching to conditional sampling. Each of these techniques requires some degree of subjectivity in the identification of coherent structures, for example, in the choice of an initial pattern in pattern matching or in the selection of a threshold when using conditional sampling. However, Lumley (1967) proposed an objective technique for the identification of coherent structures in turbulent flows, the Proper Orthogonal Decomposition (POD). The POD, also referred to as the Karhunen-Loève decomposition, selects the most energetic feature from an ensemble of (random) turbulent velocity fields. In the wall region, the first and most energetic eigenfunction of the POD captures 60% of the turbulent kinetic energy there. The eigenfunctions of the POD are referred to as empirical eigenfunctions because they are derived from data (obtained either experimentally or through simulations) rather than from the governing equations. The empirical eigenfunctions form a basis for the turbulent velocity which possesses optimal convergence: when the velocity field is expanded in terms of the eigenfunctions, any truncation of the expansion will contain the most energy among all truncations of that order.

Using the empirical eigenfunctions as a basis, a model for the dynamics of the large-scales may be constructed from the Navier-Stokes equations by means of Galerkin projection. The projection yields a coupled system of non-linear ordinary differential equations. When the equations are truncated — including only a few modes — the effect of the unresolved modes and the mean velocity profile must be accounted for. The unresolved modes exert a Reynolds stress on the large, resolved scales through the nonlinear terms. The Reynolds stress may be approximated using a gradient diffusion model with an eddy viscosity that may be varied parametrically. The mean velocity profile may be expressed as an integral of the Reynolds stress from the resolved modes. This introduces cubic terms into the model equations which globally stabilize the dynamical system which results from the low-dimensional model. These models may be studied through simulations and by use of the tools of dynamical systems theory.

The coherent structures observed in the boundary layer — the streaks and rolls — appear as fixed points in the models, and for certain values of the eddy viscosity parameter the models display intermittent dynamics with the trajectories in the phase space of the model following heteroclinic orbits connecting fixed points which correspond to the structures. The jumps in the phase space of the model correspond roughly to the bursts observed in simulations and experiments in the boundary layer. The updraft between the rolls strengthens into an ejection as energy is transferred from/to the smaller scales represented in the model. (The severe truncation in the models limits their ability to capture the energy transfer fully.) The energy transfer leads to the weakening and breakdown of the structures. Following their breakdown, the structures re-form, and the process repeats.

2   Motivation for Control

Our success in applying open-loop control to turbulent channel flow gave us confidence to take the same approach to closed-loop, feedback control with the aim of sustained drag reduction. The open-loop control (described in Lumley & Blossey (1998a)) interfered with the coherent structures near the wall, draining energy from the rolls and weakening momentum transport away from the wall. As a result, the drag was substantially reduced. Although the reduction in drag lasted much longer than the application of the control, the drag eventually rose again as the coherent structures strengthened and burst. The success of the open loop control led to two conclusions: that channel flow may be successfully controlled by focusing the control on only the large-scale structures, and that the effect of the control may last much longer than the duration of the control. This suggests an approach for applying control in a closed-loop, feedback control framework. The control may be applied as relatively short pulses whose purpose is to weaken the cross-stream coherent structures, or rolls. These pulses may be repeated when the coherent structures strengthen to prevent the bursting of these coherent structures. To be a realistic approach, the control must rely only on information available in a practical implementation. This information includes pressure and shear stress measurements at the wall, but not any information about the velocity field above the wall. Thus, our heuristic procedure of choosing the position of the forcing in the open-loop control — from observations of the high-speed streak — must be replaced by an objective method which relies, most likely, on the shear stress at the wall.

This control strategy is descended from work originating from our extended group at Cornell. A low-dimensional model was constructed for the wall region of the turbulent boundary layer by Aubry et al. (1988). Their model successfully mimicked the bursting process observed in experiments and simulations of near-wall turbulence, and was found to possess heteroclinic connections between fixed points in the phase space of the model. The fixed points — which were saddle points for realistic parameter values — represented rolls in the fluids systems. The jumps along the heteroclinic connections resembled the bursting process with a strengthening of the updraft between the rolls and a transfer of energy to the higher-wavenumber modes. If we accept the dynamical system as a reasonable model for the bursting process, the problem of control may be approached from this point of view. The first method to explore is stabilizing the fixed point: setting up linear feedback control with the aim of turning the saddle point into a stable fixed point. The stabilization of these saddle points would require sustained and substantial control input. This is not realistic in a practical setting. The difficulty of estimating the state of the coherent structures based on partial information available at the wall and the susceptibility of such a strategy to the dynamics of unmodeled modes would further complicate such an approach. In designing a control strategy for near-wall turbulence, our hope is that limited but intelligent application of our control effort would lead to efficient techniques for the reduction of drag.

Bloch & Marsden (1989) suggested another strategy for controlling the heteroclinic orbits that was more in line with the thinking outlined above. Their strategy relies on the identification of a controllable region around the stable manifold of a given saddle point. Once a trajectory enters the controllable region, the control is applied to prevent the escape of the trajectory along the unstable direction. In the absence of noise, the trajectory can be directed right to the fixed point and will sit in the neighborhood of the fixed point for all time. For a more realistic setting which includes noise, the trajectory will fall in towards the fixed point along the stable direction — with the control keeping the trajectory close to the stable manifold — until the controllable region becomes small enough that the noise may bump the trajectory out of the controllable region, leading to the escape of the trajectory along the unstable manifold. The controllable region usually takes the form of a cone which encloses the stable manifold and has its vertex at the fixed point. Therefore, the control can be expected to be most susceptible to noise near the fixed point. However, the goal of a practical control strategy is not to delay the heteroclinic jumps or bursts indefinitely, but only to increase the period between bursting events (the inter-burst time, T_B). By increasing the time between jumps/bursts, fewer turbulence-generating events will occur, less turbulence will be generated and the momentum transport away from the wall by the turbulence (skin friction drag) will be weakened.

Our strategy for control in the minimal flow unit, which contains a single set of coherent structures, is to apply the control directly to the coherent structures. Earlier control work (Coller et al. 1994a, Coller et al. 1994b) relied on an adjacent pair of vortices as an actuator for control, but the minimal flow unit is not large enough to permit the introduction of another set of structures without directly interfering with the naturally occurring coherent structures. The form of the control in the minimal flow unit will be similar to that used in our experiments with open-loop control and will be applied with a body force whose form is a Gaussian in the wall-normal direction which is largest near the wall (at $y^+ = 10$). The body force is applied to the first two spanwise Fourier modes, and its position is determined by our estimation technique which predicts the location of the rolls. The strength of the control is determined by our control algorithm which is outlined in section 4. The control is chosen to be invariant in the streamwise direction. Although the structures themselves are not uniform in the streamwise direction, streamwise invariance is a reasonable assumption in a small box like the minimal flow unit except when the structures break down. The control focuses on the prevention of the instability and bursting of the near-wall coherent structures. To this end, we will attempt to suppress the rolls when they grow strong. Weakening the rolls reduces the Reynolds stress generated by the near-wall structures and prevents the ejections which lead to instabilities and bursts. However, before we can describe the control strategy in detail, the problem of estimating the state of the flow based on measurements at the wall must be addressed.

3 Estimation

In our attempt to move from open-loop control to some form of feedback control, one key input to the control — an estimate of the state of the flow — is necessary. Our control strategy must incorporate information from the flow so that the control may adapt. The control will be applied to the flow selectively, i.e. only when the structures are tending towards bursting, in an attempt to maximize the effectiveness of the control. To this end, our estimate of the state of the flow will focus on the strength of the coherent structures. Our strategy for estimation draws from the work of Podvin & Lumley (1998), who applied the Proper Orthogonal, or Karhunen-Loève, Decomposition (POD) to velocity fields from the minimal flow unit and used the eigenfunctions of the POD to construct a low-dimensional model for the large-scale flow structures in the minimal flow unit. (For background on low-dimensional models and the POD, see Holmes et al. (1996) or Lumley & Blossey (1998*b*).)

The POD provides a complete expansion of the velocity field in terms of orthogonal eigenfunctions. The convergence of this expansion is optimal in the sense that any truncation contains, on average, the most kinetic energy of any truncation of that order. (The POD yields Fourier modes when applied to homogeneous directions — in this case, the streamwise and spanwise directions — so that the velocity field is expanded in Fourier modes in the homogeneous directions and eigenfunctions in the inhomogeneous directions.)

$$u_i(x, y, z, t) = \sum_{n=1}^{\infty} a_{k_1 k_3}^{(n)}(t) e^{2\pi i \left(\frac{k_1 x}{L_x} + \frac{k_3 z}{L_z} \right)} \phi_{i k_1 k_3}^{(n)}(y) \tag{3.1}$$

The wall shear stress can be expanded in terms of the wall-normal derivatives of the eigenfunctions.

$$\frac{\partial u_i}{\partial y}(x, y = 0, z, t) = \sum_{n=1}^{\infty} a_{k_1 k_3}^{(n)}(t) e^{2\pi i \left(\frac{k_1 x}{L_x} + \frac{k_3 z}{L_z} \right)} \frac{\partial \phi_{i k_1 k_3}^{(n)}}{\partial y}(y = 0) \tag{3.2}$$

We are interested in tracking the strength of the first two spanwise Fourier modes of the most energetic eigenfunction. These two modes combine to form the streaks and rolls that we have been talking about in our picture of turbulence generation in near-wall turbulent flow. The coefficients of these two modes, denoted by $a_{01}^{(1)}(t)$ and $a_{02}^{(n)}(t)$, can be estimated from the wall shear stress very simply if we assume that the wall shear stress in those two Fourier modes comes only from those two modes. Remember that the first eigenfunction carries the most energy among all eigenfunctions. Furthermore, the first eigenfunction in the wall region of the turbulent boundary layer carries more than half of the kinetic energy there. We truncate our eigenfunction expansion, including only the first eigenfunction, to determine the coefficients of the eigenfunctions from the shear stress. This truncation is a good approximation when the flow is dominated by the coherent structures. When the structures break down, we can expect higher eigenfunctions to play a larger role and perhaps degrade the accuracy of our estimate.

$$\frac{\partial \hat{u}_{i_{k_1 k_3}}}{\partial y}(y = 0, t) \approx a_{k_1 k_3}^{(1)}(t)\frac{d\phi_{i_{k_1 k_3}}^{(1)}}{dy}(y = 0) \tag{3.3}$$

$$a_{k_1 k_3}^{meas}(t) = \beta\left(\frac{D\hat{u}_{1_{k_1 k_3}}}{D\phi_{1_{k_1 k_3}}^{(1)}}\right)_{wall} + (1 - \beta)\left(\frac{D\hat{u}_{3_{k_1 k_3}}}{D\phi_{3_{k_1 k_3}}^{(1)}}\right)_{wall} \tag{3.4}$$

Here, $D \equiv \partial/\partial y$ and $a_{k_1 k_3}^{meas}$ refers to our instantaneous estimate ("measurement") of $a_{k_1 k_3}^{(1)}$. We drop the superscript in $a_{k_1 k_3}^{meas}$ since we only estimate the strength of a single eigenfunction for a given Fourier mode (k_1, k_3). The scalar β determines the relative weight of the spanwise and streamwise shear stress in determining the coefficients. The two modes that we want to estimate have wavenumbers $k_1 = 0$ and $k_3 = 1, 2$. The streamwise and spanwise shear stress for modes of this form (with no streamwise variation) come from the streaks and rolls, respectively. The rolls are disturbances of the form $\mathbf{u}(x, y, z) = (0, u_2(y, z), u_3(y, z))$ which do not directly affect the streamwise shear stress, except indirectly through their nonlinear interaction with the mean velocity which supplies energy to the streaks. Similarly, the streaks do not affect the spanwise shear stress. For the purposes of control, we are more interested in the rolls and choose a value of β close to zero. In fact, Podvin & Lumley (1998) found that their estimation was more successful for such values of β. The importance of spanwise shear stress in the prediction of ejections was also highlighted by Lee et al. (1997), who employed neural networks to optimize the prediction of strong ejections and sweeps in the wall region based on wall shear stress measurements.

The procedure outlined above will give us instantaneous "measurements" of the coefficients, which will likely be aliased and noisy. Our estimate comes from the application of a simple filter to the time series of instantaneous measurements. The filter takes the form of a dynamic equation:

$$\frac{d}{dt}a_{k_1 k_3}^{est} = \frac{a_{k_1 k_3}^{meas} - a_{k_1 k_3}^{est}}{T} \tag{3.5}$$

Here, T is a timescale. We find that values of T on the order of 40 wall time units are effective. The choice of T involves a balance between the smoothing of non-physical oscillations in measurements from aliased higher frequencies and the time lag inherent in choosing a large T. The performance of the estimation using an array of six rows of eight sensors each in the minimal

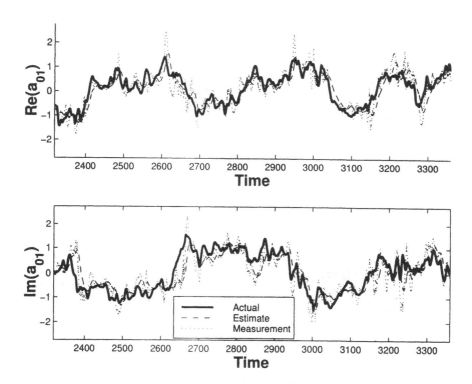

Figure 1: A comparison of our estimate of $a_{01}^{(1)}(t)$ with its actual value, computed by taking an inner product between the velocity field and the eigenfunction $\phi_{i01}^{(1)}$.

flow unit is displayed in Figure 1. The spacing of the sensors is approximately 22 wall units in the spanwise direction and 60 wall units in the streamwise direction. This array of sensors covers the entire wall in our minimal flow unit $(L_z = 184, L_z = 368)$.

For the purposes of our control, we are interested in the strength of the rolls in particular. However, in the interest of a robust estimation scheme, it is useful to include both streamwise and spanwise shear stress in the estimation routine. In fact, a value of $\beta = 0.1$ was most successful in providing an effective control for drag reduction. The inclusion of the streamwise shear stress can be seen as "contaminating" our measurement of the strength of the rolls, but the streaks (through the streamwise shear stress) do provide an indication of the past strength of the rolls, since the rolls give rise to the streaks through interaction with the mean flow. Nevertheless, the quality of our estimate suffers when the control is turned on, as can be seen in Figure 2.

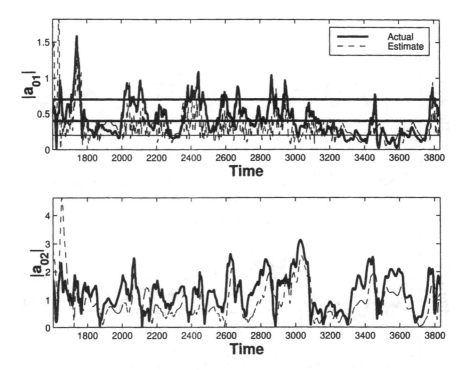

Figure 2: The estimation technique is sensitive to the control and is not as successful in predicting the strength of the first mode when the control is switched on. The horizontal lines correspond to the levels at which the control is switched on and off.

The estimate is much more sensitive to the control than the actual value of the coefficient: the control has the effect of driving the value of the estimate towards zero. After the control is switched off, the estimate increases and approaches the actual value of the coefficient until the control switches on again and drives the estimate back towards zero. The effect of the control on the estimate — driving it towards zero — apparently results from a decoupling of the shear stress at the wall, which provides the input for our estimate, from the coherent structures when the control is switched on. The causes of this will be explored in section 6, when we take a look at the effect of the control on the eigenfunctions of the POD. While the estimate is increasing and decreasing with the pulses of control in Figure 2, the shear stress and the actual value of the coefficient rise slowly but steadily. This seems to call into question the effectiveness of our control. However, if the control were switched off, the increases in shear stress and in the strength of the rolls would be much larger. In fact, turning off our control leads to a quick return of the shear stress to the levels of the uncontrolled flow. The motivation for our control strategy, which is based on a discrete switching algorithm, is described in the next section.

4 Control Algorithm

Our approach to control is simply to limit the strength of the rolls in the minimal flow unit. The rolls supply energy to the streaks through nonlinear interaction with the mean velocity profile. In this way, the rolls cause an increase in momentum transport away from the wall and drag at the wall. In addition, by strengthening the streaks, the rolls promote instability and the generation of smaller-scale turbulence which gives rise to a further increase in momentum transport and shear stress at the wall. By acting to control the rolls, we attempt to intervene and prevent the production of turbulence and an increase in shear stress at the wall. This strategy is an attempt to mimic the success of a passive technique for turbulence control, polymer drag reduction. The polymers act only at the smallest scales of the turbulence by expanding in the fluctuating strain rate field just outside of the viscous sublayer and damping out the smallest eddies there. However, their action at the smallest scales has a significant effect on the large scales: the rolls are significantly weakened in polymer drag reduced flow, while the streaks grow stronger. We take this configuration of the coherent structures as a cue and focus on using the control to weaken the rolls when they grow strong while simply ignoring the strength of the streaks. Although the streaks are likely to burst sometimes while we focus on the rolls, our control effort will be most efficiently applied to the rolls which are less energetic than the streaks. Both in the polymer drag reduced case and with our control, the flow will sometimes burst — this is nearly unavoidable — but our goal is not the complete suppression of the turbulence. The rolls and streaks are part of the turbulence, and completely suppressing them would be nearly impossible in a practical situation. Rather, our control attempts to suppress the instability of the rolls and streaks, so that this instability is triggered less often, less small-scale turbulence is generated, and momentum transport away from the wall is weakened.

   The full fluid system does not exactly possess the heteroclinic jumps of the low-dimensional models. The heteroclinic connections in the phase space of the model are partly a result of the combination of the roll and streak modes into a single mode in the dynamical system of Aubry et al. (1988). The behavior of the full fluid system does not seem to possess the simple geometry of a saddle point as the rolls and streaks give rise to a burst. It appears that the streaks do not change in strength very substantially during a burst. The rolls do strengthen in the time leading up to a burst, but the substantial signal of a burst lies in the streamwise-varying ($k_1 \neq 0$) modes which grow significantly in strength as the burst develops. (Hamilton et al. (1995) provide a nice description of the breakdown and regeneration of these structures in turbulent Couette flow.) However, for the purposes of our control, these streamwise-varying modes would be very difficult to track and become strong only when the structures have started to break down. Our control aims to prevent the breakdown of the structures and, as a result, focuses on the rolls. We attempt to limit the strength of the rolls when they appear to be growing and tending towards a burst. This control strategy draws from the work of Guckenheimer (1995) who constructed a class of hybrid (discrete switching plus linear feedback) controllers for problems with two unstable eigendirections. The linear feedback controllers have a limited domain of attraction and may be susceptible to disturbances and unmodeled dynamics. Guckenheimer proposed that the domain of attraction of these linear controllers might be increased substantially by applying discrete switching control to trajectories which escaped from the domain of attraction of the linear controllers. The amplitude and direction of this control would be piecewise constant over stretches of phase space, presumably increasing in amplitude as the trajectories moved farther away from the fixed point. The discrete control does not attempt to direct the trajectory gracefully and efficiently back to the fixed point, rather the control switches on and drives the system directly towards the linear controller's domain of attraction. When the domain of attraction is reached, the discrete control switches off and the linear feedback controller takes over. As a demonstration of his concept, Guckenheimer (1995) constructs a hybrid control strategy for the inverted double pendulum whose domain of attraction is considerably larger than that of a linear feedback controller alone.

For our fluid system, we wish to borrow from the hybrid control strategy outlined above. However, we do not have a clear-cut, low-dimensional dynamical system underlying our fluid simulations. The rolls have a stable fixed point at zero strength when the flow is laminar (since our simulations are subcritical), but the domain of attraction is apparently small and would be difficult to attain. We will not attempt to apply linear feedback control when the strength of the rolls is small.[1] However, we will apply discrete switching control to the rolls when they grow strong. Our "hybrid" strategy relies solely on the discrete switching control to limit the strength of the rolls. One point must be addressed here before describing the control explicitly: the control reacts in response to our estimate of the strength of the rolls, not to the rolls themselves. The estimate is susceptible to aliasing, a finite response time, and other difficulties, and should not be accepted without skepticism. If the estimate is varying wildly, the estimate should probably not be used to guide the control. Either the estimate is responding to higher modes which have been aliased down to the wavenumber of the coherent structures — or the coherent structures are drifting quickly and will not respond well to the ap-

---

[1]Some groups apply linear feedback control to channel flow, with feedback laws based on suboptimal control theory (Lee et al. 1998), neural networks (Lee et al. 1997), or arguments about the suppression of vorticity flux (Koumoutsakos 1997).

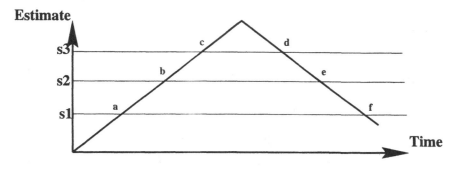

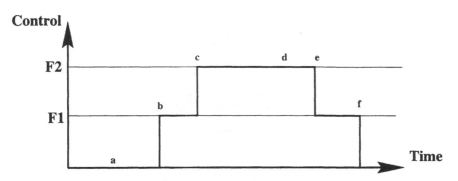

Figure 3: Our discrete switching scheme for the control switches on or up a level as the estimate passes above the upper thresholds s_2 and s_3, and switches off or down when passing below the lower thresholds s_1 and s_2.

plication of a control which has a fixed position in the spanwise direction. In such situations, the control will be switched off for a few time units until the estimate has settled down, after which time the control may be activated again.

Our control strategy involves on/off switching based on the size of our estimate of the strengths of the structures, $|a_{01}^{est}|$. We set three threshold levels: s_1, s_2 and s_3, and two control strengths F_1 and F_2. As the strength of the estimate increases from zero with the control initially turned off, the control will turn on at level F_1 when the estimate passes above s_2. If $|a_{01}^{est}|$ subsequently falls below s_1, the control is turned off. If the strength of the estimate rises above s_3, the control is increased to F_2. The control strategy is illustrated in Figure 3.

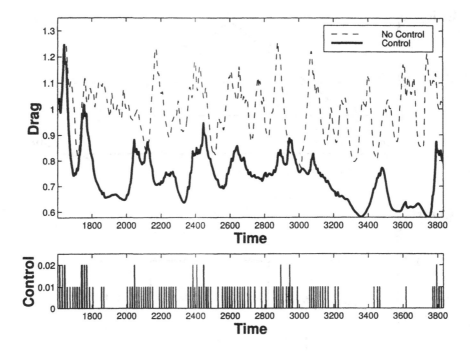

Figure 4: The streamwise shear stress at the wall as a function
of time (displayed in outer time units, i.e. tU/h). Note that the
control tends to switch on during periods of relatively high shear
stress and off when the shear stress is relatively low.

5   Results

This control strategy has been implemented in turbulent channel flow in the
minimal flow unit with two different Reynolds numbers: $R_\tau = u_\tau h/\nu = 100$
and $R_\tau = 175$. (Here, u_τ is the friction velocity, h the half-height of the channel and $\nu = \mu/\rho$ the kinematic viscosity. The friction velocity u_τ is defined by
$\rho u_\tau^2 = \mu \partial U/\partial y$ where ρ and μ are the density and dynamic viscosity, respectively, and the gradient of the mean velocity U is evaluated at the wall.) The
eigenfunctions of Podvin & Lumley (1998) which serve as the basis of our
estimation technique were produced using data from the $R_\tau = 175$ channel.
These eigenfunctions (which are only defined in the wall region, $0 \leq y^+ \leq 40$)
were then rescaled for use in the $R_\tau = 100$ channel. The eigenfunctions of the
two different channels should be very similar when scaled in wall units.[2]

The effect of the control in the $R_\tau = 100$ channel can be seen in Figures 4

---

[2]Later computations revealed that the eigenfunctions matched closely when scaled in
wall units.

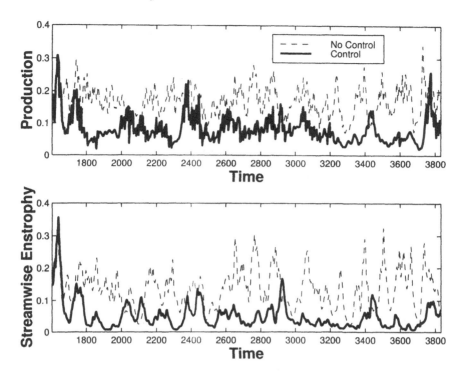

Figure 5: Turbulent production and streamwise enstrophy, both integrated over the bottom half of the channel (where the control is applied).

through 6. The control is switched on and off repeatedly while the drag is high, but remains off for extended periods while the drag is low. The coherent structures are responsible for much of the drag through their momentum transport so that the coherent structures are strong during periods of high drag and weak during periods of low drag. (The one exception to this behavior is that during a burst, the coherent structures will begin to weaken before the drag reaches its peak.) The repeated switching of the control during the periods of high drag results from the effect of the control on our estimation technique. The estimate of the strength of the coherent structures drops when the control is applied. The location of the strongest control is at $y^+ = 10$ and the control drops off as a Gaussian from that point with a half-width of 10 wall units. The coherent structures, in contrast, are strongest away from the wall.

Relative to the simulation without control, the control results in a reduction in drag of more than 20%. The bursting events, represented by peaks in the drag, are not eliminated by the control: rather the peaks are reduced in magnitude and frequency relative to those in the uncontrolled flow and the period between the peaks is increased. The reduction in drag is apparently a result of a decrease in momentum transport by the coherent structures. The structures provide most of the Reynolds stress during the quiescent periods between bursting events; therefore, their weakening will have a substantial effect on the shear stress at the wall. (A control strategy designed to enhance the activity of the coherent structures can increase the shear stress at the wall significantly.) The weakening of the structures may be seen not only indirectly through the plot of shear stress at the wall, but also more concretely through plots of turbulent production $-\left\langle uv\frac{dU}{dy} \right\rangle$ and streamwise enstrophy $\langle \omega_1^2 \rangle$. These statistics, displayed in Figure 5, were computed at regular time intervals by integrating over the bottom half of the channel:

$$\left\langle \omega_1^2 \right\rangle (t) = \frac{1}{L_x L_z} \int_{-1}^{0} \int_{0}^{L_x} \int_{0}^{L_z} \omega_1^2 \, dz \, dx \, dy \qquad (5.1)$$

Note the strength of the peaks in streamwise enstrophy without the control. The streamwise enstrophy is a good indicator of the strength of the rolls, since these are primarily composed of streamwise vorticity. There is a strong correlation between the peaks in streamwise enstrophy and peaks in drag. It appears, although not conclusively, that the peaks in enstrophy tend to lead the peaks in drag slightly, indicating perhaps that the rolls contribute to the drag substantially but that other modes may take over as the energy is transmitted from the large to the small scales.

The control is effective in suppressing both the production of turbulent kinetic energy and streamwise enstrophy as well as the shear stress at the wall. This indicates that our control has a substantial effect on the dynamics of the coherent structures near the wall (which are responsible for much of the turbulent production near the wall (Kim et al. 1971)). This further suggests that a control strategy may be successful by focusing its attention only on the large-scale structures. Other control strategies, in particular that of Koumoutsakos (1997) and Koumoutsakos (1999) which is based on canceling the vorticity flux from the wall, attempt to suppress numerous small-scale events which contribute to the generation of turbulence near the wall. These schemes attempt to influence the large-scale statistics like drag through small-scale interactions. Our control is applied only to the largest scales. The smaller scales evolve without control. Some of the schemes which focus on canceling the small-scale events induce the formation of a no-penetration layer away from the wall. Blowing and suction at the wall exactly cancel the wall-normal velocity away from the wall, creating a virtual no-penetration (but free slip) layer. The blowing and suction substantially weaken the coherent structures (as seen in a plot of streamwise enstrophy in Lee et al. (1998)) as well as reducing their impact on the shear stress at the wall. (The physics of these oppositional control schemes is discussed in Hammond et al. (1998).)

By focusing our control effort on the large-scale structures, we limit the scale of the actuators to that of the structures.[3] This may alleviate to some extent the very small scale required of the actuators and may also make the control's performance more robust with respect to the number of actuators, as long as there is at least one actuator per set of coherent structures (roughly one actuator per 100 wall units). Although this strategy may be relatively robust with respect to the spacing of the actuators, the sensor spacing seems to be more critical. To achieve the high degree of drag reduction in our simulations, we have used eight sensors across the width of our box, $L_3^+ = 160$. We envision an actual implementation to have actuators spaced by approximately 60–100 wall units and sensors by 30–50 wall units.

The work of Aubry et al. (1990) suggested that active control strategies would have a similar effect on the boundary layer to polymer drag reduction. In fact, we employed ideas based on the large-scale effect of polymer drag reduction in the design of our control strategy. As a result, we would expect our controlled flow to exhibit large-scale properties similar to those found in the polymer drag-reduced boundary layer, namely a thickening of the buffer layer — corresponding to the growth of the coherent structures — as well as the strengthening of the streamwise fluctuations and the attenuation of the spanwise fluctuations. As shown in Figure 6, the statistics do respond in this way. The streamwise fluctuations do not grow as much as in polymer drag-reduced flow; however, the cross-stream fluctuations are reduced considerably, resulting in smaller Reynolds stress across the half of the channel in which the control is applied.

Our control was also applied in a minimal flow unit with Reynolds number $R_\tau = 175$. The control succeeded in reducing the drag in this setting, but the success, a drag reduction of up to 13%, was limited when compared to the $R_\tau = 100$ channel. The effect of the control on the statistics — from the drag to the mean velocity profile and rms turbulent fluctuations — is similar to the control's effect in the $R_\tau = 100$ channel, although the impact of the control on the statistics is smaller. We attempted to increase the effectiveness of the control in the larger $R_\tau = 175$ channel by increasing the domain of the control to match that of the smaller channel (in outer rather than wall units) but found that this does not result in comparable drag reduction. We speculate that the discrepancy in the success of our control is a result of the increased number of active modes in the larger, $R_\tau = 175$, channel which adversely affects our estimation technique, since the first eigenfunction in a given wavenumber will tend to contribute less to the shear stress at the wall as the number of active modes increases. In addition, the control of only two modes may not be sufficient since more modes may contribute actively to the

---

[3]Schoppa & Hussain (1998) and Sirovich & Karlsson (1997) both have suggested and implemented (the former numerically and the latter experimentally) passive control strategies based on actuators which are larger than the scale of the coherent structures. Schopppa & Hussain suggest that their strategy interferes with the mechanism by which the coherent structures form.

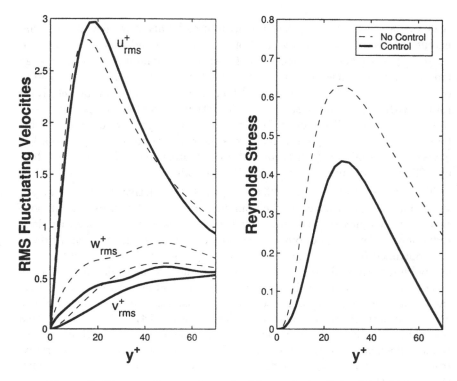

Figure 6: The rms fluctuating velocities and the Reynolds stress with and without control, normalized by their respective shear velocities.

dynamics of the flow at higher Reynolds numbers. The inclusion of dynamical information in our estimation technique in the future may alleviate some of these difficulties, helping to make our control strategy more robust and less sensitive to Reynolds number.

6   Control and the POD eigenfunctions

A few questions are raised by the formulation and implementation of our control strategy:

- If the control strategy is based on POD eigenfunctions, what happens if the shape of the modes is affected by the control?

- If the control changes the shape of the modes, how does that affect our estimation technique, which is constructed using modes from the uncontrolled flow?

We propose to address these questions by examining the Proper Orthogonal Decomposition of these flows. Changes in the spectrum of the eigenfunctions and the shapes of the modes in the different flows should yield insight into the effect of the control on the flow as well as the robustness of our control strategy.

Comparisons of the shape of the most energetic eigenfunction of the Proper Orthogonal Decomposition are employed to address these questions. Since the coherent structures form the basis of our control strategy and the first eigenfunction provides a good picture of these coherent structures, changes in the first eigenfunction as a result of the control will affect the robustness of our control strategy. The eigenvalue spectrum gives an indication of the distribution of energy among the modes. In flows where the first eigenfunction carries almost all of the turbulent kinetic energy, our control may be relatively more effective than in flows where more modes carry a significant fraction of the energy.

As validation of our eigenvalue computations, we compared the first POD eigenfunction in uncontrolled flow with those of Podvin & Lumley (1998) and found that the eigenfunctions matched closely. Both eigenfunctions were generated using data from turbulent channel flow in a minimal flow unit with Reynolds number $R_\tau = 175$. These eigenfunctions and the ones described below were defined in the wall region only, i.e. for $0 \le y^+ \le 40$. The two eigenfunctions match closely near the wall with slight variations away from the wall. The size of our statistical sample is relatively small, so the eigenfunctions are not fully converged. However, the close correspondence of the eigenfunctions, particularly their gradient at the wall (which is employed in our estimation scheme), gives us confidence in the eigenfunctions generated from our data and allows us to move forward into the comparison of eigenfunctions in different flows.

In Figure 7, the first eigenfunctions of the controlled and uncontrolled flow are compared. The shape of the eigenfunctions are similar between the two cases. The peak of the streamwise component in the controlled flow appears to be shifted away from the wall, even when both eigenfunctions are scaled according to their respective friction velocities. In addition, the cross-stream components in the controlled flow are weaker, by approximately a factor of two, than those of the uncontrolled flow. This picture corresponds to that presented by the rms fluctuating velocities in Figure 6. This is not unexpected since the first, most energetic, eigenfunction captures more than half of the kinetic energy in the wall region. The change in the gradient of the eigenfunctions at the wall will affect our estimation technique; however, these will change our estimates quantitatively but not qualitatively. Since we have chosen the weight β in our estimation scheme to be near zero, the estimate depends primarily on the spanwise components of the eigenfunctions and will behave as $a^{est} \sim \left(\frac{d\phi_3}{dy}\big|_{wall}\right)^{-1}$. Since we use the eigenfunctions from the un-

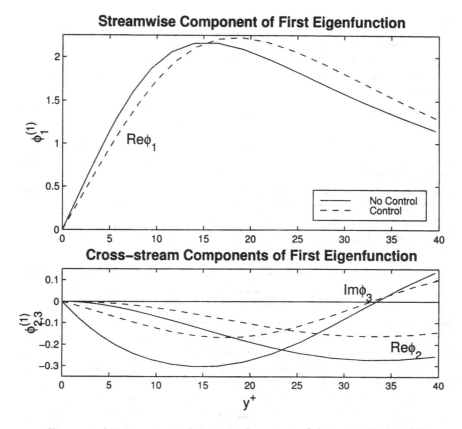

Figure 7: A comparison of the eigenfunctions of the controlled and uncontrolled flows with $R_\tau = 100$.

controlled flow (which have a larger gradient at the wall) in our estimation techniques, the strength of the first mode in the controlled flow will be systematically underestimated.

The eigenfunctions displayed in Figure 7 reflect the properties of the large-scales of the controlled flow as a whole. Difficulties with the estimation technique when the control was active suggest that another decomposition, using only velocity fields where the control was switched on, should be performed. A comparison between the eigenfunctions of the controlled flow as a whole and the controlled flow sampled only when the control was active is provided in Figures 8 and 9. The large-scale properties of the eigenfunctions are similar. Examination of the shapes of the eigenfunctions near the wall (displayed in Figure 9) reveals that the shear stress footprints of the first eigenfunction are quite different in the two cases. The control induces a flow in the span-wise direction very close to the wall, i.e. below $y^+ = 2$, which runs opposite

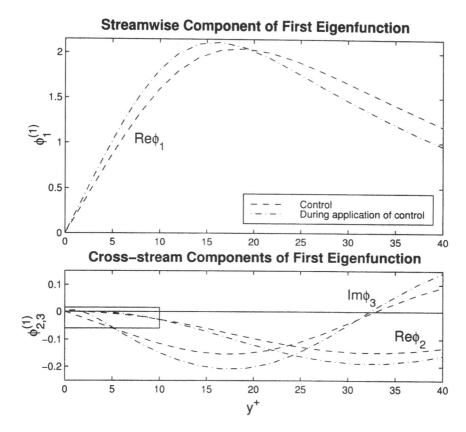

Figure 8: A comparison of the eigenfunctions of the controlled flow between eigenfunctions generated from data taken only when the control was switched on and eigenfunctions from the whole simulation (including samples both with and without the control switched on). The region enclosed in a box around the origin is expanded in Figure 9.

the outer flow in the spanwise direction. The streamwise component of the eigenfunction is not affected by the control in the same way. This may explain why our control is more effective when some of the streamwise shear stress is included in the estimate: the inclusion of the streamwise shear stress, which is relatively unaffected by the control, makes the estimation scheme more robust in the presence of control. The cross-flow near the wall is likely a result of the body force, which is 60% of its full strength at the wall. Away from the wall, the body force only acts to damp the coherent structures. However, close to the wall, the body force is sufficiently strong (relative to the background, turbulent flow) to cause a flow counter to the flow induced by the

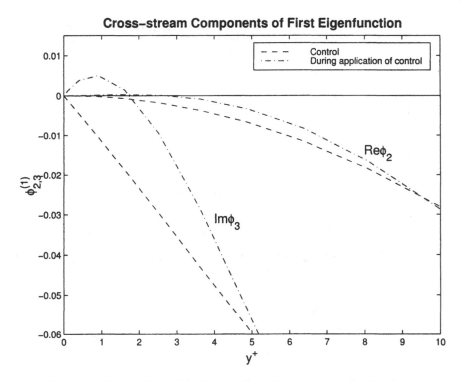

Figure 9: The application of control results in a spanwise flow very close to the wall which opposes that of the coherent structures away from the wall.

coherent structures away from the wall. In the short term, this counter-flow reverses the shear stress footprint of the coherent structures. As a result, after the control is switched on, the instantaneous estimate will be out of phase with the previous instantaneous estimates. The magnitude of the smoothed estimate will drop, and the rate of decrease will be limited only by the time scale of the filtering (and the parameter β which is the relative weighting of the streamwise and spanwise shear stress in the estimate). Over the longer term, the shear stress at the wall will become decoupled from the outer flow as the flow near the wall is driven by the body force rather than influenced by the coherent structures. In an attempt to improve the performance of our estimation technique when the control was switched on, we performed simulations with alternative forms of the body force which died away faster close to the wall, but the difficulties with the body-force-induced counter-flow near the wall remained.

Despite the systematic errors in the estimation induced by changes in the eigenfunctions, the greatest difficulty lay in the spanwise counter-flow induced near the wall when the control is switched on. This flow interferes with the shear stress footprints of the coherent structures and makes the strength of the structures away from the wall very difficult to predict. We feel, however, that the cross-flow is specific to the nature of our control — a body force — and that wall-mounted actuators based on fluid motion would not exert such strong effects very close to the wall. In addition, practical implementations of control (and subsequent simulations that we plan to undertake) would not sense and actuate the flow at the same location. The sensors will likely be located upstream of the actuators, and as a result the estimate would be relatively independent of the effect of the actuators.

7   Conclusions

Our future work with control in the simulations will include a move to settings which would be more realistic for practical implementations of control. This includes the elimination of the co-location of the sensors and actuators, which would eliminate some of the problems with our estimation during control described in section 6. We envision successive rows of sensors and actuators. The sensors would feed into an estimation algorithm which would predict the state of the flow and, after an appropriate time delay — corresponding to the time taken for the structures to be convected from the sensors downstream to the actuators — apply control with the actuators. More rows of sensors could be placed downstream of the actuators to judge the effectiveness of the control and feed back information to the estimation and control algorithms. We have found that the greatest effect of our control occurs well after the control is applied, so that we might expect that the rows of actuators could be spaced well apart, perhaps by a few thousand wall units, so that the effect of one actuator could be felt by the flow before encountering another actuator. In such a setting, the POD-based, low-dimensional models could play a larger role in control, predicting the evolution of the flow from the sensor to the actuator and possibly improving the estimation of the flow. The models might also contribute to the formulation of the control if the flow is found to reside near fixed points of the models, where the dynamics might be accurately modeled by a linearization about these fixed points.

We have implemented an active control strategy in simulations of turbulent channel flow in the minimal flow unit. Estimates of the strength of the two most energetic empirical eigenfunctions of the Proper Orthogonal Decomposition are generated from measurements of the shear stress at the wall. The control is applied using a discrete switching technique to limit the influence of inaccuracies in the estimation when the control is switched on. A body force, which is localized near the wall, is added to the wall-normal momentum equation and serves as the actuator in the simulations. Drag reductions on the order of 20% are achieved in a minimal flow unit with $R_\tau = 100$.

8 Acknowledgments

This work has been supported in part by the Air Force Office of Scientific Research through Contract No. F49620-96-1-0329; in part by Contracts Nos. F49620-92-J-0287 and F49620-97-1-0203 jointly funded by the US Air Force Office of Scientific Research (Control and Aerospace Programs) and the US Office of Naval Research; in part by Grants No. F49620-92-J-0038, F49620-99-1-0012 and F49620-96-1-0329, funded by the US Air Force Office of Scientific Research (Aerospace Program); in part by Grant Number CTS-9613948 funded by the U. S. National Science Foundation; and in part by the Physical Oceanography Programs of the US National Science Foundation (Contract No. OCE-901 7882) and the US Office of Naval Research (Grant No. N00014-92-J-1547). PNB was supported in part by a National Science Foundation Graduate Research Fellowship.

References

Aubry, N., Holmes, P., Lumley, J. L. & Stone, E. 1988. The dynamics of coherent structures in the wall region of a turbulent boundary layer. *J. Fluid Mech.* 192: 115–173

Aubry, N., Lumley, J. L. & Holmes, P. 1990. The effect of modeled drag reduction in the wall region. *Theo. Comput. Fluid Dyn.* 1: 229–248

Bloch, A. M. & Marsden, J. E. 1989. Controlling homoclinic orbits. *Theo. Comput. Fluid. Dyn.* 1: 179–190

Coller, B. D., Holmes, P. & Lumley, J. L. 1994*a*. Controlling noisy heteroclinic cycles. *Physica D* 72: 135–160

Coller, B. D., Holmes, P. & Lumley, J. L. 1994*b*. Interaction of adjacent bursts in the wall region. *Phys. Fluids* 6(2): 954–961

Guckenheimer, J. 1995. A robust hybridization stabilization strategy for equilibria. *IEEE Trans. Automatic Control* 40(2): 321–326

Hamilton, J. M., Kim, J. & Waleffe, F. 1995. Regeneration mechanisms of near-wall turbulence structures. *J. Fluid Mech.* 287: 317–348

Hammond, E. P., Bewley, T. R. & Moin, P. 1998. Observed mechanisms for turbulence attenuation and enhancement in opposition-controlled wall-bounded flows. *Phys. Fluids* 10(9): 2421–2423

Holmes, P. J., Berkooz, G. & Lumley, J. L. 1996. *Turbulence, Coherent Structures, Dynamical Systems and Symmetry*, New York/Cambridge: Cambridge University Press

Kim, H. T., Kline, S. J. & Reynolds, W. C. 1971. The production of turbulence near a smooth wall in a turbulent boundary layer. *J. Fluid Mech.* 50: 133–160

Koumoutsakos, P. 1997. Active control of vortex-wall interactions. *Phys. Fluids* 9(12): 3808–3816

Koumoutsakos, P. 1999. Vorticity flux control for a turbulent channel flow. *Phys. Fluids* 11(2): 248–250

Lee, C., Kim, J., Babcock, D. & Goodman, R. 1997. Application of neural networks to turbulence control for drag reduction. *Phys. Fluids* 9(6): 1740–1747

Lee, C., Kim, J. & Choi, H. 1998. Suboptimal control of turbulent channel flow for drag reduction. *J. Fluid Mech.* 358: 245–258

Lumley, J. L. 1967. The structure of inhomogeneous turbulence. In *Atmospheric Turbulence and Wave Propagation*, ed. A. M. Yaglom & V. I. Tatarski, Moscow: Nauka, pp. 166–176

Lumley, J. L. & Blossey, P. N. 1998a. Control of turbulence. *Ann. Rev. Fluid Mech.* 30: 311–327

Lumley, J. L. & Blossey, P. N. 1998b. The low dimensional approach to turbulence. In *Modeling Complex Turbulent Flows*, ed. M. D. Salas, J. Hefner & L. Sakell, Kluwer Academic Press. to appear

Lumley, J. L. & Kubo, I. 1985, Turbulent drag reduction by polymer additives: a survey. In *The Influence of Polymer Additives on Velocity and Temperature Fields*, ed. B. Gampert, New York: Springer-Verlag, pp. 3–24

Podvin, B. & Lumley, J. L. 1998. A low dimensional approach for the minimal flow unit. *J. Fluid Mech.* 362: 121–155

Schoppa, W. & Hussain, F. 1998. A large-scale control strategy for drag reduction in turbulent boundary layers. *Phys. Fluids* 10(5): 1049–1051

Sirovich, L. & Karlsson, S. 1997. Turbulent drag reduction by passive means. *Nature* 388: 753–755

Sil'nikov Chaos in Fluid Flows

T. Mullin, A. Juel and T. Peacock

Abstract

We review the results of experimental studies of Sil'nikov chaos in three different fluid flows. These are Taylor–Couette flow between rotating cylinders, side-wall convection in a liquid metal flow with an applied magnetic field and electrohydrodynamic convection in a nematic liquid crystal. In each case the observed low-dimensional dynamics is found to be organised by codimension-2 bifurcations. The results illustrate that progress can be made using ideas from dynamical systems theory to understand complicated fluid flows of practical relevance which are beyond computation. It is also suggested that contact can be made between the full equations of motion and low-dimensional models provided physically relevant symmetries are taken into account.

1 Introduction

It has long been recognised that insight into complicated fluid motion may be gained by studying the stability of the corresponding laminar flow and the transition to disorder as a parameter is varied. Since studies of this type can be carried out both theoretically and experimentally, it provides an opportunity for comparison between experimental, theoretical and numerical results. This powerful combination can often be insightful in making progress with one of the most challenging and difficult problems of modern physics, fluid flow turbulence.

A cornerstone of hydrodynamic stability theory was laid by Taylor (1923) who used a combined theoretical and experimental approach to good effect in studying the stability of flow between concentric rotating cylinders. He was interested in the effect of viscosity on the Rayleigh stability criterion for rotating flows. He obtained striking agreement between analytical and experimental results for the limits of stability of the base flow as a function of the rotation rates of the inner and outer cylinders. Since that time, the so-called 'Taylor–Couette' problem has been studied extensively in a number of contexts and has become a classic example of the transition to disorder in a closed fluid flow. Reviews of many of the papers on the topic can be found in DiPrima & Swinney (1981). More recently it has been studied as an example of a fluid flow where progress can be made in understanding complicated motions using ideas from low-dimensional dynamical systems and chaos. A review of this work is provided by Mullin & Kobine (1996).

Modern research on the transition to low-dimensional chaos in fluid systems is often centred on the celebrated work of Gollub & Swinney (1975). Their results provided the first indications that the onset of disorder occurs after a finite sequence of bifurcations leading to a strange attractor and chaos. This is more in accord with the ideas of Ruelle and Takens than the Landau–Hopf model, where turbulence is modelled dynamically as a complicated quasiperiodic motion (Joseph (1985)). Research since that time has shown that, as in any nonlinear system, there are many routes to chaos in Taylor–Couette flow. These include period doubling (Pfister (1985)), quasiperiodicity (Brandstater & Swinney (1987)), homoclinicity (Mullin & Price (1989)) and intermittency (Price & Mullin (1991)).

One way of making progress in these difficult problems is to explore the possibility that apparently different sources of chaotic dynamics are in some sense connected and organised by the underpinning solution structure. In particular, it is known in dynamical systems theory that codimension-2 organising centres can form when paths of qualitatively different bifurcations intersect in a control parameter plane (Guckenheimer & Holmes (1986)). For example, if a path of pitchfork and one of Hopf bifurcations meet then this local interaction can produce dynamical structure which is global. This is important since it suggests that local structures which can be analysed mathematically can provide predictions about global dynamical behaviour. Thus apparently different types of low-dimensional dynamical behaviour can be linked. Moreover, it offers the possibility of providing a link between the Navier–Stokes equations and low-dimensional models.

In the following sections we shall present evidence for low-dimensional chaos in the Taylor–Couette problem. The nature of the dynamics described is shown to be consistent with a model problem investigated by Sil'nikov (1965). In addition, we present recent results from two other experiments, concerning side-wall convection in a liquid metal which has a magnetic field applied and electrohydrodynamic convection in a liquid crystal flow. For both these problems we show that some of the complicated dynamical behaviour can be understood in terms of low-dimensional models qualitatively similar to those used for Taylor–Couette flow, implying a degree of universality for Sil'nikov dynamics.

2 Taylor–Couette flow

Taylor–Couette flow is concerned with fluid motion on apparently simple boundary conditions between concentric rotating cylinders. Here, we will discuss results from studies of a restricted version of the problem where only

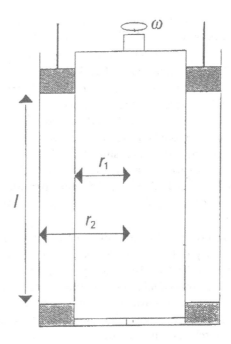

Figure 1. A schematic of the Taylor–Couette apparatus.

---

the inner cylinder rotates and the outer is stationary. Even in this version there exists a wealth of challenging and interesting flow phenomena which test the limits of observation and calculation.

A schematic diagram of a typical experimental arrangement is given in figure 1. There are three control parameters which govern the experimental flow. The principal dynamical parameter is the *Reynolds number*, Re, which is defined as

$$Re = \frac{r_1 \omega d}{\nu}, \tag{1}$$

where r_1 is the radius of the inner cylinder, ω is the angular rotation speed of that cylinder, d is the width of the gap between the cylinders and ν is the kinematic viscosity of the fluid. There are also two geometrical parameters. The *aspect ratio* Γ is defined as

$$\Gamma = \frac{l}{d}, \tag{2}$$

where l is the distance between the two horizontal end plates. The *radius ratio* η is defined as

$$\eta = \frac{r_1}{r_2}, \tag{3}$$

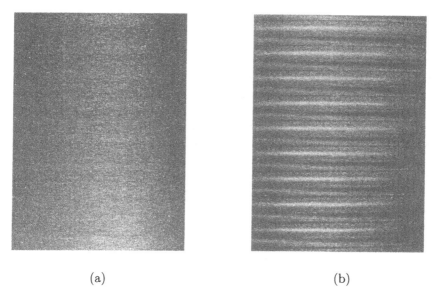

(a)   (b)

Figure 2. (a) Front view of Taylor–Couette flow when Re is below
the range at which cells appear.(b) Front view of flow visualisation
of Taylor cells.

where r_2 is the radius of the stationary outer cylinder. In practice the radius
ratio is a constant for any particular experimental configuration and qualitative changes in the state of the flow are usually brought about through
variation of the Reynolds number at fixed values of the aspect ratio.

At low rotation rates the flow in any experiment is more or less featureless,
as shown in Figure 2a but there is always some three-dimensional motion at
the ends. When Re is increased, cells form along the length of the cylinder spreading inwards from the non-rotating ends. A typical visualisation of
a cellular flow is presented in Figure 2b. The cellular structure comprises
secondary circulations superposed on the main azimuthal flow, and this is
readily made visible using anisotropic flow visualisation particles. Observations in apparatus with aspect ratios greater than 10 suggest that the onset
of cells occurs at a bifurcation to which a critical Reynolds number can be
ascribed. However, it was first pointed out by Benjamin (1978) that the
situation is more subtle than a simple bifurcation.

The finite size of the domain in the experiment is qualitatively different to
the situation in most mathematical and numerical models of the flow, where
periodicity is assumed along the cylinders. Most importantly, the onset of

cells is described in the models by a pitchfork bifurcation, but this is not what is observed in experiments. Casual observation suggests that the effect of the ends is simply to soften the bifurcation as in any physical system where small natural imperfections will disconnect the bifurcation. However, the 'imperfection' of the ends is $O(1)$ and so the pitchfork is effectively destroyed. This is readily demonstrated by observing the lower limit of stability of the disconnected part of the bifurcation. It has been shown by Cliffe *et al.* (1992) that the fold or limit point is at least a factor of two in excess of the onset of cells for small radius ratios. It becomes even greater in the limit of a narrow gap between the cylinders, which is the case considered by Taylor and in most subsequent studies. This is an important feature of the problem since it removes a crucial element of our codimension-2 organising centre, the pitchfork bifurcation.

Experimental evidence of Lorenzen & Mullin (1985) suggests that the disconnection of the pitchfork bifurcation from cellular flow is equally big in large aspect ratio systems. In other words, performing experiments with long cylinders will not minimise end-effects. Indeed, if long cylinders are used in an experiment then the solution set can become immeasurably large, as discussed by Benjamin & Mullin (1982). Practical difficulties associated with this large multiplicity, such as long term transient behaviour and the effects of external perturbations, will dominate. If we wish to study individual bifurcation sequences to examine codimension-2 points then an essential step is to restrict the physical size of the problem and thus reduce the rich multiplicity found at large aspect ratios. It could be argued that restricting the number of spatial modes, by virtue of limiting the size of the experiment, will enforce low-dimensional behaviour and produce special types of dynamical motion which are of limited scope. However, it has been shown by Mullin (1993a) that both high- and low-dimensional dynamics can co-exist in a small enclosed flow so that the infinite-dimensional aspect of the Navier–Stokes equations is retained in spatially restricted flows.

Research on the Taylor–Couette problem has shown that pitchfork bifurcations do exist in the solution set, as discussed by Mullin & Cliffe (1986). These arise because of the reflectional symmetry of the flow domain about the horizontal mid-plane. Cellular flows that appear with quasistatic increase of the Reynolds number from zero are always symmetric about the mid-plane. However, it is found both experimentally and computationally that these initially symmetric flows typically can become asymmetric in some regions of parameter space. When this occurs, cells in one half of the domain grow at the expense of cells in the other half.

A perfect symmetric pitchfork bifurcation is a mathematical idealisation. In any physical problem there will always be small imperfections which break the perfect Z_2 symmetry of the domain, and this results in the disconnection of one of the asymmetric solution branches. The disconnection is observed to be on the order of a few per cent, which is presumably because the geometrical imperfections are only small perturbations away from the ideal case. Thus, the elements of the pitchfork bifurcation remain locally complete. One state evolves smoothly and the other terminates nearby in a saddle-node or fold bifurcation. This is in complete contrast to the case for the onset of cells where the bifurcation is effectively destroyed. Thus it is extremely important to recognise the relevant physical symmetries of a problem and to treat those associated with mathematical models with care.

Upon increasing the Re, the first appearance of time-dependent motion after the onset of cells is normally a travelling wave. This was first investigated in detail by Coles (1965). Davey *et al.* (1968) applied weakly-nonlinear stability theory to the case of infinitely long cylinders to show that steady cellular flow becomes unstable to a travelling wave by means of a Hopf bifurcation. Unlike the pitchfork bifurcation, the Hopf bifurcation is structurally stable to perturbations, and thus it has the same form in the infinite-cylinder model as in the experiment. This appears to be true in practice despite the small amount of forcing that arises from imperfections such as lack of perfect circularity of the two cylinders.

Having established that both symmetry-breaking and Hopf bifurcations are observable in Taylor–Couette flow we now have the ingredients for codimension-2 organising centres for chaotic dynamics. The fact that the steady-state bifurcation is not from a trivial flow makes theoretical studies of such phenomena extremely complicated. Nevertheless, normal form analysis could, in principle, allow progress to be made in producing low-order models from the Navier–Stokes equations at the codimension-2 point.

We now give an example of chaotic dynamics in a particular realisation of the Taylor–Couette problem, where the presence of an organising centre in parameter space appears to be the crucial factor. The boundary conditions in this case are slightly different to those that are normally applied. It is still the case that there are a stationary outer cylinder and a rotating concentric inner cylinder. However, the end plates which bound the domain this time rotate with the inner cylinder instead of being stationary. The example of the interaction of Hopf and saddle-node bifurcations is provided by Mullin *et al.* (1989), who studied the mechanism for production of quasiperiodicity and irregularity in a flow which consists of four Taylor cells. Numerical bifurcation techniques were applied to discretised versions of the Navier–Stokes equations with the appropriate boundary conditions. These calculations showed that the initially symmetric four-cell flow undergoes a symmetry-breaking bifurcation as the Reynolds number is increased. It is a more complicated bifurcation than the one discussed previously, since it is subcritical at the point of bifurcation. The unstable asymmetric branches which arise at the bifurcation

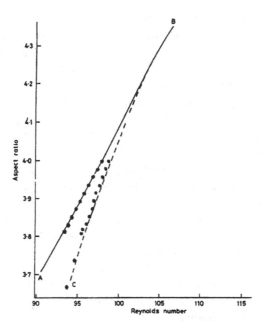

Figure 3. Bifurcation set near a codimension-2 point with four cells. AB is a path of saddle-nodes and BC is the locus of Hopf bifurcations.

point turn round at saddle-node bifurcations at a lower value of Re than that associated with the symmetry-breaking bifurcation. Thus stable asymmetric four-cell states are formed at the saddle-node or limit point.

If the system is initialised such that the flow is on the steady asymmetric state just above the saddle-node point then the computations show that an increase of Re gives rise to a Hopf bifurcation, where a travelling wave with azimuthal wavenumber $m = 1$ is created. It is found that as the aspect ratio is increased from $\Gamma \approx 3.7$, the critical Reynolds numbers for the saddle-node and Hopf bifurcations approach each other and become coincident at $Re = 106.2$, $\Gamma = 4.34$. This is shown in Figure 3, where the numerically determined loci of the saddle-node (AB) and Hopf bifurcations (BC) are plotted together with some experimental points. Point B therefore represents the location of a codimension-2 bifurcation in this particular problem, involving the simultaneous occurrence of two structurally stable codimension-1 bifurcations. The experiments are very difficult to perform in this parameter regime and hence it was not possible to get close to the codimension-2 point in practice. Nevertheless, we have been able to establish the solution structure close to the organising centre and would therefore expect to observe robust global dynamics nearby.

From the experimental and numerical results we can infer that a low-order model of the following type ought to apply in the neighbourhood B.

$$
\begin{aligned}
\dot{X} &= XZ - WY, \\
\dot{Y} &= XW + YZ, \\
\dot{Z} &= P + Z - \frac{1}{3}Z^3 - (X^2 + Y^2)(1 + QX + EZ),
\end{aligned}
$$

where $W = 10$, $E = 0.5$, $Q = 0.7$, and P is the bifurcation parameter. The parameter values have been chosen empirically as discussed below.

The experimental flow in the neighbourhood of B consists of two different frequencies and so, dynamically, the motion takes place on a 2-torus. The first frequency corresponds to a travelling tilt wave of wavenumber one, which breaks the continuous $SO(2)$ symmetry of the annular domain. The second frequency is lower and is a wave of wavenumber zero which is in phase at all points around the domain. A time-series of a measured velocity component in this state contains discrete packets of the high frequency separated by relatively long periods of approximately quiescent behaviour. The quiescent periods are equally spaced when the flow is regular but when the motion is chaotic, their lengths show considerable variation.

In Figure 4 we show a comparison between experimental observations and results from numerical integrations of the model equations. On the left-hand side the figure contains attractors reconstructed from experimental time series using the technique of phase portrait reconstruction, as discussed by Broomhead & Jones (1989). For comparison, corresponding attractors from the model are presented on the right-hand side. If the Reynolds number is in excess of 133.0 the motion is quasiperiodic and the attractor is a 2-torus. Reducing the Re to 132.7 then the attractor shown in Figure 4a is obtained. There are still two frequencies present but they have become commensurate at a ratio 6:1 and the motion then takes place on a complicated limit cycle. This feature can clearly be seen in Figure 4b which is reconstructed from data obtained by numerically integrating the model equations.

The virtue of investigating the experimental dynamics in this geometrical way is that it allows one to identify possible low-dimensional mechanisms that are responsible for controlling the observed behaviour. In this particular case, it would appear that a torus is approaching homoclinicity to an unstable fixed point at the left-hand end of the central core of trajectories. A high concentration of data points at this end of the core indicates that the flow

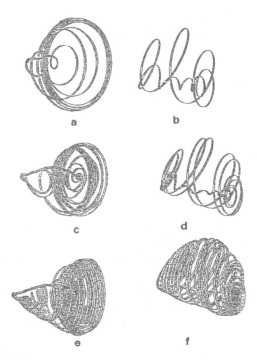

Figure 4. A period doubling sequence to chaos on a near homoclinic torus. The left-hand attractors were reconstructed from experimental time-series and the right-hand set are from the model.

spends a relatively long time in this region of phase space. This feature is
more noticeable in Figures 4c and 4e. Animation of the trajectories shows
that the fixed point is weakly attracting along the nearly one-dimensional
core. However it is also seen to be strongly unstable with respect to spiralling
motion in a plane orthogonal to the core. This is indicative of near homoclinic
motion on a torus which is of the same type as that investigated in a particular
normal form model by Langford (1984).

The type of dynamical motion described is commonly referred to as the
Sil'nikov mechanism for the onset of chaos. A detailed description of this
low-dimensional mechanism can be found in Guckenheimer & Holmes (1986).
It involves motion in phase space that passes close to a saddle point which
has a stable and an unstable manifold associated with it. In the particular
case studied by Sil'nikov (1965) the saddle point is such that the motion close
to the fixed point on one of the manifolds is a spiral on a sheet and the other

is a line orientated normal to the sheet. Once a trajectory has been ejected along the unstable manifold, it is then reinjected via the stable manifold. A bifurcation parameter controls this reinjection process, and it is possible for an orbit to become homoclinic to the saddle point.

The data sets shown in Figure 4 suggest that the dynamics of the flow are close to conditions of homoclinicity, since the period of motion around the attractor is approximately ninety to one hundred times the rotation period of the inner cylinder. When the Reynolds number was then reduced to 129.6, the locked torus was observed to period-double. The model is three-dimensional and so the locking process is an essential part of the transition to chaos. However, period doubling is a generic feature of Sil'nikov orbits, as shown by Glendinning & Sparrow (1984). It was also shown by Healey *et al.* (1991) to have relevance to a nonlinear electronic oscillator, where the underpinning solution structure provided an organising centre for the observed dynamics.

Further reduction in Re to 127.9 produces chaotic motion. Hence, these observations represent a strong indication for a connection between the Navier–Stokes equation of fluid motion and properties of finite-dimensional dynamical systems of a nontrivial kind. However, there remain outstanding problems before definite statements can be made. The normal form used in the model has not been systematically derived from the Navier–Stokes equations using Lyapunov–Schmidt reduction techniques, and so there are coefficients in the model which have not yet been completely identified. Some of them have been specified from a reduction of the discretised Navier–Stokes equations used in the numerical studies but they involve combinations of the two control parameters Re and Γ (Cliffe (1999)). In addition, empirical information from the experimental data can be used to estimate the order of magnitude and the signs of some coefficients but their exact values remain unknown. In any case, the model is strictly only valid at the codimension-2 point and all of the observations discussed above were taken a finite distance away from the multiple bifurcation point.

The role played by the continuous azimuthal symmetry of the Taylor–Couette problem in predisposing the system to display low-dimensional behaviour has been investigated. The approach was to replace the stationary circular outer cylinder of the original geometry with one of non-circular cross-section, thereby breaking the continuous azimuthal symmetry. In this system pure travelling waves are no longer permitted, and it is therefore of interest to see what forms of dynamical behaviour occur instead. Kobine & Mullin (1994) chose the outer boundary to have a square cross-section, thereby breaking the continuous azimuthal symmetry. Mathematically, this results in the continuous $SO(2)$ symmetry group of the original problem being replaced with a discrete Z_4 group. The robustness of steady vortices in such configurations was first pointed out by Terada & Hattori (1926). Dynamical phenomena

Figure 5. A reconstructed attractor from experimental time-series
measured between a rotating round cylinder in a square outer.
$\Gamma = 1.25, Re = 776$.

were investigated by Kobine & Mullin (1994) in the regime of very small aspect ratios ($\Gamma \sim 1$), where the flow typically consists of just one cell. The
onset of time-dependence with increasing Reynolds number was shown to
have all the characteristics required of a Hopf bifurcation. Loci of several different codimension-1 bifurcations, including Hopf bifurcations, were mapped
out experimentally in the control plane formed by aspect ratio and Reynolds
number. Amongst those, a line of secondary Hopf bifurcations and a line of
periodic folds (the time-dependent equivalent of a saddle-node bifurcation)
were found to meet, forming a codimension-2 bifurcation. At parameter values close to this point dynamical behaviour was found that gave rise to the
phase portrait shown in Figure 5. The qualitative nature of Sil'nikov dynamics is displayed clearly. However, unlike the standard Taylor–Couette
problem where cellular structure is evident on average at all supercritical
Reynolds numbers, the square box flow is featureless for $Re \approx 1200$.

3   Side-wall convection in liquid gallium

The flow we consider here is convection driven by a horizontal temperature
gradient applied to a liquid metal, gallium. The problem originates in a
horizontal Bridgman crystal growth geometry where a crucible containing a

crystal melt is slowly removed from a furnace. A horizontal temperature gradient is therefore established between the melt and the solid which gives rise to buoyancy-driven convection. In most practical situations the motion so induced is found to be highly disordered or even turbulent. It is known that these flows can produce irregular distributions of dopant called striations in the crystallised host material, and this inhomogeneity is undesirable if one wants to grow good quality semiconductor crystals. Hence there is considerable interest in methods of suppressing or controlling fluid convection in this system. One method of controlling the convection is to apply an external magnetic field, as reviewed by Series & Hurle (1991), which induces an electromotive field that can be non-uniform in regions in the melt. Thus electrical currents can flow, and these interact with the applied magnetic field to damp the convective motion.

The pioneering experimental work on this problem was carried out by Hurle (1966) and Hurle et al. (1974). They observe thermal oscillations in the flow and show that a magnetic field applied orthogonally to the main convective flow can be used to damp the time-dependence. However, a more recent investigation by McKell et al. (1990) reports that an applied magnetic field may also promote instabilities, including low-dimensional chaos. Hence there is clearly a need for an in-depth study of the interaction between the flowing conductor and the applied magnetic field to progress our understanding at a fundamental level.

The three main parameters which govern the MHD flow are the Grashof number, $Gr = \beta\gamma gh^3/\nu^2$, the Prandtl number, $Pr = \nu/\kappa$, and the Hartmann number, $Ha = B_0 h\sqrt{\sigma/\nu\rho}$. The various terms in these control parameters, β, g, h, ν, κ, B_0 and ρ, refer to the coefficient of thermal expansivity, the acceleration of gravity, the height of the enclosure, the kinematic viscosity, the thermal diffusivity, the applied magnetic induction and the density respectively. $\gamma = (T_h - T_c)/A_x$ is the applied temperature difference where T_h refers to the temperature at the hot end, T_c is the temperature at the cold end and A_x is the horizontal length scale. Gr gives a measure of the relative importance of buoyancy to viscous forces and corresponds to the non-dimensional temperature gradient. Ha is a ratio between the Lorentz and viscous forces and is proportional to the strength of the imposed magnetic field. The third relevant quantity, the Prandtl number, Pr, is the ratio of viscous to thermal diffusivities. It is approximately zero for gallium, which is a liquid metal above 29.8°C, and hence heat transport principally takes place by conduction. However, as pointed out by Braunsfurth & Mullin (1996) the qualitative form of observed oscillatory motion can depend sensitively on its precise value. Thus it is important that a finite value of Pr is used in any model of the flow, rather than the mathematically convenient, but singular, limit of zero.

A schematic diagram of the apparatus is shown in Figure 6. It consisted of

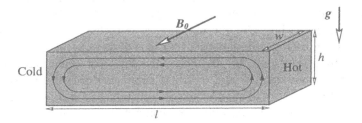

Figure 6. Schematic of the gallium experiment with applied magnetic field. Hot fluid rises at the end marked hot and falls at
the cold end. The magnetic field is applied orthogonally to the
convective circulation.

---

an insulating rectangular channel which held the liquid gallium between two
thermally conducting molybdenum plates. The temperature of the end walls
was set using turbulent silicon oil baths, whose temperature was held constant
to within $\pm 0.05°$C by commercial temperature controllers. The experiment
was further enclosed in an air cabinet whose temperature was held constant at
$32 \pm 0.5°$C. The channel containing the gallium was thermally insulating and
was made from pyrophelite, whose thermal conductivity is 29 times smaller
than that of gallium. A non-conducting ceramic lid was fitted on top of the
gallium to form the upper boundary. The channel was thermally isolated
from its surroundings using a thick layer of insulating material.

The channel containing the gallium was centred between the 4-inch pole
pieces of an electromagnet, allowing steady fields of up to 1250 G to be applied using a stabilised DC power supply. Hartmann numbers in the range 0
to 64 could thereby be achieved. The magnetic field was found to be uniform
within $\pm 0.3\%$ of its mean value across the region occupied by the sample of
gallium. Temperatures were measured with type K insulated thermocouples
of diameter 0.25mm, which could be either accurately positioned within the
melt using a micromanipulator, or embedded within the side-walls. The precision achieved in the measurement of relative temperatures was better than
$\pm 0.01°$C.

Singly periodic oscillations were observed for $Ha = 0$ and found to occur
via a supercritical Hopf bifurcation at $Gr = 4.6 \times 10^4$. This was also found
to be the case with magnetic fields up to $Ha = 0.9$. For fields in excess

of this value the onset of oscillations was complicated and often involved direct transition to irregular motion which involved a hysteretic transition. Details can be found in Juel (1997). When the Hartmann number was set to $Ha = 4.7$, the first onset of time-dependent flow was hysteretic and gave rise to a quasiperiodic motion. The time-series then contained two frequencies. One was 7.6×10^{-2}Hz, and was found to be almost independent of Gr. The other comprised a large amplitude modulation whose frequency approached zero monotonically with increase of Gr.

These observations point to the existence of complicated dynamics in the form of a torus which becomes homoclinic to an unstable fixed point. As before, this dynamical phenomenon is remarkable as it combines local instability in the region of the fixed point with global stability as the periodic orbit reinjects the solution into the unstable region. It is often associated with the presence of chaotic dynamics in nearby regions of parameter space, and is of the same Sil'nikov type as discussed for Taylor–Couette flow. An example of this homoclinic behaviour is shown in the phase portrait presented in Figure 7. A time-series sampled for $Gr = 7.37 \times 10^4$, $Pr = 1.97 \times 10^{-2}$ and $Ha = 4.7$ was used to reconstruct the attractor. The trajectories spiral out rapidly from the unstable fixed point near one end of the attractor and are subsequently reinjected into the central region along the weakly attracting manifold along the axis.

4   Electrohydrodynamic convection in a nematic liquid crystal

A nematic is a complex fluid composed of rod- or disc-like molecules which possess orientational but not positional order. It has a viscosity which is typically five times that of water and so it can flow freely as a bulk fluid. However, complicated flow properties arise from the coupling between translational and orientational motions of the molecules. In typical liquid crystal device applications the fluid layers are less than the thickness of a human hair. The fluid molecules can be given a preferred orientation by treating the bounding surfaces which is referred to as the director n. It is this alignment which gives the fluid strongly anisotropic material and optical properties and the latter make them useful as optical display materials.

When an electric field is applied across a thin layer of nematic material between a pair of conducting electrodes cellular convection can occur above a critical field strength. It originates from the motion of free ions in the fluid which move preferentially due to the anisotropic conductivity and permittivity. The onset of cellular convection in a nematic liquid crystal is commonly referred to as electroconvection (Kramer & Pesch (1996)) and, in many ways,

Figure 7. Reconstructed phase portrait measured at $Pr = 1.97 \times 10^{-2}, Gr = 7.37 \times 10^4$ and $Ha = 4.7$.

---

is analogous to Bénard convection in heated fluid flayers. Associated with the convection is a distortion of the alignment of the material which is manifest as a spatially varying refractive index. This enables flow to be observed directly as an intensity pattern formed by the focusing and defocusing of transmitted light.

Experimental investigations of electroconvection are often concerned with large aspect ratio systems containing many convection cells. Under these conditions it is assumed that one can describe the onset of cellular convection, and subsequent transitions, using general arguments about pattern formation in nonequilibrium systems. As in the Taylor–Couette problem discussed above, the large number of states available makes it very difficult to investigate low-dimensional dynamics in any detail. The potential for such behaviour is particularily obscured in liquid crystal flows because thermal fluctuations are important in these small scale flows and can cause the system to jump between the many coexisting states. However, small aspect ratio electro-convection experiments can exhibit low-dimensional behaviour as shown by Tsuchiya *et al.* (1988). Despite this observation, relatively little work has been done to explore the possibility of organised dynamics within origin of this complex time-dependency.

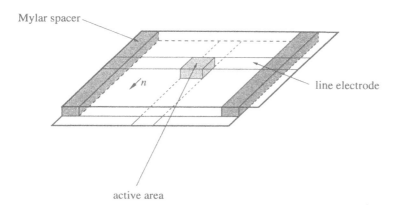

Figure 8. Schematic of the liquid crystal experiment. The director
n is indicated. The cell was full of material but only the shaded
region between the two electrodes was active.

The liquid crystal cell comprised a $46 \pm 1.0\mu$m thick layer of nematic liquid
crystal BDH–17886 sandwiched between two optically flat glass plates, and
a schematic of the cell is presented in Figure 8. An indium–tin oxide line
electrode of thickness $\sim 185\mu$m was etched onto the inner surface of each
plate and the arrangement was such that, when viewed from above, the lines
overlapped at right angles. This created an active region of aspect ratio
4:4:1 to which an electric field could be applied. The largest dimension of
the active area was equivalent to only 10^5 molecular lengths and comparable
to the width of a human hair. On such a small scale inherent microscopic
fluctuations are known to cause large variation in continuum quantities such
as the director (the unit vector field describing the macroscopic average of
molecular orientation), as discussed by De Gennes & Prost (1995). Alignment
of the material, which was parallel to the lower electrode, was obtained using a
rubbed layer of Poly-Vinyl Alcohol (PVA) spin coated on top of the electrodes.
The cell was mounted on a microscope translation stage and maintained at a
constant temperature of $32.0 \pm 0.02°$C. Applied AC voltages were of the order
of 10V rms at ~ 600 Hz with a long term stability of better than 0.5%. Light
transmitted through the cell was imaged using a CCD camera and analysed
using a computer system.

An image of eight-cell flow in the liquid crystal cell is presented in Figure

Figure 9. A top view of an eight-cell steady state of electrohydrodynamic convection where the bright lines indicate upward
convection.

---

9. Within a given parameter range this flow was primary, i.e. it smoothly
evolved from the undisturbed nematic as the voltage, v, was continuously
increased or the frequency, f, was continuously reduced. In neighbouring
regions of parameter space the primary flow comprised six or ten convection
cells.

Systematic variation of v and f was used to measure a two-dimensional experimental bifurcation set. The steady eight-cell flow was found to undergo
a mixed-mode oscillation between eight and ten cells. An attractor reconstructed from mixed-mode flow is presented in Figure 10. Here trajectories
are ejected vertically from the region surrounding an unstable fixed point
and then spiral back in. Consistent with the abstract model of Sil'nikov,
the period of mixed-mode oscillation was found to increase montonically by
an order of magnitude as control parameters were varied and a transition to
seemingly chaotic dynamics was observed.

5   Conclusions

The ideas discussed in this review have proved to be very successful in understanding complicated fluid flows from three diverse problems even when
the steady states of the governing equations are at the limits of calculation.
The same qualitative low-dimensional dynamics have been identified in each
case which gives credence to idea that organising centres play a central role
in understanding the global dynamics of fluid systems. Outstanding challenges remain for a formal connection between the Navier–Stokes equations

Figure 10. Reconstructed attractor for a nematic liquid crystal flow oberved near the interaction of eight and ten cells.

Figure 11. Flow visualisation of 'turbulent flow' in square box apparatus at $\Gamma = 4.0$, $Re = 1230$.

and low-dimensional models but the experimental evidence for robust chaotic dynamics is now overwhelming and the ideas can be used with confidence to make progress with flows of both scientific and practical interest. However, we finish with a photograph in Figure 11 of a disordered flow at a modest Reynolds number where there is no evidence for low-dimensional chaos. It serves to remind us that exciting scientific challenges remain.

References

Benjamin, T.B. (1978) Bifurcation phenomena in steady flows of a viscous fluid. *Proc. R. Soc. Lond.* **A359**, 1–26, 27–43.

Benjamin, T.B. & Mullin, T. (1982) Notes on the multiplicity of flows in the Taylor–Couette experiment. *J. Fluid Mech.* **121**, 219–230.

Braunsfurth, M.G., & Mullin, T. (1996) An experimental study of oscillatory convection in liquid gallium. *J. Fluid Mech.* **327**, 199–219.

Brandstater, A. & Swinney, H.L. (1987) Strange attractors in weakly turbulent Couette–Taylor flow. *Phys. Rev.* **A35**, 2207–2220.

Broomhead, D.S. & Jones, R. (1989) Time-series analysis. *Proc. R. Soc. Lond.* **A423**, 103–121.

Cliffe, K.A. (1999) Private communication.

Cliffe, K.A., Kobine, J.J. & Mullin, T. (1992) The role of anomalous modes in Taylor–Couette flow. *Proc. R. Soc. Lond.* **A439**, 341–357.

Coles, D. (1965) Transition in circular Couette flow. *J. Fluid Mech.* **21**, 385–425.

Davey, A., DiPrima, R.C. & Stuart, J.T. (1968) On the instability of Taylor vortices. *J. Fluid Mech.* **31**, 17–52.

DiPrima, R. & Swinney, H.L. (1981) In *Hydrodynamic Instabilities and the Transition to Turbulence*, Springer.

De Gennes, P.G. & Prost, J. (1995) *The Physics of Liquid Crystals*, Oxford University Press.

Glendinning, P. & Sparrow, C. (1984) Local and global behaviour near homoclinic orbits. *J. Stat. Phys.* **35**, 645–696.

Gollub, J.P. & Swinney, H.L. (1975) Onset of turbulence in a rotating fluid. *Phys. Rev. Lett.* **35**, 927–930.

Guckenheimer, J. & Holmes, P. (1983) *Nonlinear Oscillations, Dynamical Systems and Bifurcations of Vector Fields*, Springer Applied Mathematical Sciences **42**.

Healey, J.J., Broomhead, D.S., Cliffe, K.A., Jones, K.A. & Mullin, T. (1991) The origins of chaos in a modified Van der Pol oscillator, *Physica D* **48**, 322–339.

Hurle, D.T.J. (1966) Temperature oscillations in molten metals and their relationship to growth striae in melt-grown crystals. *Phil. Mag.* **13**, 305–310.

Hurle, D.T.J., Jakeman, E., & Johnson, C.P. (1974) Convective temperature oscillations in molten gallium. *J. Fluid Mech.* **64**, 565–576.

Joseph, D.D. (1985) In *Hydrodynamic Instabilities and the Transition to Turbulence*, Springer.

Juel, A. (1997) D. Phil. Thesis, Oxford University.

Kobine, J.J. & Mullin, T. (1994) Low-dimensional bifurcation phenomena in Taylor-Couette flow with discrete azimuthal symmetry. *J. Fluid Mech.* **275**, 379–405.

Kramer, L. & Pesch, W. (1996) In *Pattern Formation in Liquid Crystals*, Springe.

Langford, W.F., (1984) Numerical studies of torus bifurcations. *Int. Ser. Num. Math.* **70**, 285–93.

Lorenzen, A. & Mullin, T. (1985) Anomalous modes and finite-length effects in Taylor–Couette flow. *Phys. Rev. E* **31**, 3463–3465.

McKell, K.E., Broomhead, D.S., Jones, R., & Hurle, D.T.J. (1990) Torus doubling in convecting molten gallium. *Europhys. Lett.* **12**, 513–518.

Mullin, T. (1993a) Disordered fluid motion in a small closed system. *Physica D* **62**, 192–201.

Mullin, T. (1993b) *The Nature of Chaos*, Oxford University Press.

Mullin, T. & Cliffe, K.A. (1986) In *Nonlinear Phenomena and Chaos*, Adam Hilger.

Mullin, T. & Kobine, J.J. (1996) In *Nonlinear Mathematics and its Applications*, Cambridge University Press.

Mullin, T. & Price, T.J. (1989) An experimental observation of chaos arising from the interaction of steady and time-dependent flows. *Nature* **340**, 294–296.

Mullin, T., Tavener, S.J. & Cliffe, K.A. (1989) An experimental and numerical study of a codimension-2 bifurcation in a rotating annulus. *Europhys. Lett.* **8**, 251–256.

Pfister, G. (1985) Deterministic chaos in rotational Taylor–Couette flow. In *Flow of Real Fluids* (ed. G.E.A. Meier & F. Obermeier). *Springer Lecture Notes in Physics,* **235**, 199–210.

Price T.J. & Mullin T. (1991) An experimental observation of a new type of intermittency, *Physica D* **48**, 29–52.

Series, R.W. & Hurle, D.T.J. (1991) The use of magnetic fields in semiconductor crystal growth. *J. Crystal Growth* **113**, 305–328.

Sil'nikov, L.P. (1965) A case of the existence of a denumerable set of periodic motions. *Sov. Math. Dokl.* **6**, 163–166.

Taylor, G.I. (1923) Stability of a viscous liquid contained between two rotating cylinders. *Phil. Trans. R. Soc. Lond.* **A223**, 289–343.

Terada, T. & Hattori, K. (1926) Some experiments on the motions of fluids. Part 4. Formation of vortices by rotating disc, sphere or cylinder. *Rep. Tokyo Univ. Aeronaut. Res. Inst.* **2**, 287–326.

Tsuchiya, Y., Horie, S. & Itakura, H. (1988) Coherent pattern oscillations (OS2) of Williams domains in nematic liquid crystal. *J. Phys. Soc. Jap.* **57**, 669–670.

Phase Turbulence and Heteroclinic Cycles

F.H. Busse, O. Brausch, M. Jaletzky and W. Pesch

Abstract

Various cases of phase turbulence in convection layers heated from below are reviewed. In several cases a close connection between the onset of phase turbulence and the existence of a heteroclinic orbit in a reduced system of equations can be found. New results are presented for phase turbulence in the case of centrifugally driven convection in a rotating cylindrical annulus.

1 Introduction

Phase turbulence is observed in many extended fluid systems that are characterized by supercritical or only weakly subcritical bifurcations from the uniform basic static state under steady external conditions. A large number of bifurcating solutions usually exists in the neighborhood of the critical value of the control parameter and the mathematical problem can be considered as an unfolding from a bifurcation point of infinite codimension. In the absence of a variational principle guaranteeing a unique asymptotic state the competing modes often give rise to a spatio-temporally complex state. In particular the phases of the flow at a given location appear to vary in a chaotic fashion.

The standard methods for analyzing pattern forming instabilities apply to large aspect-ratio systems, which can be idealized as infinitely extended. Therefore a description in terms of Fourier modes in dependence on two-dimensional wave vectors k is natural. In particular near onset of convection, i.e. when the main control parameter R, such as the Rayleigh number in Rayleigh–Bénard convection, is slightly beyond its critical value, almost perfect periodic patterns like rolls (characterized by a single wave vector), but also squares (two distinct wave vectors) or hexagons can be obtained. Besides the nearly periodic pattern, experiments often exhibit persistent spatio-temporal dynamics. Snapshots of the patterns in this case show local roll patches corresponding to wave vectors varying with respect to their direction while the absolute values of the wave vectors are nearly constant. The discontinuities in the wave vector field like grain boundaries or immersed point defects (dislocations) trigger the dynamics of the patterns. The notion 'phase turbulence' or 'weak turbulence' has been introduced for such states in distinct contrast to more fully developed turbulence.

Properties of Turbulence

Chaotic time dependence	Decay of spatial correlations	Broad wavenumber spectrum	Inertial range, fractal structure

Dynamical Systems
few degrees of freedom,
e.g. convection in a box

Phase Turbulence
many degrees of freedom, nearly degenerate bifurcation,
e.g. convection in extended layers with rotation

Classical Turbulence
Interaction with boundaries, e.g. turbulent pipe and
channel flow; thermal convection at high Rayleigh numbers

Asymptotic Turbulence
Inertial range scaling, e.g. atmospheric flows and other high Reynolds number systems

Table 1

Phase turbulence is thus a phenomenon that exhibits certain properties of fluid turbulence while others are missing. The reduced complexity of phase turbulent systems can therefore provide examples in which certain aspects of fluid turbulence can be studied in a relatively simple setting. Table 1 may serve to illustrate the place of phase turbulence within the general field of fluid turbulence.

One way of elucidating the mechanism of phase turbulence is the exploration of the stability of periodic patterns in the R-k-space. In fact, in the most important and representative cases phase turbulence originates from the nonexistence of stable periodic states immediately above onset. Historically the first example has been convection in a rotating Rayleigh–Bénard layer. Küppers and Lortz (1969) realized that all steady solutions describing convection flows in a horizontal fluid layer heated from below and rotating about a vertical axis are unstable when the rotation parameter Ω exceeds a critical value. They concluded that some kind of turbulent motion must be realized as a result. In later studies a close connection of the time-dependent states with a heteroclinic orbit was recognized (Busse and Clever, 1979a) and simple models based on this idea (Busse, 1984) could explain the experimental observations (Busse and Heikes, 1980; Heikes and Busse, 1980) quite well. Since then the system of rotating convection has become a favored example for the study of spatio-temporal chaos in pattern forming systems. Both experimental investigations (Zhong *et al.*, 1991; Zhong and Ecke, 1992; Bodenschatz *et al.*, 1992; Hu *et al.*, 1995, 1997, 1998) and numerical simulations (Tu and Cross, 1992; Fantz *et al.*, 1992; Neufeld and Friedrich, 1995; Millán-Rodriguez *et al.*, 1995; Pesch, 1996; Ponty *et al.*, 1997) have been employed to study various aspects of this phase turbulent system.

In the meantime other systems exhibiting phase turbulence were found. Zippelius and Siggia (1982) showed that Rayleigh–Bénard convection in a non-rotating layer with stress-free boundaries also exhibits the property that none of the existing infinite steady solutions is stable for sufficiently low Prandtl numbers and as a consequence phase turbulence must be expected. The mathematical analysis of Zippelius and Siggia was based on incorrect assumptions, but a later, more general analysis of Busse and Bolton (1984) confirmed that all solutions are indeed unstable albeit through different mechanisms of instability. The resulting phase turbulence has been studied in various papers (Busse, 1986; Busse and Sieber, 1991; Busse *et al.*, 1992; Xi *et al.*, 1997). But, unfortunately, there exists no possibility for a direct comparison with experiments since stress-free boundaries can be realized only for fluid layers with large Prandtl numbers (Goldstein and Graham, 1969).

Another system in which spatio-temporal chaos is easily observable is electroconvection in a layer of nematic liquid crystals. For experimental and theoretical studies of this system we refer to a recent review by Pesch and Behn (1998). Of special interest are cases where the primary bifurcation from the spatially homogeneous state occurs in the form of an oscillatory instability giving rise to traveling waves at threshold. It can happen that none of these wave states is stable owing to the Benjamin–Feir instability resulting in spatio-temporally disordered patterns (Dangelmayr and Kramer, 1998). Less well understood are the situations where phase turbulence cannot be related to an instability such as spiral defect chaos which has attracted much attention in recent years (Morris *et al.*, 1993).

It is common to make a distinction between phase turbulence and defect turbulence. The latter is characterized by the persistent spontaneous creation and extinction of defects which are defined as points where the phase jumps by 2π while the amplitude vanishes in an otherwise smooth pattern. But since experimentally realized phase turbulence is always associated with certain kinds of defects and grain boundaries, it is difficult to sustain the distinction for physical systems. The two ideal cases of spatio-temporal chaos generated solely by defects corresponding to phase singularities and the spatio-temporal chaos associated with entirely smoothly varying phase can be regarded as extreme cases of various possibilities for turbulence close to the critical value of the control parameter.

In the following we shall first discuss briefly the cases of variational and non-variational dynamics in Rayleigh–Bénard layers and then focus in the third section on the role of heteroclinic orbits. In the fourth section the newly

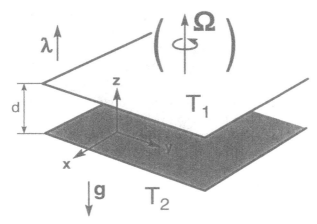

Fig. 1: Sketch of a horizontal fluid layer heated from below and possibly rotating about a vertical axis.

studied case of a Rayleigh–Bénard layer with 'horizontal' axis of rotation will be presented in which heteroclinic cycles also play a role. The paper closes with some concluding remarks in section 5.

2 Variational and Non-Variational Dynamics in Rayleigh–Bénard Convection

We consider a horizontal, infinitely extended fluid layer of height d heated from below as shown in figure 1. Because of the horizontal isotropy the onset of convection corresponds to a bifurcation of infinite codimension. The general solution of the linearized basic equations (see, for example, Busse, 1978) can be written in the form

$$w = f(z, \alpha) \sum_{n=-N}^{N} C_n \exp\{i\boldsymbol{k}_n \cdot \boldsymbol{r}\} \qquad (2.1a)$$

where the conventions

$$C_n = C_{-n}^{+}, \quad \boldsymbol{k}_{-n} = -\boldsymbol{k}_n, \quad |\boldsymbol{k}_n| = \alpha, \quad \boldsymbol{k}_n \cdot \boldsymbol{\lambda} = 0 \quad \text{for all } n \qquad (2.1b)$$

have been used and where a Cartesian system of coordinates (x, y, z) with z in the vertical direction has been assumed; $\boldsymbol{\lambda}$ is the vertical unit vector and C^{+} denotes the complex conjugate of C. Expression (2.1a) is given for the vertical velocity component. Analogous expressions hold for the other variables such as the temperature field. A special role is played by the regular distributions of $\boldsymbol{k}$-vectors for which all angles φ between neighboring vectors $\boldsymbol{k}_n$ assume the same value, $\varphi = \pi/N$. Patterns which are periodic in the plane are obtained in the special cases $N = 1, 2, 3$ which correspond to rolls (or stripes), squares and hexagons, respectively. In the general case N may tend to infinity in which case functions of the form (2.1) correspond to the class of almost periodic functions in the x-y-plane with fixed wavenumber.

The control parameter of the problem is the Rayleigh number R which is proportional to the applied temperature difference between bottom and top of the layer and which assumes a minimum as a function of wavenumber α for $\alpha_c = 3.116/d$ and $\alpha_c = \pi/d\sqrt{2}$ in the cases of no-slip and stress-free conditions at the boundaries, respectively. The corresponding critical values of the Rayleigh number are $R_c = 1707.76$ and $27\pi^4/4$. A solution of the form (2.1) is also obtained when the layer is rotating about a vertical axis, since it is possible to design experiments such that the centrifugal force is negligible while the Coriolis force plays an important role.

The arbitrary choice of the amplitudes C_n becomes restricted when the nonlinear problem is considered. For example, $|C_1|^2 = |C_2|^2 = \cdots = |C_N|^2$ must be satisfied for steady solutions corresponding to regular distributions of the vectors $\boldsymbol{k}_n$. These constraints on the coefficients C_n are obtained from the solvability conditions in the cubic order of the expansion of the basic equation in powers of the amplitude of convection. More general constraints are obtained when a time dependence of the coefficients on a long time scale is admitted. The solvability conditions give rise to evolution equations for the amplitudes $C_n(t)$ of the form

$$\frac{d}{dt}C_j^+ = (R - R_0)KC_j^+ + \beta \sum_{n,m=-N}^{N} C_n C_m \delta(\boldsymbol{k}_j + \boldsymbol{k}_n + \boldsymbol{k}_m)$$

$$- \left[\sum_{n=1}^{N} |C_n|^2 A(\boldsymbol{k}_j \cdot \boldsymbol{k}_n) + \Omega E(\boldsymbol{k}_j \cdot \boldsymbol{k}_n) \; \boldsymbol{\lambda} \cdot \boldsymbol{k}_j \times \boldsymbol{k}_n \right] C_j^+$$

$$+iM\nabla\psi \times \boldsymbol{\lambda} \cdot \boldsymbol{k}_j C_j^+ \qquad \text{for} \quad j = -N, \ldots, N, \quad (2.2a)$$

$$\left(\frac{\partial}{\partial t} - \frac{\partial^2}{\partial x^2} - \frac{\partial^2}{\partial y^2} \right) \psi = \sum_{r,p} \hat{q}_{rp} C_r C_p \exp\{i(\boldsymbol{k}_r + \boldsymbol{k}_p) \cdot \boldsymbol{r}\}. \qquad (2.2b)$$

Here K, β and M are constants while A and E are functions of $\boldsymbol{k}_i \cdot \boldsymbol{k}_n$ and β provides a measure for deviations from the Boussinesq approximation, i.e. for the temperature dependence of the viscosity etc., which tend to favor hexagonal convection cells. The terms underlined by a solid line enter the problem only in the case of stress-free boundary conditions in which case a z-independent mean flow described by a slowly varying stream function $\psi(x, y, t)$ can be generated. The term proportional to Ω and underlined by a dashed line enters the problem only in the case of a rotating layer where $\Omega = \Omega_D d^2/\nu$ is the rotation parameter made dimensionless with the thickness

d of the layer and the kinematic viscosity ν of the fluid. Ω_D is the dimensional angular velocity of rotation. Because the reflection symmetry of the basic equations with respect to vertical planes is lost in a rotating system, terms proportional to $\boldsymbol{\lambda} \cdot \boldsymbol{k}_j \times \boldsymbol{k}_n$ enter in addition to those depending only on inner products $\boldsymbol{k}_j \cdot \boldsymbol{k}_n$ between k-vectors. Without the underlined terms the right hand side of expressions (2.2a) can be written as derivatives of a Lyapunov functional $F(C_{-N}, \ldots, C_N)$,

$$\frac{d}{dt}C_j^+ = -\frac{\partial}{\partial C_j}F(C_{-N}, \ldots, C_N) \qquad \text{for } j = -N, \ldots, N \qquad (2.3a)$$

with

$$F(C_{-N}, \ldots, C_N)$$
$$\equiv -\frac{1}{2}(R - R_0)K \sum_{j=-N}^{N} |C_j|^2 - \frac{1}{3}\beta \sum_{j,n,m} C_j C_n C_m \delta(\boldsymbol{k}_j + \boldsymbol{k}_n + \boldsymbol{k}_m)$$
$$+\frac{1}{4}\sum_{n,j} A(\boldsymbol{k}_j \cdot \boldsymbol{k}_n)|C_n|^2|C_j|^2. \qquad (2.3b)$$

The variational dynamics expressed by these equations guarantees the existence of at least one stable stationary solution of the problem corresponding to a minimum of the functional F. But, of course, there may be more than one attractor corresponding to more than one local minimum of F, as happens in the case of the competition between rolls and hexagonal convection cells (Busse, 1967).

The property (2.3) disappears in problems of Rayleigh–Bénard convection with rotation about a vertical axis or with stress-free boundaries and indeed it can be demonstrated that all steady solutions are unstable if either Ω is sufficiently large (Küppers and Lortz, 1969) in the former case or the Prandtl number P is less than 0.543 in the case of stress-free boundaries (Busse and Bolton, 1984; see also Mielke, 1997). In other cases of Rayleigh–Bénard convection the steady attracting flow also does not remain stable as the Rayleigh number R is increased much beyond the critical value due to higher order terms which can no longer be neglected in equations (2.2). But except for the two special cases mentioned above there always exists a region close to the critical value of the Rayleigh number where the variational dynamics described by equations (2.3) is applicable.

3 Phase Turbulence in a Convection Layer Rotating about a Vertical Axis

In the case of a rotating layer as shown in figure 1 the phase turbulence assumes a particularly simple form and the connection to a heteroclinic cycle

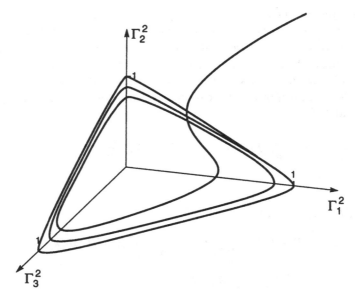

Fig. 2: Sketch of a trajectory in the phase space spanned by the amplitudes $\Gamma_i^2, i = 1, 2, 3$, which approaches the heteroclinic orbit connecting the fixed points $\Gamma_i^2 = \delta_{ij}$ with $j = 1, 2, 3$.

is most evident. As shown by Küppers and Lortz (1969), the growing disturbance of steady rolls, which is the only stable steady solution in the case $\beta = 0$, $\Omega < \Omega_c$, assumes the form of rolls oriented with an angle of about $60°$ with respect to the given rolls when Ω exceeds Ω_c. Because of this property it is sufficient to use a set of three rolls for a simple model of the time dependence of convection (Busse and Clever, 1979a). The corresponding evolution equations are

$$\frac{d}{d\tau}\Gamma_1 = (1 - \Gamma_1^2 - \xi\Gamma_2^2 - \gamma\Gamma_3^2)\Gamma_1,$$

$$\frac{d}{d\tau}\Gamma_2 = (1 - \Gamma_2^2 - \xi\Gamma_3^2 - \gamma\Gamma_1^2)\Gamma_2, \tag{3.1}$$

$$\frac{d}{d\tau}\Gamma_3 = (1 - \Gamma_3^2 - \xi\Gamma_1^2 - \gamma\Gamma_2^2)\Gamma_3,$$

where the variables Γ_n are rescaled versions of the amplitudes C_n and τ is identical with the time t except for a constant factor. Without losing generality we may assume real variables Γ_n. The system (3.1) of equations has eight fixed points all of which are unstable for $\gamma + \xi > 2$ when either $\gamma < 1$ or $\xi < 1$ holds. The same system (3.1) of equations was first used in the context of population biology by May and Leonard (1975) and some mathematical properties are discussed in their paper. ¿From arbitrary initial conditions the trajectory in the space spanned by coordinates $\Gamma_1^2, \Gamma_2^2, \Gamma_3^2$ approaches a heteroclinic cycle in the plane $\Gamma_1^2 + \Gamma_2^2 + \Gamma_3^2 = 1$ as indicated in figure 2.

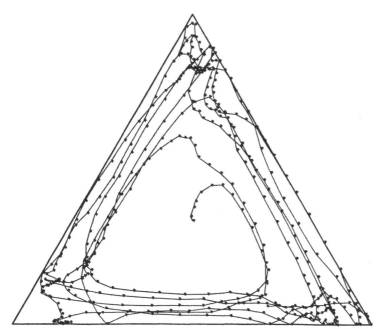

Fig. 3: Trajectory of the statistical limit cycle described by equations (3.1) with superimposed noise (see Busse, 1984) in the triangular domain given by the three corners $\Gamma_i^2 = \delta_{ij}, j = 1, 2, 3$. The trajectory starts at the unstable fixpoint $\Gamma_1 = \Gamma_2 = \Gamma_3$ in the center of the triangle. The stars are placed at equal time intervals.

As the heteroclinic cycle is approached the unphysical feature of an ever increasing period becomes apparent. In realistic situations the interaction of patches of rolls with varying orientations will create disturbances which will prevent the ultimate approach to the heteroclinic orbit. The effect of the disturbances can be modeled through the addition of noise in the system (3.1) of equations as was done by Busse (1984). A typical picture of the conversion of the heteroclinic cycle into a 'statistical limit cycle' is shown in figure 3. The average frequency of this cycle is proportional to

$$(R - R_c)(1 - \gamma)(\log(1/\eta))^{-1}$$

where $R - R_c$ is the excess of the Rayleigh number over its critical value, $1 - \gamma$ is proportional to the growth rate of the Küppers–Lortz instability in the case $\gamma < 1$ and η is a typical amplitude of the disturbances generated by white noise.

The statistical limit cycle concept is useful for the understanding of the local replacement of rolls by other rolls differing in their orientation by approximately $60°$ as seen in experiments with high Prandtl number fluids. An

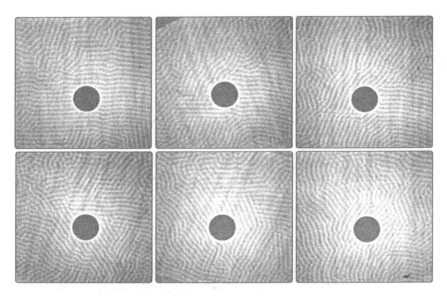

Fig. 4: Phase turbulent convection induced by the Küppers–Lortz instability in a horizontal layer of methyl alcohol of thickness $d =$ 3.3mm rotating about a vertical axis. The shadowgraph pictures have been taken 2 minutes apart in the clockwise sequence (upper row left to right, then lower row right to left). The central circle originates from the cooling water channel at the top of the layer and does not interfere with the pattern dynamics. Rolls tend to be replaced by other rolls turned counterclockwise by about 60°. (For details see Heikes and Busse, 1980; Busse and Heikes, 1980.)

example from the work of Heikes and Busse (1980) is shown in figure 4. The disturbance amplitude η can be interpreted as the influence of neighboring patches and η^{-1} thus is roughly proportional to the size of the patches which increases as the critical Rayleigh number is approached. Through the inclusion of gradient terms in the system (3.1) non-local effects can be taken into account (Tu and Cross, 1992) and a statistical limit cycle can be found without the inclusion of noise. In the case of rotating convection layers with Prandtl numbers of the order of unity the angle between rolls and the most strongly growing disturbance rolls becomes smaller than 60° and the dynamics can no longer be described by as simple a system as (3.1). In spite of a smaller angle of the strongest growing disturbance a predominance of rolls differing by 60° in orientation can still be noticed in the experimental measurements (Hu *et al.*, 1995, 1998).

Heteroclinic cycles also play a role in the other case of phase turbulence

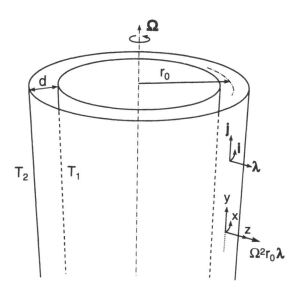

Fig. 5: Geometrical configuration of an annular convection layer with 'horizontal' axis of rotation.

occurring at the critical value of the Rayleigh number. A discussion of the dynamical features leading to phase turbulence in a convection layer with stress-free boundaries has been given by Busse *et al.* (1992). But, as has already been mentioned, a comparison between theory and experiment does not seem to be possible in this case.

4 Phase Turbulence in a Convection Layer with Horizontal Axis of Rotation

Another way of realizing buoyancy driven motions is to use the centrifugal force in place of gravity. By cooling an inner rotating cylinder and heating an outer co-rotating one a Rayleigh–Bénard convection layer is realized with the Rayleigh number given by

$$R = \frac{\gamma(T_2 - T_1)\Omega_D^2(r_1 + r_2)d^3}{2\nu\kappa}, \qquad (4.1a)$$

where γ is the thermal expansivity, T_2 and T_1 are the temperatures at which the outer and inner cylindrical boundaries are kept and r_2 and r_1 are the corresponding radii, d is the gap width, $d = r_2 - r_1$, Ω_D is the rotation rate, ν is the kinematic viscosity and κ is the thermal diffusivity. A sketch of the geometrical configuration is shown in figure 5. There are two additional di-

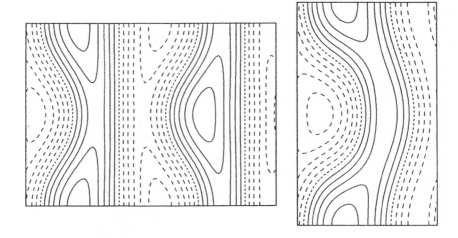

Fig. 6: Hexarolls (left) for $R = 2300$ with $\boldsymbol{k}_1 = (3.117, 0)$, $\boldsymbol{k}_{2,3} = (-1.559, \pm 2.2)$ and knot convection (right) for $R = 2400$ with $\boldsymbol{k}_1 = (3.117, 0)$, $\boldsymbol{k}_2(0, 2.2)$. $P = 1000, \Omega = 20$ in both cases. Solid (dashed) lines indicate positive (negative) isotherms in the midplane $z = 0$ of the fluid layer.

mensionless parameters, the rotation parameter Ω and the Prandtl number P,

$$\Omega = \Omega_D d^2 / \nu, \qquad P = \frac{\nu}{\kappa}, \tag{4.1b}$$

if we restrict our attention to the small gap limit, $d \ll r_1$. The idea is to minimize the effect of the Coriolis force by keeping Ω small while making the centrifugal force sufficiently large that it exceeds gravity by a good margin. This constraint can be satisfied for laboratory experiments through the use of high Prandtl number fluids such as highly viscous silicone oils.

On the theoretical side the assumption of a small Ω permits the consideration of the problem as an unfolding from the bifurcation with infinite codimension in the isotropic limit of $\Omega = 0$. We can thus use the formulation of the problem (2.1), (2.2) and restricting ourselves to the case of rigid boundaries with $\beta = 0$ we arrive at the equations

$$\frac{d}{dt} C_l^+ = (R - R_c - \tau^2 (\boldsymbol{k}_l \cdot \boldsymbol{j})^2 \alpha^{-2}) K C_l^+ - \sum_{n=1}^{N} |C_n|^2 A(\boldsymbol{k}_l \cdot \boldsymbol{k}_n) C_l^+$$

$$+ i\Omega \sum_{n,m=-N}^{N} \delta(\boldsymbol{k}_l + \boldsymbol{k}_n + \boldsymbol{k}_m) B C_n C_m (\boldsymbol{k}_n \times \boldsymbol{k}_m \cdot \boldsymbol{\lambda}) \boldsymbol{j} \cdot \boldsymbol{k}_n, \tag{4.2}$$

where B is a constant as long as the absolute values of the $\boldsymbol{k}$-vectors are constant. Equations (4.2) admit the familiar roll solutions corresponding to

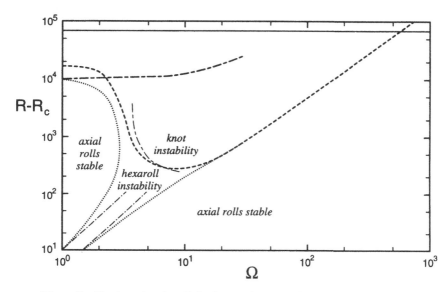

Fig. 7: Regions in the R-Ω-plane where axial rolls are stable or unstable with respect to the indicated instabilities for $P = 7$.

$N = 1$. Among these the axially oriented rolls corresponding to $\boldsymbol{k}_1 \cdot \boldsymbol{j} = 0$ are preferred since the Ω-dependence vanishes for these rolls. The unusual last term of equations (4.2) gives rise to new instabilities and new forms of three-dimensional convection flows. An example of the latter are the hexarolls which are shown in figure 6. They have been derived by Auer *et al.* (1995) in the case $N = 3$ with $\boldsymbol{k}_1 \cdot \boldsymbol{j} = 0$ and $\boldsymbol{k}_1 + \boldsymbol{k}_2 + \boldsymbol{k}_3 = 0$. Another form of three-dimensional convection is knot convection. This type of convection can be observed in a non-rotating Rayleigh–Bénard layer (Busse and Clever, 1979b) and has also been investigated numerically (Clever and Busse, 1989). In the present case of a convection layer with horizontal axis of rotation, knot convection occurs at much lower values of the Rayleigh number so that the analysis of the weakly nonlinear limit applies. Auer *et al.* (1995) have shown that the description of knot convection based on equations (4.2) in the case $N = 2$ agrees well with the numerical solutions of the full equations. A plot of knot convection is also given in figure 6. A diagram indicating the regions of the R-Ω-plane where axial rolls are stable and where they are unstable with respect to the hexaroll and knot instabilities is given in figure 7.

The hexaroll instability leads to the evolution of stable steady hexarolls only in a small fraction of the unstable region. For small values of $R - R_c$ the hexarolls are also unstable and a nearly heteroclinic cycle results (Busse and Clever, 1999). It manifests itself in the form of the long period oscillation shown in figure 8. Axial rolls described by $\boldsymbol{k}_1$ become unstable with respect

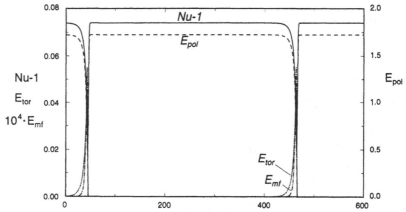

Fig. 8: Nearly heteroclinic cycle between axial rolls 180° out of phase with the brief changeover induced by the hexaroll instability. The energies E_{pol}, E_{tor} and E_{mf} of the poloidal and toroidal components of the velocity field and of the mean flow have been plotted as functions of time. Also shown is the Nusselt number Nu in dependence on t. Hexaroll convection is characterized by a finite toroidal component and by a finite mean x-component of the velocity field both of which vanish in the case of axial rolls.

to the hexaroll instability, the amplitudes C_2, C_3 corresponding to k_2, k_3 with $k_1 + k_2 + k_3 = 0$ grow while C_1 decays to zero. But a new steady state can not be attained. Instead C_1 changes sign and axial rolls 180° out of phase with the original one become established until after many thermal diffusion times the hexaroll disturbances grow again. The switch over can be seen in the time sequence of plots of figure 9. For the analysis of convection it is convenient to use the decomposition of the velocity field u into poloidal and toroidal components, and into its mean component U which represents the average of u over the x-y-plane,

$$u = \nabla \times (\nabla\varphi \times \lambda) + \nabla\psi \times \lambda + U, \qquad (4.3)$$

where the condition can be imposed that the x-y-average of the functions φ and ψ vanishes. Axial rolls are characterized by the property $\psi \equiv 0$ while all other solutions exhibit a toroidal function ψ proportional to Ω. The results are essentially independent of the Prandtl number P as soon as P exceeds order 10. The time scale of all dynamic processes is the thermal diffusion time, d^2/κ, where κ is the thermal diffusivity.

Considerable efforts have been expended to realize the interesting dynamics of a convection layer with parallel axis of rotation in the laboratory (Jaletzky and Busse, 1998; Jaletzky, 1999). A sketch of the apparatus is given in those papers. It represents a realization of the configuration of figure 5 with

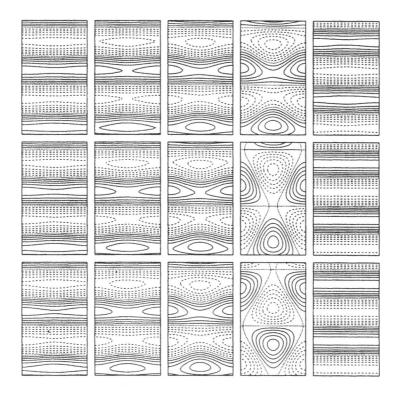

Fig. 9: Time sequence of plots $\Delta t = 4$ apart ($\Delta t = 2$ for the last five intervals), from top to bottom then left to right, in the transition regime when axial rolls become unstable to the hexaroll instability and convection returns to axial rolls $180°$ out of phase with the original ones. The plots show lines of constant temperature in the midplane of the layer for $R = 1800, \Omega = 5, P = 7$. The wavenumber in the x-direction (upwards) is $2.7d^{-1}$ and in the y-direction (towards the right) it is $1.559d^{-1}$ (after Busse and Clever, 1999.)

vertical axis such that laboratory gravity does not give rise to an oscillatory force. It has been possible to generate both hexaroll convection and knot convection, as shown in figure 10. Through the use of highly viscous fluids it is possible to achieve low values of Ω, while the centrifugal acceleration exceeds that of gravity by as much as a factor of 10. Thermochromatic liquid crystals embedded in thin plastic sheets and attached to the inner cylinder have been used for visualization. As shown in the photographs of figure 10, hexaroll convection and knot convection can be observed in accordance with the theory. At higher Rayleigh numbers oblique rolls predominate which have not been considered in the theoretical analysis of Auer *et al.* (1995). Unfortunately, it has not yet been possible to observe experimentally the

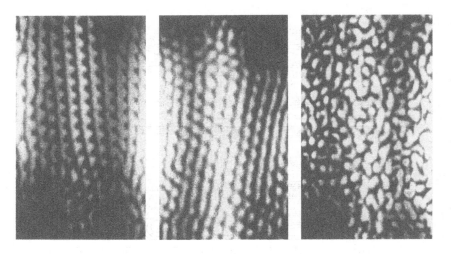

Fig. 10: Hexaroll convection at $R = 2300, \Omega = 10$ (middle), knot convection at $R = 4880, \Omega = 24$ (left) and phase turbulent convection at $R = 3970, \Omega = 14$ (right) observed in a rotating cylindrical annulus as sketched in figure 5. $P \approx 400$ and rotation is clockwise in all three cases.

phase turbulence arising from the existence of the nearly heteroclinic cycle shown in figures 8 and 9. Since this cycle seems to be confined to a region of less than 10% above the critical value of the Rayleigh number it can not easily be visualized. The amplitude of the temperature variations induced by convection is too weak in this regime to induce a visible change of color of the liquid crystals.

It is possible, however, to carry out numerical simulations based on computer code developed by Pesch (1996). An example of these simulations is shown in figure 11 where the intermittent appearance of hexaroll convection is clearly visible. It can also be seen that the phase of the two-dimensional axial rolls changes by 180° as they reappear after the hexaroll episode. The numerical simulations can be run for extended periods in time and the statistical properties of the phase turbulence can be analyzed as functions of the parameters of the problem. As must be expected on the basis of the analysis of the spatially periodic convection displayed in figures 8 and 9 the phase turbulent convection becomes purely time periodic in the case of low aspect ratio layers when the coherence length of patches of hexaroll convection becomes comparable to the horizontal periodicity interval.

The numerical simulation can also be employed to describe the phase turbulent convection at higher Rayleigh numbers as seen in the third picture of figure 10. A typical time sequence of the spatio-temporally chaotic state is

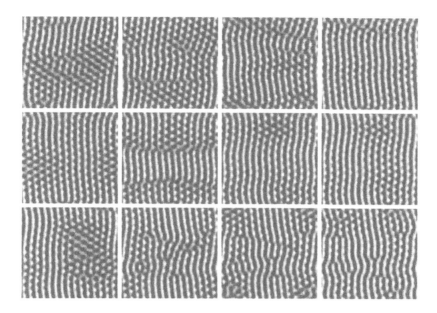

Fig. 11: Time sequence of plots (left to right, then top to bottom) with $\Delta t = 14$ showing the normal velocity in the midplane of the layer in the case of phase turbulence induced by the hexaroll instability of axial rolls in the case $R = 1814, \Omega = 5, P = 10$. The shift of the rolls by $180°$ after the hexaroll interlude is clearly visible.

shown in figure 12 which resembles the experimental observation in figure 10. It is of interest to note that an oscillation in which the axial roll component switches phase by $180°$ is still noticeable. A full period corresponds to about 1.6 thermal time units so that after four pictures in figure 12 vertical line segments have switched from black to white or vice versa. A quantitative measure of this chaotic switching phenomenon can be obtained from the correlation between patterns at the times t_0 and $t_0 + t$. As a typical example the correlation averaged over t_0 for the pattern of the vertical velocity has been plotted in figure 13. The period evident in this figure appears to be a remnant of the heteroclinic cycle. Attempts to measure the correlation experimentally are under way. Preliminary results appear to be consistent with the numerical results.

5 Concluding Remarks

In this article we have focused on systems where the phenomenon of phase turbulence is related to heteroclinic cycles. It is an open question whether similar mechanisms operate in other situations as well. The absence of a Lyapunov

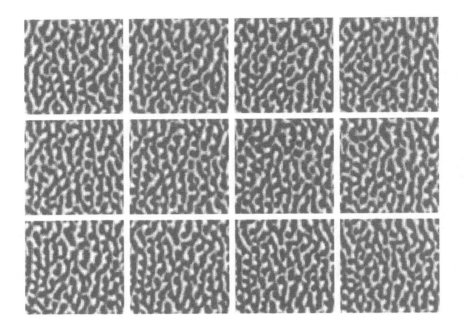

Fig. 12: Time sequence of plots (left to right, top to bottom) with $\Delta t = 0.2$ showing the normal velocity in the midplane of the layer in the case $R = 3440, \Omega = 15, P = 10$. The phase of the axial roll component still appears to switch by $180°$ about every fourth plot.

potential is a necessary condition for complex spatio-temporal dynamics, but not a sufficient one. High Prandtl number convection in a non-rotating layer, for instance, shows little time dependence even at 10 times the critical value of the Rayleigh number.

In the case of convection with stress-free boundaries a Lyapunov potential does not exist owing to the presence of the mean-flow mode. This mode can be understood as a Goldstone mode originating from the spontaneously broken (continuous) Galilean invariance. Since such modes are only weakly damped they can easily be excited by any small perturbations in the patterns, which are then typically reinforced. Phase turbulence in liquid crystals is closely related to this mechanism. Here a Goldstone mode with respect to the orientational degrees of freedom comes into play (Rossberg *et al.*, 1996).

Even less well understood is the phenomenon of spiral-defect chaos in convection layers with Prandtl numbers of the order of unity. It seems that here a kind of wave vector frustration is important. The different building blocks of the patterns (spirals, grain boundaries etc.) select different wave vectors and the system is unable to reconcile those into a stationary periodic pattern

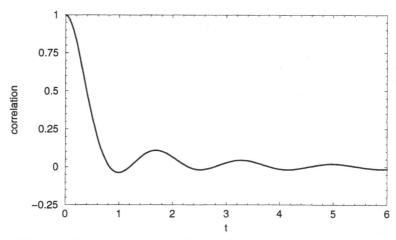

Fig. 13: Correlation function $\int w(t_0)w(t_0+t)dxdy$ averaged over t_0 in the range $0 < t_0 < 400$ as a function of t, for the parameters of figure 12; w denotes the normal component of the velocity at the midplane of the layer.

(Cakmur *et al.*, 1997).

References

Auer, M., Busse, F.H., and Clever, R.M. (1995) 'Three-dimensional convection driven by centrifugal buoyancy', *J. Fluid Mech.* **301**, 371–382.

Bodenschatz, E., Cannell, D.S., de Bruyn, J.R., Ecke, R., Hu, Y.-C., Lerman, K. and Ahlers, G. (1992) 'Experiments on three systems with non-variational aspects', *Physica D* **61**, 77–93.

Busse, F.H. (1967) 'The stability of finite amplitude cellular convection and its relation to an extremum principle', *J. Fluid Mech.* **30**, 625–649.

Busse, F.H. (1978) 'Nonlinear properties of convection', *Rep. Progress in Physics* **41**, 1929–1967.

Busse, F.H. (1984) 'Transition to turbulence via the statistical limit cycle route', in *Turbulence and Chaotic Phenomena in Fluids*, T. Tatsumi, ed., Elsevier, pp. 197–202.

Busse, F.H. (1986) 'Phase-turbulence in convection near threshold', *Contemp. Math.* **56**, 1–8.

Busse, F.H. and Bolton, E.W. (1984) 'Instabilities of convection rolls with stress-free boundaries near threshold', *J. Fluid Mech.* **146**, 115–125.

Busse, F.H. and Clever, R.M. (1979a) 'Nonstationary convection in a rotating system', in *Recent Developments in Theoretical and Experimental Fluid Mechanics*, U. Müller, K.G. Roesner and B. Schmidt, eds., Springer, pp. 376–385.

Busse, F.H. and Clever, R.M. (1979b) 'Instabilities of convection rolls in a fluid of moderate Prandtl number', *J. Fluid Mech.* **91**, 319–335.

Busse, F.H. and Clever, R.M. (1999) 'Heteroclinic cycles and phase turbulence', in *Pattern Formation in Continuous and Coupled Systems*, M. Golubitsky, D. Luss and S. Strogatz, eds., IMA Volumes in Mathematics and its Applications, vol. 115, Springer, pp. 25–32.

Busse, F.H. and Heikes, K.E. (1980) 'Convection in a rotating layer: a simple case of turbulence', *Science* **208**, 173–175.

Busse, F.H., Kropp, M. and Zaks, M. (1992) 'Spatio-temporal structures in phase-turbulent convection', *Physica D* **61**, 94–105.

Busse, F.H. and Sieber, M. (1991) 'Regular and chaotic patterns of Rayleigh–Bénard convection', in *Bifurcation and Chaos: Analysis, Algorithms, Applications*, R. Seydel, F.W. Schneider, T. Küppers and H. Troger, eds., Birkhäuser, pp. 75–88.

Cakmur, R.V., Egolf, D.A., Plapp, B.B. and Bodenschatz, E. (1997) 'Bistability and competition of spatiotemporal chaotic and fixed point attractors in Rayleigh–Bénard convection', *Phys. Rev. Lett.* **79**, 1853–1856.

Clever, R.M. and Busse, F.H. (1989) 'Three-dimensional knot convection in a layer heated from below', *J. Fluid Mech.* **198**, 345–363.

Dangelmayr, G. and Kramer, L. (1998) 'Mathematical tools for pattern formation', in *Evolution of Spontaneous Structures in Dissipative Continuous Systems*, F.H. Busse and S.C. Müller, eds., Springer Lecture Notes in Physics, vol. M55, pp. 1–85.

Fantz, M., Friedrich, R., Bestehorn, M. and Haken, H. (1992) 'Pattern formation in rotating Bénard convection', *Physica D* **61**, 147–154.

Goldstein, R.J. and Graham, D.J. (1969) 'Stability of a horizontal fluid layer with zero shear boundaries', *Phys. Fluids* **12**, 1133–1137.

Heikes, K.E. and Busse, F.H. (1980) 'Weakly nonlinear turbulence in a rotating convection layer', *Ann. NY Acad. Sci.* **357**, 28–36 .

Hu, Y., Ecke, R.E. and Ahlers, G. (1995) 'Time and length scales in rotating Rayleigh–Bénard convection', *Phys. Rev. Lett.* **74**, 5040–5043.

Hu, Y., Ecke, R.E. and Ahlers, G. (1997) 'Convection under rotation for Prandtl numbers near 1: linear stability, wavenumber selection and pattern dynamics', *Phys. Rev. E* **55**, 6928–6949.

Hu, Y., Pesch, W., Ahlers, G. and Ecke, R.E. (1998) 'Convection under rotation for Prandtl numbers near 1: Küppers–Lortz instability', *Phys. Rev. E* **58**, 5821–5833.

Jaletzky, M. (1999) 'Über die Stabilität von thermisch getriebenen Strömungen im rotierenden konzentrischen Ringspalt', Dissertation, University of Bayreuth.

Jaletzky, M. and Busse, F.H. (1998) 'New patterns of convection driven by centrifugal buoyancy', *ZAMM* **78**, S525–S526.

Küppers, G. and Lortz, D. (1969) 'Transition from laminar convection to thermal turbulence in a rotating fluid layer', *J. Fluid Mech.* **35**, 609–620.

May, R.M. and Leonard, W.J. (1975) 'Nonlinear aspects of competition between three species', *SIAM J. Appl. Math.* **29**, 243–253.

Mielke, A. (1997) 'Mathematical analysis of sideband instabilities with application to Rayleigh–Bénard convection', *J. Nonlinear Sci.* **7**, 57–99.

Millán-Rodriguez, J., Bestehorn, M., Pérez-Garcia, C., Friedrich, R. and Neufeld, M. (1995) 'Defect motion in rotating fluids', *Phys. Rev. Lett.* **74**, 530–533.

Morris, S.W., Bodenschatz, E., Cannell, D.S. and Ahlers, G. (1993) 'Spiral defect chaos in large aspect ratio Rayleigh–Bénard convection', *Phys. Rev. Lett.* **71**, 2026–2029.

Neufeld, M. and Friedrich, R. (1995) 'Statistical properties of the heat transport in a model of rotating Bénard convection', *Phys. Rev. E* **51**, 2038–2045.

Pesch, W. (1996) 'Complex spatio-temporal convection patterns', *CHAOS* **6**, 348–357.

Pesch, W. and Behn, U. (1998) 'Electrohydrodynamic convection in nematics', in *Evolution of Spontaneous Structures in Dissipative Continuous Systems*, F.H. Busse and S.C. Müller, eds., Springer Lecture Notes in Physics, vol. M55, pp. 335–383.

Ponty, Y., Passot, T. and Sulem, P.L. (1997) 'Chaos and structures in rotating convection at finite Prandtl number', *Phys. Rev. Lett.* **79**, 71–74.

Rossberg, A.G., Hertrich, A., Kramer, L. and Pesch, W. (1996) 'Weakly nonlinear theory of pattern-forming systems with spontaneously broken isotropy', *Phys. Rev. Lett.* **76**, 4729–4732.

Tu, Y. and Cross, M.C. (1992) 'Chaotic domain structure in rotating convection', *Phys. Rev. Lett.* **69**, 2515–2518.

Xi, H.-W., Li, X.-J. and Gunton, J.D. (1997) 'Direct transition to spatiotemporal chaos in low Prandtl number fluids', *Phys. Rev. Lett.* **78**, 1046–1049.

Zhong, F., Ecke, R. and Steinberg, V. (1991) 'Rotating Rayleigh–Bénard convection: Küppers–Lortz transition', *Physica D* **51**, 596–607.

Zhong, F. and Ecke, R.E. (1992) 'Pattern dynamics and heat transport in rotating Rayleigh–Bénard convection', *CHAOS* **2**, 163.

Zippelius, A. and Siggia, E.D. (1982) 'Disappearance of stable convection between free-slip boundaries', *Phys. Rev. A* **26**, 1788–1790

An ODE Approach for the Enstrophy of a Class of 3D Euler Flows

Koji Ohkitani

Abstract

We investigate the time evolution of the total enstrophy of 3D Euler flows by introducing a systematic method for tracing it. We are particularly interested in the Taylor–Green vortex.

(i) We establish that the enstrophy is governed by an ODE locally around the initial state. We argue that the domain where the ODE is defined should be extended, as long as the enstrophy remains regular. For the conventional choice of initial condition, we classify the functional form of the ODE for some typical kinds of evolution.

(ii) Using the results of Brachet *et al.* (1983) we study the analytic structure of the ODE numerically. It is found that if the enstrophy blows up in finite time t_*, then it should do so as $(t_* - t)^{-\alpha}$ with $\alpha \geq 1$. This result is based on a Taylor series expansion in a dependent variable, in contrast to a conventional series expansion in time.

The current approach is illustrated by the inviscid Burgers equation. In this case, without using its exact solution we obtain the correct form of the ODE. The ODE is remarkably simple, that is, the underlying analytic function turns out to be a simple polynomial.

1 Introduction

Vortex stretching in fluids generally increases as the total enstrophy grows in time. However, how fast it grows in time is not known theoretically.

To study the enstrophy growth, Taylor and Green (1937) considered a special kind of periodic flow, which develops from a simple initial velocity field

$$\boldsymbol{u} = \begin{pmatrix} \cos x \sin y \sin z \\ -\sin x \cos y \sin z \\ 0 \end{pmatrix}. \tag{1}$$

They considered the viscous case and later other people used it to investigate possible formation of singularities in inviscid flows. Taylor and Green examined evolution of this flow by a series expansion in time. In particular they computed the enstrophy up to $O(t^4)$. Morf *et al.* (1980) computed

64

numerically the Taylor coefficients up to $O(t^{40})$ for the Euler case and interpreted their result to be suggestive of the presence of a singularity. However, when Brachet *et al.* (1983) extended the analysis up to $O(t^{80})$, their previous conclusion was not confirmed. Some detailed analyses on the basis of the pseudo-spectral computations have been reported by Brachet *et al.* (1992) and Kerr (1993). Nevertheless, long-time behaviour of the enstrophy is not sufficiently well understood. The evolution of the Taylor–Green vortex turned out to be extremely complicated despite the simplicity of the initial condition. We note the same kind of analysis has been made using another kind of periodic flows by Pelz and Gulak (1997).

   In this paper, we introduce a different approach to the Taylor–Green problem. Our aim here is to provide a basis on which the problem of singularity (or regularity) is discussed and analysed.

2   The inviscid Burgers equation

2.1   Test problem

In order to illustrate 'the ODE approach', we consider the 1D inviscid Burgers equation

$$\frac{\partial u}{\partial t} + u\frac{\partial u}{\partial x} = 0, \tag{2}$$

under a periodic boundary condition in $[0, 2\pi]$. We consider an initial condition

$$u_0(x) = -A\sin x, \tag{3}$$

where A is the initial amplitude of velocity.

   As the exact solution of this equation is well known, we describe briefly what are needed for our purpose. Introducing $w(a, t) = -\partial u/\partial x$, the evolution equation takes a simple form

$$\frac{Dw}{Dt} = w^2 \tag{4}$$

in the Lagrangian frame of reference, see for example, Frisch (1983). The position x of a fluid particle which is located at a at time $t = 0$ is given by $x = a + tu_0(a)$. If we define 'enstrophy' by

$$Q(t) = \frac{1}{2}\left\langle w^2(x, t)\right\rangle, \tag{5}$$

where the angle brackets denote a spatial average over a spatial period, that is, $\langle\ \rangle = \frac{1}{2\pi}\int dx$, we obtain after integration

$$Q(t) = \frac{2Q_0}{\sqrt{(1 - 4Q_0 t^2)}(1 + \sqrt{(1 - 4Q_0 t^2)})}. \tag{6}$$

Here $Q_0 = A^2/4$ is the initial value of the enstrophy. For example, for the case of $A = 1$ the solution breaks down at $t_* = 1$ and the enstrophy diverges simultaneously as $(t_* - t)^{-1/2}$.

Now, suppose that we know neither the exact solution nor the singular behaviour of the enstrophy. Suppose that we are given a finite number of Taylor coefficients of the enstrophy which is parametrised by A. The problem we will consider in this paper is how to constrain the time evolution of the enstrophy out of this information.

2.2   The ODE approach

Because the enstrophy is an even function of t for the initial condition, it is convenient to introduce $\xi \equiv t^2$ as a new independent variable. We thus regard the enstrophy

$$Q = Q(\xi, A) \tag{7}$$

as a function of ξ and A. By working successively, we can obtain the Taylor expansion in ξ as follows:

$$Q = \frac{1}{4}A^2 + \frac{3}{16}A^4\xi + \frac{5}{32}A^6\xi^2 + \frac{35}{256}A^8\xi^3 + \frac{63}{512}A^{10}\xi^4 + \cdots . \tag{8}$$

By differentiating (7) w.r.t. ξ successively with fixed A we write

$$\frac{dQ}{d\xi} = Q_\xi(\xi, A) \tag{9}$$

and

$$\frac{d^2Q}{d\xi^2} = Q_{\xi\,\xi}(\xi, A), \tag{10}$$

where the subscripts denote differentiation. We note that the Jacobian determinant between $(Q, Q_\xi) \leftrightarrow (\xi, A)$ is not zero initially,

$$\frac{\partial(Q, Q_\xi)}{\partial(\xi, A)} = -\frac{1}{64}A^7 \neq 0, \tag{11}$$

therefore by the implicit function theorem we can solve (7) and (9) locally for $\xi = \phi(Q, Q_\xi)$ and $A = \psi(Q, Q_\xi)$ using differentiable functions ϕ, ψ. Plugging them into (10) we may write an ODE of the form

$$\frac{d^2Q}{d\xi^2} = f(Q, Q_\xi). \tag{12}$$

So far f is unknown. But some properties of f follow from a few Taylor coefficients

$$Q(0) = \tfrac{1}{4}A^2,$$

$$Q_\xi(0) = \tfrac{3}{16}A^4 = 3Q(0)^2, \tag{13}$$

$$Q_{\xi\,\xi}(0) = \tfrac{5}{16}A^6 = 20Q(0)^3.$$

Here the far right expression are obtained by eliminating A using the first equation. It is necessary to use A to keep track of dependence on $Q(0)$.

First, because (12) is autonomous and A is arbitrary, f should satisfy the following *identity*:

$$f(Q, 3Q^2) = 20Q^3, \quad \forall Q \geq 0. \tag{14}$$

Second, because the enstrophy involves the dimension of time only, f should also satisfy the *similarity* relationship

$$f(\alpha Q, \alpha^2 Q_\xi) = \alpha^3 f(Q, Q_\xi), \quad \forall \alpha \geq 0. \tag{15}$$

Putting $\alpha = 1/Q$ in (15), we have

$$\frac{d^2 Q}{d\xi^2} = Q^3 f\left(1, \frac{Q_\xi}{Q^2}\right). \tag{16}$$

Thus introducing a new function $F(R) \equiv f(1, R)$ and a new variable $R \equiv Q_\xi / Q^2$, the ODE can be reduced to a second-order autonomous ODE of the form

$$\frac{1}{Q^3} \frac{d^2 Q}{d\xi^2} = F\left(\frac{Q_\xi}{Q^2}\right). \tag{17}$$

By (14), $F(R)$ satisfies $F(3) = 20$. By higher-order Taylor coefficients we can determine derivatives of $F(R)$ w.r.t. R. For example, by differentiating (17) w.r.t. ξ using the chain rule we find

$$\frac{d}{d\xi}\left(\frac{Q_{\xi\xi}}{Q^3}\right) = \frac{dF}{dR} \frac{d}{d\xi}\left(\frac{Q_\xi}{Q^2}\right). \tag{18}$$

Putting $\xi = 0$, i.e. $R = 3$, we find $F'(3) = 15$. Working successively we find

$$F''(3) = 6, \quad F'''(3) = 0, \quad F''''(3) = 0, \ldots. \tag{19}$$

It should be noted that all the derivatives of $F(R)$ at order equal to or higher than the third appear to vanish. Assuming $F(R)$ is an analytic function, we deduce

$$F(R) = 3R^2 - 3R + 2. \tag{20}$$

Actually, we can confirm that this is an exact expression by checking it against (6). In this sense we obtain a complete closure for the enstrophy of the Burgers solution associated with this class of initial condition.

Strictly speaking, by a successive method, we can deduce neither that all the higher derivatives actually vanish nor that the function $F(R)$ is analytic w.r.t. R. Nevertheless, the above procedure is a systematic method which can lead to an exact ODE. In other words, it is not easy to predict a correct formula

$$Q(\xi, A) = \frac{1}{2\xi} \sum_{n=1}^{\infty} \frac{(2n-1)!!}{(2n)!!}(A^2\xi)^n, \tag{21}$$

solely by observing the first few terms of the series. In contrast, by the above method it is possible to predict (21) easily. As far as this case is concerned, the Taylor expansion w.r.t. a dependent variable does a better job than that w.r.t. an independent variable in predicting the correct evolution.

Having obtained $F(R)$, the problem is now reduced to that of an ODE. One may solve it numerically or analytically. But as can be verified, for a finite-time blow-up to form it is sufficient that $F(R) > c$ (> 0) for some positive c.

3   3D Euler equations

In order to apply the above method to 3D Euler equations, we need an initial condition which has some arbitrary parameters. Here we consider a velocity field

$$\boldsymbol{u} = \begin{pmatrix} A \cos x \sin y \sin z \\ B \sin x \cos y \sin z \\ C \sin x \sin y \cos z \end{pmatrix}. \tag{22}$$

The corresponding vorticity field is

$$\boldsymbol{\omega} = \begin{pmatrix} (C - B) \sin x \cos y \cos z \\ (A - C) \cos x \sin y \cos z \\ (B - A) \cos x \cos y \sin z \end{pmatrix}. \tag{23}$$

¿From the incompressibility condition we have $A + B + C = 0$, therefore we can eliminate C as $C = -A - B$. We mainly consider the case $B = -A$, which was originally chosen by Taylor and Green (1937) and which has been extensively studied numerically by Brachet *et al.* (1983, 1992).

3.1   The ODE approach

Because Q is an even function of t for the initial condition, we write it in terms of $\xi \equiv t^2$ as follows:[1].

$$Q = \frac{3}{8}A^2 + \frac{5}{128}A^4\xi + \frac{25}{4224}A^6\frac{\xi^2}{2!} - \frac{157317}{188334080}A^8\frac{\xi^3}{3!} + \cdots. \tag{24}$$

The higher-order Taylor coefficients were determined by an iteration scheme. The calculation is straightforward but we need symbolic manipulation to handle them (see the Appendix) because the algebra involved is formidable.

Regarding Q as a continuously differentiable function of ξ and A, we write

$$Q = Q(\xi, A). \tag{25}$$

---

[1]We use the same notations for Q, ϕ, ψ as we used in the previous section. Needless to say, they represent different functions

Differentiating with respect to ξ with fixed A, we have

$$\frac{\partial Q}{\partial \xi} = Q_\xi(\xi, A), \tag{26}$$

and

$$\frac{\partial^2 Q}{\partial \xi^2} = Q_{\xi\xi}(\xi, A). \tag{27}$$

Noting that

$$\frac{\partial(Q, Q_\xi)}{\partial(\xi, A)} = \frac{75}{45056} A^7 \neq 0, \tag{28}$$

by the implicit function theorem, we can solve (25) and (26) for ξ and A at least locally: $\xi = \phi(Q, Q_\xi)$, $A = \psi(Q, Q_\xi)$ with some differentiable functions ϕ and ψ. Plugging them into (27) we write

$$\frac{d^2 Q}{d\xi^2} = f(Q, Q_\xi). \tag{29}$$

So far f is unknown, except that it is differentiable. Note we have at the initial instant of time

$$
\begin{aligned}
Q(0) &= \tfrac{3}{8} A^2, \\
Q_\xi(0) &= \tfrac{5}{128} A^4 = \tfrac{5}{18} Q(0)^2, \\
Q_{\xi\xi}(0) &= \tfrac{25}{4224} A^6 = \tfrac{100}{891} Q(0)^3.
\end{aligned}
\tag{30}
$$

Now, since f is autonomous and A is arbitrary, we have the following

identity:

$$f\left(Q, \frac{5}{18} Q^2\right) = \frac{100}{891} Q^3, \quad \forall Q \geq 0. \tag{31}$$

Moreover, since Q involves only the dimension of time we also have the

similarity

$$f(\alpha Q, \alpha^2 Q_\xi) = \alpha^3 f(Q, Q_\xi), \quad \forall \alpha \geq 0. \tag{32}$$

Putting $\alpha = 1/Q$ in (32) and defining $F(R) = f(1, R)$ and a non-dimensional variable $R = Q_\xi/Q^2$ we write

$$\frac{1}{Q^3} \frac{d^2 Q}{d\xi^2} = F\left(\frac{Q_\xi}{Q^2}\right). \tag{33}$$

The problem is reduced to an ODE defined by a function $F(R) \equiv f(1, R)$ of R. In section 4.1 we see numerically that $F(R)$ is analytic w.r.t. R near a point determined by the initial condition. By (30) we have

$$F\left(\frac{5}{18}\right) = \frac{100}{891}. \tag{34}$$

Differentiating (33) w.r.t. ξ we can obtain

$$F'\left(\frac{5}{18}\right) = \frac{1516951}{470250}. \tag{35}$$

In principle, it is possible to obtain $F''\left(\frac{5}{18}\right)$, $F'''\left(\frac{5}{18}\right)$, ... by successive differentiation. Unlike the case of Burgers' equation, $F(R)$ is not a polynomial in R according to an analysis using data up to $O(\xi^{38})$.

3.2 Description of long-time behaviour by the ODE

By the implicit function theorem, we have shown that a differentiable function $F(R)$ exists such that the evolution of Q is governed by an ODE of the form (32). At this stage, the existence of a continuously differentiable function $F(R)$ is guaranteed only in a local region (interval) near the initial state.

We consider a case where there is a singular point on the boundary of the interval and the variable R approaches the singular point from inside. If $F(R)$ becomes unbounded in this limit, then there is a blow-up at finite time, whether or not R reaches the singular point in finite time or not. On the other hand, if $F(R)$ remains bounded, there may not be a blow-up, even though the function $F(R)$ becomes non-differentiable. (For example, consider $F(R) \propto (R - c_1)^{1/3} + c_2$ where c_1, $c_2 > 0$ are constants.) Thus, long-time evolution of the enstrophy can be described by an ODE with a continuous function $F(R)$, beyond the interval of R guaranteed by the implicit function theorem.

If the enstrophy remains finite for all time, the interval where the extended function $F(R)$ is defined should include the immediate neighbourhood of the origin $R = 0$, because otherwise we have $Q_\xi/Q^2 > c$ for some $c > 0$ all the time and the enstrophy would become infinitely large while R stays within this range.

Below we will see some typical kinds of evolution specifically.

3.3 Typical kinds of evolution

Before examining the coefficients of $F(R)$, it is useful to consider some typical cases of the time evolution of Q in the ODE (33).

First we consider as a typical evolution which is regular for all time an

- exponential growth

$$Q(\xi) \propto \exp(\beta\sqrt{\xi}), \quad \beta > 0. \tag{36}$$

In this case we have

$$F(R) \to R^2 \text{ as } R \to 0. \tag{37}$$

On the other hand, we can consider as a typical evolution which blows up in finite time an

- algebraic blow-up

$$Q(\xi) \propto \frac{1}{(\xi_* - \xi)^m}, \quad m > 0 \ (m \neq 1), \ \xi_* > 0. \tag{38}$$

In this case we find

$$R = \frac{Q_\xi}{Q^2} \propto (\xi_* - \xi)^{m-1} \tag{39}$$

and

$$F(R) = \frac{Q_{\xi\xi}}{Q^3} \propto (\xi_* - \xi)^{2(m-1)}. \tag{40}$$

According as $m > 1 \ (< 1)$, both R and $F(R) \to 0 \ (\to \infty)$.

For a particular initial condition (22) with $B = -A$, in fact R decreases in time as shown by numerical simulations (not shown here). Therefore we are interested in the case $m > 1$, where we find

$$F(R) \to \left(1 + \frac{1}{m}\right) R^2 \quad \text{as } R \to 0. \tag{41}$$

In these cases, what matters is the behaviour of $F(R)$ near the origin $R = 0$. We can classify possible kinds of evolution using $C \equiv \lim_{R \to 0} \frac{F(R)}{R^2} \ (> 0)$.

- $C > 1 \Rightarrow$ blow-up in finite-time,

- $0 < C \leq 1 \Rightarrow$ regular for all time.
  The case $C = 1$ includes an exponential growth. For $0 < C < 1$ we have $Q(\xi) \propto \xi^{1/(1-C)}$, a long-time regular behaviour of an algebraic form.

4   Structure of the function $F(R)$

4.1   The Domb–Sykes plot

We consider the Taylor expansion of $F(R)$ w.r.t. a dependent variable R,

$$F(R) = \sum_{n=0}^{\infty} A_n \left(R - \frac{5}{18}\right)^n. \tag{42}$$

Applying the iteration procedure described in section 3.1 and using data presented in Brachet *et al.* (1983), we obtain A_n numerically for $n = 0, \ldots, 38$.

To examine the analytic structure of $F(R)$, we try the usual Domb–Sykes plot for the coefficients A_n, that is,

$$\frac{A_{n+1}}{A_n} \to \frac{1}{R_*} \left(1 + \frac{\gamma - 1}{n} \right) \quad \text{as } n \to \infty. \tag{43}$$

In Figure 1 we show the Domb–Sykes plot together with a fitted straight line, a linear function of $1/n$. A least-squares fit was done on the interval $15 \le n \le 38$.

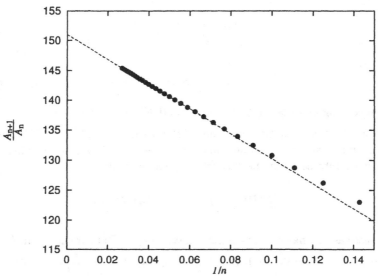

Figure 1: The Domb–Sykes plot.

We find a small radius of convergence $R_* \approx \dfrac{1}{151.01}$ and an exponent $\gamma \approx -0.376$. We identify a singular point in the function $F(R)$,

$$F(R) \approx \left(\frac{5}{18} + R_* - R \right)^{0.376} + \text{(regular terms)}. \tag{44}$$

Noting that $1/3 < 0.376 < 1/2$, we deduce the time evolution of the enstrophy is constrained by

$$R < \frac{5}{18} + R_*. \tag{45}$$

This does not mean that the enstrophy blows up in finite time, since it places an upper bound on R rather than a lower bound. Still, there follows from (45) a constraint on the strength of a possible singularity in the enstrophy.[2]

---

[2] By a theorem due to Beale *et al.* (1984), it is the maximum vorticity which controls the breakdown of smooth solutions. On physical grounds, we expect that the total enstrophy should also go singular, if there is singularity formation

That is, if the total enstrophy blows up like

$$Q(t) \approx \frac{1}{(t_* - t)^\alpha}, \tag{46}$$

then α should be constrained by $\alpha \geq 1$, because if $\alpha < 1$, then $R \to \infty$ as $t \to t_*$, which contradicts (45). Actually the form of singularity for $Q(t)$ once proposed in Morf *et al.* (1980) had the exponent $\alpha \approx 0.8$. As mentioned in the Introduction, this conclusion was rejected in Brachet *et al.* (1983), consistent with the present analysis.

4.2   Euler's transformation

As mentioned above R starts to decrease in time. So the relevant region is $R < 5/18$ rather than $R > 5/18$ where a singular point (44) lies. In the relevant region $R < 5/18$, the series (42) is an alternating series; its terms change their signs consecutively. It is useful to employ the classical Euler transformation to improve convergence of the series. We introduce a new variable $r \equiv -\frac{1}{R_*}\left(R - \frac{5}{18}\right)$ and rewrite $F(R)$ as

$$F(R) = \sum_{n=0}^{\infty} (-1)^n A_n R_*^n r^n = \sum_{n=0}^{\infty} a_n r^n, \tag{47}$$

where $a_n = (-1)^n A_n R_*^n$. By applying Euler's transformation we transform it as

$$F(R) = \frac{1}{1+r}\left(a_0 + \zeta \delta a_0 + \zeta^2 \delta^2 a_0 + \cdots\right), \tag{48}$$

where $\zeta \equiv \frac{r}{1+r}$ and $\delta a_n = a_n - a_{n+1}$, $\delta^2 a_n = \delta a_n - \delta a_{n+1}, \ldots$ for $n = 0, 1, 2, \ldots$

In Figure 2 we plot $F(R)$ as a function of R. Two curves of different orders of approximation are shown. If the enstrophy growth is exponential in time (or, iterated exponentials in time) the curve should hit the origin, like the short-dashed curve. $F(R) = R^2$. At the order $n = 38$ currently available, the value of $F(R = 0)$ is positive. It is not difficult to deduce the presence of a singularity if $F(0)$ is really positive. However, as the series (48) has not yet sufficiently well converged, we refrain from concluding whether $F(0)$ is actually positive or not.

We can solve the ODE defined by an approximate function $F(R)$ using coefficients A_n up to $n = 38$. In Figure 3 we compare the approximate solution with a result of pseudo-spectral computation by Brachet *et al.* (1992). As expected, the approximation grows faster than the direct simulation in the later stage. This means the value of $F(0)$ observed in Figure 2 should decrease as we increase the number of terms in (38). (At this order of approximation, the enstrophy becomes singular at $t = 4.48$.)

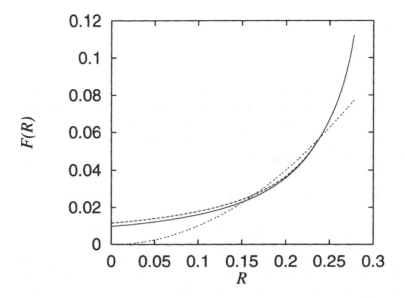

Figure 2: The function $F(R)$. The solid line denotes a result using data $\delta^n a_0$ up to $n = 38$ and the dashed line up to $n = 32$. The short-dashed line denotes $F(R) = R^2$.

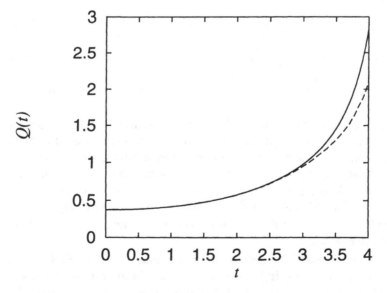

Figure 3: A comparison of the approximate enstrophy (solid) with that of a result of direct simulation (dashed) by Brachet *et al.* (1992).

5   Summary

We have introduced a method of tracing the enstrophy for 3D Euler equations, which we have called an ODE approach. Applying it first to the Burgers equation, we show the function defining the ODE is a simple polynomial and 'complete closure' is obtained. In this sense the presence of a singularity is deduced systematically.

For the 3D Euler equations, such a nice truncation has not yet been found. That is, we checked that $F(R)$ is neither a polynomial nor a rational function, using data up to $O(\xi^{40})$. It may be worthwhile to search for other forms of the ODE. While we do not know if a singularity exists or not, a constraint is obtained upon the strength of singularity in the time evolution of the enstrophy. It may be useful to perform the present analysis using numerical data of higher order in the Taylor coefficients.

Acknowledgements

Part of this work was done while the author was attending a research programme *Turbulence*, held during January 4–July 2, 1999, at the Isaac Newton Institute for Mathematical Sciences, University of Cambridge. He thanks the Isaac Newton Institute for providing him with an excellent environment and pleasant hospitality. He is indebted to J. Brasseur, P.G. Drazin, B.J. Guerts, R. Kerswell, S. Nazarenko, M. Oberlack, A. Pumir, J.C. Vassilicos and X. Wu for helpful discussion during his stay in the UK. He also thanks A. Craik, H. Okamoto and M. Yamada for helpful comments in the earlier stage of this work.

Appendix. More general case

Retaining both A and B, we can compute Taylor coefficients of the total enstrophy up to $O(\xi^4)$ or $O(t^8)$:

$$Q(0) = c_1(A^2 + AB + B^2),$$

$$\frac{dQ}{d\xi}(0) = c_2(A^2 + AB + B^2)^2,$$

$$\frac{d^2Q}{d\xi^2}(0) = c_3(A^2 + AB + B^2)^3 - c_4 A^2 B^2(A + B)^2,$$

$$\frac{d^3Q}{d\xi^3}(0) = -c_5(A^2 + AB + B^2)^4 + c_6 A^2 B^2(A + B)^2(A^2 + AB + B^2),$$

$$\frac{d^4Q}{d\xi^4}(0) = c_7(A^2 + AB + B^2)^5 - c_8 A^2 B^2(A + B)^2(A^2 + AB + B^2)^2,$$

$$(A1)$$

where the constants are exactly given by

$$c_1 = \frac{3}{8}, \quad c_2 = \frac{5}{128}, \quad c_3 = \frac{25}{4224}, \quad c_4 = \frac{117}{11264},$$

$$c_5 = \frac{157317}{188334080}, \quad c_6 = \frac{2373909}{188334080}, \tag{A2}$$

$$c_7 = \frac{729080006648669}{995224325652480000}, \quad c_8 = \frac{7518303285665369}{746418244239360000}.$$

It is remarkable that these results are polynomials in the particular combinations of A and B, that is, $A^2 + AB + B^2$ and $A^2 B^2 (A + B)^2$. For the special case of $B = -A$, we can go up to $O(\xi^6)$ or $O(t^{12})$:

$$\frac{d^5 Q}{d\xi^5}(0) = \frac{-5780547259290435869616386599637}{999803976530379276035004825600000},$$

$$\frac{d^6 Q}{d\xi^6}(0) = \frac{4224525674636860970625319448473967116656094222957}{94815493796762160140075522693701495496638464000000000}.$$

To perform the Domb–Sykes plot, we have used the numerical data (Table 5) of Brachet *et al.* (1983) at order equal to or higher than $O(t^{14})$.

References

Beale, J.T, Kato, T., Majda, A. (1984) 'Remarks on the breakdown of smooth solutions for the 3-D Euler equations', *Commun. Math. Phys.* **94**, 61–66.

Brachet, M.E., Meiron, D.I., Orszag, S.A., Nickel, B.G., Morf, R.H., Frisch, U. (1983) 'Small-scale structure of the Taylor–Green vortex', *J. Fluid Mech.* **130**, 411–452.

Brachet, M.E., Meneguzzi, M., Vincent, A., Politano, H., Sulem, S.L. (1992) 'Numerical evidence of smooth self-similar dynamics and possibility of subsequent collapse for three-dimensional ideal flows', *Phys. Fluids A* **4**, 2845–2854.

Frisch, U. (1983) 'Fully developed turbulence, singularities and intermittency', in *Chaotic Behaviour of Deterministic Systems*, edited by Iooss, G., Helleman, R., Stora, R., North-Holland, Amsterdam, 665–704 .

Kerr, R.M. (1993) 'The role of singularities in turbulence', in *Unstable and Turbulent Motion of Fluids*, edited by Kida, S., World Scientific, Singapore, 102–112.

Morf, R.H., Orszag, S.A., Frisch, U. (1980) 'Spontaneous singularity in three-dimensional inviscid incompressible flow', *Phys. Rev. Lett.* **44**, 572–575.

Pelz, R.B., Gulak, Y. (1997) 'Evidence for a real-time singularity in hydrodynamics from time series analysis', *Phys. Rev. Lett.* **79**, 4998–5001.

Taylor, G.I., Green, A.E. (1937) 'Mechanism of the production of small eddies from large ones', *Proc. Roy. Soc. London A* **158**, 499–521.

Scale Separation and Regularity of the Navier–Stokes Equations

C.R. Doering and J.D. Gibbon

1 Introduction

Intermittency in dynamical systems and turbulence is the name given to phenomena which display localized fluctuations in time and/or space (Batchelor and Townsend (1949), Kuo and Corrsin (1971), Sreenivasan (1985)). It is most easily characterized by probability distributions in terms of the relative likelihood of large, rare events. In turbulence the relevant events are usually identified as localized fluctuations in velocity differences or energy dissipation (Sreenivasan and Meneveau (1988)). These phenomena are identified quantitatively via anomalous scaling of appropriate moments and the associated separation of (velocity or dissipation rate) scales in the vicinity of strong intermittent fluctuations.

For example, moments of velocity differences $\delta u = \hat{\mathbf{r}} \cdot (\mathbf{u}(\mathbf{x} + \mathbf{r}) - \mathbf{u}(\mathbf{x}))$ in high Reynolds number, incompressible, homogeneous, isotropic turbulence are observed to behave according to (Frisch (1995))

$$\langle \delta u^p \rangle \sim \varepsilon^{p/3} L_0^{(p/3 - \zeta_p)} r^{\zeta_p} \tag{1.1}$$

where $\hat{\mathbf{r}} = \mathbf{r}/r$, $r = |\mathbf{r}|$ is in the inertial range, L_0 is the integral scale in the flow, and ε is the energy dissipation rate per unit mass. The average $\langle \cdot \rangle$ is usually taken to mean an ensemble or time average. The classical scaling, where there is a single velocity scale at length scale r, corresponds to $\zeta_p = p/3$ in which case

$$\langle \delta u^p \rangle \sim \varepsilon^{p/3} r^{p/3}. \tag{1.2}$$

Hölder's inequality ensures ζ_p is a concave function of p, for even p (Frisch (1995)). Apart from the exact result that $\zeta_3 = 1$ (Kolmogorov (1941)), these scalings have never been deduced rigorously from the Navier–Stokes equations although some rigorous relations between scaling exponents have been established (Constantin and Fefferman (1994)). Experimental evidence suggests that the ζ_p all fall on a monotonically increasing concave curve (see the references cited in Frisch (1995)). The deviation of ζ_p from $\frac{p}{3}$ is referred to as anomalous scaling and is taken as an indication of relatively larger fluctuations at smaller scales associated with the phenomenon of intermittency.

To elucidate the connection between anomalous scaling and scale separation, let us define $U_3 = (\varepsilon r)^{1/3}$ and the 'p-velocity' scale for $p > 3$ according to

$$U_p = \left(\frac{\langle \delta u^p \rangle}{U_3^3} \right)^{\frac{1}{p-3}}. \qquad (1.3)$$

The scaling in formula (1.1) then implies (in the spirit of extended self-similarity discussed by Benzi et al (1993))

$$U_p \sim U_3^{\frac{\zeta_p - 1}{p/3 - 1}} U_0^{\frac{p/3 - \zeta_p}{p/3 - 1}} \qquad (1.4)$$

where $U_0 = (\varepsilon L_0)^{1/3}$ is a large scale velocity. Because ζ_p is monotonically increasing from $\zeta_3 = 1$ and yet bounded above by $\frac{p}{3}$, the exponent of U_3 above satisfies $0 < \frac{\zeta_p - 1}{p/3 - 1} \leq 1$. Hence all the $U_p \to 0$ as $r \to 0$ and $U_3 \to 0$. The essence of anomalous scaling and intermittency, however, is that these p-velocity scales separate from each other in this limit. Higher p-velocity scales will be dominated by rare but large fluctuations, i.e., by extreme values in the probability distribution. So depending on how one chooses to measure the velocity, i.e., depending on the choice of p, different (velocity) scales emerge. This is most easily seen by looking at the dimensionless ratio of sequential p-velocity scales:

$$\frac{U_{p+1}}{U_p} \sim \left(\frac{U_p}{U_0} \right)^{\gamma_p} \qquad (1.5)$$

where

$$\gamma_p = \left(\frac{\zeta_{p+1} - 1}{(p+1) - 3} \right) \left(\frac{p - 3}{\zeta_p - 1} \right) - 1. \qquad (1.6)$$

For the classical scaling $\zeta_p = \frac{p}{3}$, each $\gamma_p = 0$ and the scales are all the same. On the other hand, the concavity of ζ_p as a function of p for $p > 3$ means precisely that $\gamma_p < 0$ and that the ratio diverges as r descends to smaller and smaller scales in the inertial range and $U_p \to 0$. We reiterate that these are purely phenomenological observations to date; no rigorous predictions of such anomalous scaling exponents have ever been derived from the Navier–Stokes equations.

In this paper we want to consider a different hierarchy of moments and investigate the implications of a separation of scales – should it occur – for questions of the regularity of solutions to the Navier–Stokes equations. The moments we consider are not of velocity differences; instead we focus on the 2nth moments of the energy spectrum in wavenumber space as defined by second moments of nth derivatives of the velocity field. These wavenumber moments correspond to small length scales which may be deep in the dissipation range. We do not consider averages in a turbulent steady state, but rather the time evolution of these length scales as constrained by the Navier–Stokes equations. The question addressed here is how these small length

scales behave when a singular event occurs in the Navier–Stokes equations –
if such a singular event ever occurs.

2   Scale separation in Navier–Stokes dynamics

We begin in the context of decaying turbulence, with the incompressible 3D
Navier–Stokes equations on a periodic domain $\Omega = [0, L]^3$:

$$\mathbf{u}_t + \mathbf{u} \cdot \nabla \mathbf{u} + \nabla p = \nu \Delta \mathbf{u}, \qquad (2.1)$$

where $\nabla \cdot \mathbf{u} = 0$, with the divergence-free initial condition $\mathbf{u}(\mathbf{x}, 0) = \mathbf{u}_0(\mathbf{x})$.
Here are some previously established facts (Constantin and Foias (1988),
Temam (1988), Doering and Gibbon (1995)): For small enough values of
the dimensionless quantity

$$I_0 = \frac{\|\nabla \mathbf{u}_0\|_2^2 L}{\nu^2}, \qquad (2.2)$$

where $\| \cdot \|_2$ denotes the L^2 norm on Ω, we can be assured of the existence
of a unique smooth (C^∞ in fact) solution for $0 < t < \infty$. This condition is
effectively a restriction to small Reynolds numbers in the initial data (strictly
speaking, to small R_λ, the Reynolds number based on the Taylor microscale,
in the initial data) which effectively rules out the possibility of turbulence.
For larger values of I_0, after a finite time t_* that depends on ν, L and the
magnitude of I_0, we may only assert the existence of weak, not necessarily
unique, solutions (Leray (1934)). Weak solutions have finite kinetic energy
$\frac{1}{2}\|\mathbf{u}(\cdot, t)\|_2^2$ at all times, but may exhibit isolated integrable singularities in the
instantaneous bulk energy dissipation rate $\nu\|\nabla \mathbf{u}(\cdot, t)\|_2^2$. Such singularities, *if*
they occur (which remains an open question), are responsible for our inability
to establish the uniqueness of solutions. Moreover, singularities of this sort
would violate the assumption of separation of scales invoked in the derivation
of the Navier–Stokes equations from the Boltzmann equations, raising the
question of the validity of these equations in the presence of such violently
turbulent dynamics. From a more practical standpoint, our inability to establish regularity for high Reynolds number flows represents an obstruction
to the derivation of rigorous bounds on turbulent fluctuations.

Singular events would be characterized by the vanishing of small length
scales in the flow. Indeed, a divergence in the instantaneous energy dissi-
pation rate means that the Fourier energy spectrum $E(\mathbf{k}, t) = |\hat{\mathbf{u}}(\mathbf{k}, t)|^2$ for
the solution has no effective high wavenumber cutoff. In terms of moments,
singular events correspond to the divergence of the second and all higher even
moments of $E(\mathbf{k}, t)$. During the initial period $0 < t < t_*$ when the solution is

known to be smooth, it is very smooth indeed, as all the moments of $E(\mathbf{k}, t)$ exist.

The moments of the energy spectrum are given by the squares of the L^2 norms of derivatives of the velocity vector field, and they define small length scales. We denote

$$H_n(t) = \int_\Omega |\nabla^n \mathbf{u}|^2 \, d^3 x \qquad (2.3)$$

where $|\nabla^n \mathbf{u}|^2$ means $|\Delta^{\frac{n}{2}} \mathbf{u}|^2$ for n even and $|\Delta^m \nabla \mathbf{u}|^2$ for $n = 2m + 1$ odd. For all $n \geq 1$, the ratio

$$\kappa_n(t) = \left(\frac{H_n}{H_0}\right)^{\frac{1}{2n}} = \left(\frac{\sum_{\mathbf{k}} k^{2n} |\hat{\mathbf{u}}|^2}{\sum_{\mathbf{k}} |\hat{\mathbf{u}}|^2}\right)^{\frac{1}{2n}}, \qquad (2.4)$$

defines a wavenumber corresponding to a length scale κ_n^{-1} (Foias $et\ al$ (1981), Doering and Gibbon (1995)). In particular, $\kappa_1^{-1} = \left(\frac{H_1}{H_0}\right)^{-\frac{1}{2}}$ is the familiar Taylor microscale. Hölder's inequality implies the ordering $\kappa_1 \leq \kappa_2 \leq \kappa_3 \ldots$, so the corresponding length scales $\kappa_1^{-1} \geq \kappa_2^{-1} \geq \kappa_3^{-1} \ldots$ are a decreasing sequence. This fact is not limited to κ_ns defined by solutions of the Navier-Stokes equations but is true for any velocity vector field on this domain. We do not generally expect this inequality to be saturated in practice; the κ_n are all exactly equal only if the energy is concentrated in a single wavenumber shell. They can be asymptotically equal as $n \to \infty$ only if there is a strict high wavenumber cutoff in the spectrum. Here are some previously established facts about the κ_n (Doering and Gibbon (1995)):

1. Singularities in the flow occur when κ_1 diverges, sending all the length scales (κ_n^{-1}) to zero.

2. No singularity can occur if any one of the κ_n, for $n \geq 1$, stays finite. Hence at any instant of time $t > 0$, either the κ_n are all finite or, if they diverge, they diverge together.

3. If they do diverge then the divergence is integrable in time (Gibbon (1996)).

We know that singularities must be reasonably localized in space because $\|\mathbf{u}(\cdot, t)\|_2$ remains finite and uniformly bounded, so they must have a broad-band spectrum in wavenumber space (Caffarelli $et\ al$ (1982)). If there is a single vanishing cutoff length scale associated with a singularity then all the wavenumbers diverge together, in which case the ratios $\frac{\kappa_{n+1}}{\kappa_n} = O(1)$ as $\kappa_n \to \infty$.

The central result of this paper is contained in the following hierarchy of coupled differential inequalities for the evolution of the $\kappa_n(t)$ for $n \geq 3$:

$$n \frac{d\kappa_n}{dt} \leq -\nu \left[\left(\frac{\kappa_{n+1}}{\kappa_n}\right)^{2n+2} - 1\right] \kappa_n^3 + c_n \, \kappa_n^{7/2} \|\mathbf{u}(\cdot, t)\|_2 \qquad (2.5)$$

where the c_n are absolute constants. These differential inequalities are derived directly from the Navier–Stokes equations (see the appendix) and are valid for weak as well as strong solutions. For any initial condition with finite I_0, all the κ_n are finite for $t > 0$ up until at least t_*. The restriction to $n \geq 3$ is a technical necessity in the derivation to allow us to 'close' the hierarchy to the extent that we have, involving only $\|\mathbf{u}(\cdot, t)\|_2$ which we know is a monotonically decreasing (and hence *a priori* bounded) function of time.

The ordering of the wavenumbers due to Hölder's inequality ensures that the first term on the right hand side of (2.5) originating in the viscous term in the Navier–Stokes equations is not positive. The superlinear growth of the second term $\sim \kappa_n^{7/2}$, however, allows the possibility of a finite time divergence of $\kappa_n(t)$. The broadband nature of any potential singularity implies that we can confidently assert that $\frac{\kappa_{n+1}}{\kappa_n} > 1$ (strictly) as $\kappa_n \geq \infty$, so the first term from the viscosity is likely to have even more damping power than is rigorously justified by the Hölder limit. However, the fact that $\frac{7}{2} > 3$ means that it needs to be significantly more effective if a large fluctuation is to be arrested before it leads to a divergence.

Sufficient small length scale separation (in a sense to be described precisely below) associated with potential singular events could elevate the viscous damping term in (2.5) to a more prominent, or even dominant, role in the process. By scale separation we mean the emergence of a spectrum of small length scales in the temporal vicinity of a violent fluctuation. That is, we consider the possibility that there is not just one small length scale but rather a set of distinct scales associated with the collection of the κ_n.

In analogy with the observed phenomena of intermittency and anomalous scaling for moments of velocity differences, and for the purposes of illustration, let us suppose that in a singular event each $\kappa_n(t)$ diverges with its *own* exponent β_n with respect to $\kappa_3(t)$, the lowest one valid in (2.5). That is, consider the implications of

$$\kappa_n \sim L_0^{-1+\beta_n} \kappa_3^{\beta_n} \tag{2.6}$$

as $\kappa_3 \to \infty$ (we've used the integral scale L_0 in (2.6) but any suitable bounded length scale will do). The $\beta_n \geq 1$ would then necessarily be an increasing sequence of numbers starting from $\beta_3 = 1$ and the ratio of sequential wavenumbers would scale as

$$\frac{\kappa_{n+1}}{\kappa_n} \sim L_0^{\beta_{n+1}-\beta_n} \kappa_3^{\beta_{n+1}-\beta_n} \sim (\kappa_n L_0)^{\mu_n} \tag{2.7}$$

with exponents

$$\mu_n = \frac{\beta_{n+1}}{\beta_n} - 1 > 0. \tag{2.8}$$

Under these circumstances the viscous term in (2.5) has quantitatively more power; now the nonlinear growth $\sim \kappa_n^{7/2}$ has to compete with dissipation $\sim \kappa_n^{3+2(n+1)\mu_n}$.

The hierarchy of differential inequalities in (2.5) implies that if such anomalous scaling in the κ_n occurs, then there is an upper bound on all the μ_n exponents if a true singularity actually occurs. For if

$$\mu_n > \frac{1}{4(n+1)} \tag{2.9}$$

for any $n \geq 3$, then during the development of a large fluctuation of κ_n, the magnitude of the damping term will eventually exceed the nonlinear growth so $d\kappa_n/dt$ reverses sign and no singularity is possible. The somewhat surprising conclusion is that a sufficiently divergent spectrum of small scales (sufficient 'anomalous' scaling) will regularize Navier–Stokes dynamics.

Strict algebraic scaling is not at all essential for the mechanism we are describing here. A scale separation of this sort for the κ_n in potentially singular or truly singular events is similar in spirit, and complementary in nature, to anomalous scaling for moments of velocity differences. In the inertial range scaling of velocity differences it is the presence of ever larger velocity amplitude fluctuations at ever smaller scales that is reflected in the scaling exponents. Scale separation of the κ_n would likewise be a consequence of the presence of relatively larger velocity fluctuations at ever smaller scales deep in the dissipation range in the temporal vicinity of (near) singular events. It is interesting that the Navier–Stokes equations place a limit on the magnitude of such fluctuations if solutions ever exhibit true singularities. We remark that the dissipative effect of scale separation, i.e., the first term $-\nu\kappa_{n+1}^{2n+2}\kappa_n^{-2n+1}$ on the right hand side in formula (2.5), is *exact*. The rigorous (although perhaps not perfectly sharp) estimates used in the derivation of (2.5) involved only the other terms.

3 Conclusion

These observations lead naturally to suggestions for theoretical, computational and experimental investigations. It would be interesting to evaluate the κ_n for exact solutions, such as Lundgren's spiral solution (Lundgren (1982)), to see how the spectrum behaves. Another possibility is that the κ_n could be evaluated for numerical computations of potentially singular Euler flows (Kerr (1993)). Sufficient separation of scales in the Euler solutions would suggest that such structures would be extinguished in the presence of (normal Newtonian) viscosity. A further possibility would be to test the consistency of singularity formation scenarios in terms of the energy distribution. For example the log-normal energy spectrum

$$E(k,t) = \frac{K(t)}{\sqrt{2\pi\sigma(t)^2}} \frac{1}{k} e^{-\frac{1}{2}\left(\frac{\log(kL_0)}{\sigma(t)}\right)^2} \tag{3.1}$$

has total energy $K(t)$. If the parameter $\sigma(t)$ were to diverge to infinity, then this energy would be broadly dispersed into high wavenumbers. But the spectrum of wavenumber moments for this distribution is

$$\kappa_n(t) = L_0^{-1} e^{n\sigma(t)^2} \tag{3.2}$$

so that $\beta_n = n/3$, $\frac{\kappa_{n+1}}{\kappa_n} = (L_0\kappa_n)^{\frac{1}{n}}$ and

$$\mu = \frac{1}{n} > \frac{1}{4(n+1)}. \tag{3.3}$$

Hence a singularity of this form is impossible.

The κ_n are immediately accessible from direct numerical simulations of turbulent flows. This would provide the most straightforward investigation of the question of scale separation (and the possibility of scaling and/or anomalous scaling) for the κ_n, but it is not clear how important the practical limitations on the Reynolds number will be. It is also natural to investigate these questions in the context of shell models (Gledzer (1973) and Ohkitani and Yamada (1987)) which, although somewhat artificial, can nevertheless capture many quantitative features of energy transport among length scales (Jensen *et al* (1991), Kadanoff (1998)). Most intriguingly, velocity derivatives in a turbulent flow can be measured. Although in practice it may be difficult to separate spatial from temporal averaging, some related moments of the derivatives may be extracted to probe the spectrum of small scales in the turbulence corresponding to the κ_n. The open question of regularity of solutions of the Navier–Stokes equations remains an outstanding mathematical challenge. It is an exciting prospect that experimental observations may shed new light on this problem.

Acknowledgements We thank P. Constantin, R. Kerr and K. Sreenivasan for constructive and incisive comments on this paper. This work was supported in part by a grant from the US National Science Foundation, and was completed while both authors were resident at the Isaac Newton Institute in Cambridge during the 1999 Turbulence Programme. JDG thanks the Newton Institute for their financial support during part of his stay there.

A   Appendix

The differential inequality for κ_n displayed in (2.5) can easily be found using two ladder inequalities for H_n (Doering and Gibbon (1995)),

$$\frac{1}{2}\dot{H}_n \leq -\nu H_{n+1} + c_{n,r}^{(1)} \|\nabla\mathbf{u}\|_\infty H_n, \tag{A.1}$$

$$\frac{1}{2}\dot{H}_n \geq -\nu H_{n+1} - c_{n,r}^{(1)} \|\nabla\mathbf{u}\|_\infty H_n, \tag{A.2}$$

where the $c_{n,r}^{(1)}$ is an absolute constant depending only on n and r, $\|\nabla u\|_\infty$ means the supremum over space of all derivatives of all components of the velocity field. Firstly we extend our definition of the κ_n to

$$\kappa_{n,r} = \left(\frac{H_n}{H_r}\right)^{\frac{1}{2(n-r)}} \tag{A.3}$$

and then use the inequalities in (A.1) and (A.2) on the H_n and H_r to obtain

$$(n-r)\dot{\kappa}_{n,r} \leq -\nu\left[\left(\frac{\kappa_{n+1,r}}{\kappa_{n,r}}\right)^{2(n-r)+2}\kappa_{n,r}^3 - \kappa_{n,r}\kappa_{r+1,r}^2\right] + c_{n,r}^{(2)}\|\nabla u\|_\infty\kappa_{n,r}, \tag{A.4}$$

where $c_{n,r}^{(2)}$ is another absolute constant. To estimate $\|\nabla u\|_\infty$ in terms of $\kappa_{n,r}$ we use a Gagliardo–Nirenberg inequality valid for $n \geq 3$ (see for example Doering and Gibbon (1995)),

$$\|\nabla u\|_\infty \leq c_{n,r}^{(3)}\kappa_{n,0}^{5/2}\|u\|_2. \tag{A.5}$$

Then we note two facts: $\kappa_{r+1,r} \leq \kappa_{n,r}$ and $\kappa_{n,0} \leq \kappa_{n,r}$ for $r < n$. Hence (A.4) becomes

$$(n-r)\dot{\kappa}_{n,r} \leq -\nu\left[\left(\frac{\kappa_{n+1,r}}{\kappa_{n,r}}\right)^{2(n-r)+2} - 1\right]\kappa_{n,r}^3 + c_{n,r}^{(4)}\kappa_{n,r}^{7/2}\|u\|_2. \tag{A.6}$$

Finally we set $r = 0$ and drop the second label and write $\kappa_{n,0} \equiv \kappa_n$ and $c_{n,r}^{(4)} \equiv c_n$ to obtain (2.5).

To discuss the Navier–Stokes equations with an additive forcing function $f(x)$ to extend the result to a (potentially) turbulent steady state we replace the H_n with

$$F_n(t) = \int_\Omega \left(|\nabla^n u|^2 + |\nabla^n u_f|^2\right) dV \tag{A.7}$$

where $u_f = L^2\nu^{-1}f(x)$. Then the F_n obey the same equations as the H_n in (A.1) and (A.2) with the respective addition and subtraction of an extra term $\nu\lambda_0^{-2}F_n$. The length λ_0 is defined by $\lambda_0^{-2} = L^{-2} + \lambda_f^{-2}$ where $\lambda_f \leq L$ is the smallest scale cutoff in $f(x)$ where L is the domain length. The H_n in the κ_n become F_n and the only change to equation (2.5) is the addition of a term $\nu\lambda_0^{-2}\kappa_n$ which does not change the argument or the observation concerning anomalous scaling and regularity.

References

Batchelor G.K. and Townsend A.A. (1949), 'The nature of turbulent fluid motion at large wave-numbers', *Proc. Royal Soc. Lond.* **A199**, 238–255.

Benzi R., Ciliberto S., Tripiccione C., Baudet C., Massaioli C. and Succi S. (1993), 'Extended self-similarity in turbulent flow', *Phys. Rev. E* **48**, R29–32.

Caffarelli L., Kohn R. and Nirenberg L. (1982), 'Partial regularity of suitable weak solutions of the Navier–Stokes equations', *Commun. Pure Appl. Math.* **35**, 771–831.

Constantin P. and Fefferman Ch. (1994), 'Scaling exponents in fluid turbulence: some analytic results', *Nonlinearity* **7**, 41–58.

Constantin P. and Foias C. (1988), *Navier–Stokes Equations* (University of Chicago Press).

Doering C.R. and Gibbon J.D. (1995), *Applied Analysis of the Navier–Stokes Equations* (Cambridge University Press).

Foias C., Guillopé C. and Temam R. (1981), 'New a priori estimates for the Navier–Stokes equations in dimension 3', *Commun. P.D.E.* **6**, 329–359.

Frisch U. (1995), *Turbulence: the Legacy of A.N. Kolmogorov* (Cambridge University Press).

Gibbon J.D. (1996), 'A voyage around the Navier–Stokes equations', *Physica D* **92**, 133–139.

Gledzer E.B. (1973), 'System of hydrodynamic type admitting 2 quadratic integrals of motion', *Sov. Phys. Dokl.* **18**, 216–217.

Jensen M., Paladin G. and Vulpiani A. (1991), 'Intermittency in a cascade model for three-dimensional turbulence', *Phys. Rev. A* **43**, 798–805.

Kadanoff L.P. (1998), 'A cascade model of turbulence', in *Advances in Turbulence*, pp. 201–202, ed. U. Frisch (Kluwer Academic Publishers).

Kerr R. (1993), 'Evidence for a singularity of the three-dimensional, incompressible Euler equations', *Phys. Fluids A* **5**, 1725–1746.

Kolmogorov A.N. (1941), 'Dissipation of energy in locally isotropic turbulence', *Dokl. Akad. Nauk SSSR,* **32(1)**, 16–18. (Reprinted in *Proc. Royal Soc. Lond.* **A434**, 15–17, (1991)).

Kuo A. Y.-S. and Corrsin S. (1971), 'Experiments on internal intermittency and fine-structure distribution functions in fully turbulent fluid', *J. Fluid Mech.* **50**, 285–319.

Leray J. (1934), 'Essai sur le mouvement d'un liquide visqueux emplissant l'espace', *Acta Math.* **63**, 193–248.

Lundgren T. (1982), 'Strained spiral vortex model for turbulent fine structures', *Phys. Fluids* **25**, 2193–2203.

Ohkitani K. and Yamada M. (1987), 'Lyapunov spectrum of a chaotic model of three-dimensional turbulence', *J. Phys. Soc. Japan* **56**, 4210–4213.

Sreenivasan K.R. (1985), 'On the fine-scale intermittency of turbulence',*J. Fluid Mech.* **151**, 81–103.

Sreenivasan K.R. and Meneveau C. (1988), 'Singularities of the equations of fluid motion', *Phys. Rev. A* **38**, 6287–6295.

Temam R. (1988), *Infinite Dimensional Dynamical Systems in Mechanics and Physics,* **68**, Springer series in Applied Mathematical Sciences (Springer-Verlag).

Turbulent Advection and Breakdown of the Lagrangian Flow

Krzysztof Gawędzki

1   Introduction

There are traditionally two different ways to describe fluids in motion. The Eulerian description studies the time evolution of fluid velocity in fixed points of the physical space. The Lagrangian description, in contrast, follows the time evolution of velocity along the fluid particle trajectories obeying the ordinary differential equation

$$\frac{d\mathbf{x}}{dt} \;=\; \mathbf{v}(t, \mathbf{x}). \tag{1.1}$$

One of the earliest quantitative laws of fully developed turbulence concerns the average behavior of such **Lagrangian trajectories**. Summarizing available observations, e.g. those of balloon probes released in the atmosphere, L.F. Richardson (1926) noted that the rate of growth of the mean square separation between two Lagrangian trajectories is proportional to the four-thirds power of the distance, instead of being distance-independent, as in the Brownian diffusion. In other words, the Richardson **dispersion law** states that, in mean,

$$\frac{d\rho^2}{dt} \;\propto\; \rho^{4/3}, \tag{1.2}$$

where $\rho(t)$ is the time t distance between two Lagrangian trajectories $\mathbf{x}(t)$ and $\mathbf{x}'(t)$ satisfying the ordinary differential equation (1.1). The mean-field type relation (1.2) is compatible with the Kolmogorov (1941) theory stating that the velocity differences $(\mathbf{v}(t, \mathbf{x}) - \mathbf{v}(t, \mathbf{x}'))$ scale in mean as $\rho^{1/3}$. Indeed, it follows from Eq. (1.1) that

$$\frac{d\rho^2}{dt} \;=\; 2\,(\mathbf{x} - \mathbf{x}') \cdot (\mathbf{v}(\mathbf{x}) - \mathbf{v}(\mathbf{x}')).$$

The Richardson law deserves, however, a more detailed discussion since it implies a rather surprising behavior of the Lagrangian trajectories in fully turbulent flows. This behavior and its influence on transport properties of flows are the subjects of this article.

2 Breakdown of the deterministic Lagrangian flow

First, let us note that the differential equation (1.2) is solved by

$$\rho^{2/3} = \rho_0^{2/3} + C t. \tag{2.1}$$

In the limit when the initial (i.e. time zero) separation ρ_0 tends to zero, this reduces to the relation

$$\rho^2 \propto t^3. \tag{2.2}$$

Of course, some care is needed since the law (1.2) was observed for separations in the inertial range, i.e. for $\rho \gg \eta$, where η is the scale where viscous dissipation sets in. In the limit of large Reynolds numbers $Re \to \infty$ (i.e. in fully developed turbulence), however, η tends to zero and the growth (2.2) should hold for arbitrarily short separations. In the turbulent atmosphere, for example, the viscous scale η is of the order of a millimeter so that neglecting it when we look at scales of meters or even kilometers is a reasonable approximation.

Different ways of averaging (means of powers versus powers of means) may change the proportionality factors or even, slightly, the functional form of the phenomenological dispersion law (1.2) (e.g. the power of ρ). In the observations, the averages were taken over pairs of trajectories and different turbulent velocity fields. However, since we expect the trajectory separation to have a self-averaging property, the dispersion law (2.2) should already hold in a fixed typical velocity realization after averaging over initial positions and/or times of release of two very close trajectories. Hence the Richardson law allows us to draw a surprising conclusion that at $Re = \infty$

infinitesimally close Lagrangian trajectories separate in finite time.

What is this strange behavior of trajectories? It is not common among the solutions of ordinary differential equations encountered elsewhere in physics. For example, in integrable dynamical systems the distance between two trajectories

$$\rho \simeq \mathcal{O}(1)$$

(that is what makes our planetary system stable in the 2-body approximation). In dissipative systems,

$$\rho \simeq \mathcal{O}(e^{-\lambda t})$$

(which is another stabilizing effect in our direct environment). Finally, in the chaotic dynamical systems (e.g. those describing many flows at moderately large Reynolds numbers)

$$\rho \simeq \mathcal{O}(e^{\lambda t}), \tag{2.3}$$

with the Lyapunov exponent $\lambda > 0$ signaling sensitive dependence on initial conditions. Even the last case, however, with the nearby trajectories separating fast, is quite different from the behavior (2.2). Indeed, for the exponential separation (2.3), infinitesimally close trajectories keep shadowing each other and never separate to a finite distance: chaos and fully developed turbulence are quite different phenomena despite undeniable successes of the theory of chaotic dynamical systems in describing the onset of turbulence at intermediate Reynolds numbers, see e.g. Ruelle (1991).

In fact the **explosive separation** (2.2) has a quite dramatic consequence: it means that, when $Re \rightarrow \infty$, the very concept of the Lagrangian trajectory determined by its 1-time position breaks down. Indeed, suppose that such trajectories exist in almost all realizations of the velocity field and, as random functions of velocity, depend continuously on the initial positions in any reasonable sense. Then the expectation values of powers of the time t distance between trajectories would tend to zero as the initial trajectory separation is taken to zero, in contradiction with the behavior (2.1). The breakdown of the deterministic Lagrangian flow at $Re \rightarrow \infty$ is a consequence of a breakdown of the theorem about the existence and uniqueness of solutions of the ordinary differential equation (1.1). The theorem requires $\mathbf{v}(t, \mathbf{x})$ to be Lipschitz in $\mathbf{x}$, i.e. $|\mathbf{v}(t, \mathbf{x}) - \mathbf{v}(t, \mathbf{x}')| \sim |\mathbf{x} - \mathbf{x}'|$ for small $|\mathbf{x} - \mathbf{x}'|$, whereas at $Re = \infty$ the velocities are only Hölder-continuous: $|\mathbf{v}(t, \mathbf{x}) - \mathbf{v}(t, \mathbf{x}')| \sim |\mathbf{x} - \mathbf{x}'|^\alpha$ with the exponent $\alpha < 1$ ($\alpha \simeq \frac{1}{3}$). Is then the Lagrangian description of the fluid breaking down completely at $Re = \infty$?

We expect that for typical $Re = \infty$ velocities one may still maintain a probabilistic description of Lagrangian trajectories, with such objects as the probability distribution function (PDF) $P^{t,s}(\mathbf{x}, \mathbf{y}|\mathbf{v})$ of the time s position $\mathbf{y}$ of the trajectory passing at time t through a point $\mathbf{x}$ continuing to make sense. For a regular velocity, the trajectories are deterministic and

$$P^{t,s}(\mathbf{x}, \mathbf{y}|\mathbf{v}) = \delta(\mathbf{y} - \mathbf{x}_{t,\mathbf{x}}(s)), \qquad (2.4)$$

where $\mathbf{x}_{t,\mathbf{x}}(s)$ denotes the solution of the differential equation (1.1) (with running time s) passing at time t through $\mathbf{x}$ and $P^{t,s}(\mathbf{x}, \mathbf{y}|\mathbf{v})$ is the solution of the first-order partial differential equation

$$(\partial_t + \mathbf{v}(t, \mathbf{x}) \cdot \nabla_\mathbf{x}) \, P^{t,s}(\mathbf{x}, \mathbf{y}|\mathbf{v}) = 0, \qquad (2.5)$$

$$(\partial_s - \nabla_\mathbf{y} \cdot \mathbf{v}(s, \mathbf{y})) \, P^{t,s}(\mathbf{x}, \mathbf{y}|\mathbf{v}) = 0, \qquad (2.6)$$

with the condition $P^{s,s}(\mathbf{x}, \mathbf{y}|\mathbf{v}) = \delta(\mathbf{x} - \mathbf{y})$. It is the kernel of the evolution operator $P^{t,s}(\mathbf{v})$ which may be written as a time-ordered exponential

$$P^{t,s}(\mathbf{v}) = \mathcal{T} \, e^{-\int_s^t d\sigma \, \mathbf{v}(\sigma) \cdot \nabla}$$
$$= \sum_{n=0}^{\infty} (-1)^n \int_s^t d\sigma_n \ldots \int_s^{\sigma_2} d\sigma_1 \, \mathbf{v}(\sigma_n) \cdot \nabla \ldots \mathbf{v}(\sigma_1) \cdot \nabla.$$

Suppose now that the trajectory equation (1.1) is perturbed by adding a small white noise to its right hand side, i.e. by replacing (1.1) by

$$\frac{d\mathbf{x}}{ds} = \mathbf{v}(s, \mathbf{x}) + \sqrt{2\kappa}\,\frac{d\beta(s)}{ds}, \tag{2.7}$$

where $\beta(s)$ is a d-dimensional Brownian motion and $\kappa > 0$. The stochastic differential equation (2.7) defines a Markov process which may be characterized by the transition probabilities $P_\kappa^{t,s}(\mathbf{x}, \mathbf{y}|\mathbf{v})$ satisfying the linear equation[1]

$$\left(\partial_t + \mathbf{v}(t, \mathbf{x}) \cdot \nabla_{\mathbf{x}} - \kappa\nabla_{\mathbf{x}}^2\right) P_\kappa^{t,s}(\mathbf{x}, \mathbf{y}|\mathbf{v}) = 0,$$

with the initial condition $P_\kappa^{s,s}(\mathbf{x}, \mathbf{y}|\mathbf{v}) = \delta(\mathbf{x} - \mathbf{y})$. In the operator language,

$$\begin{aligned}
P_\kappa^{t,s}(\mathbf{v}) &= \mathcal{T}\,e^{\int_s^t[-\mathbf{v}(\sigma)\cdot\nabla + \kappa\nabla^2]\,d\sigma} \\
&= \sum_{n=0}^\infty (-1)^n \int_s^t d\sigma_n \cdots \int_s^{\sigma_2} d\sigma_1\; e^{\kappa(t-\sigma_n)\nabla^2}\,\mathbf{v}(\sigma_n)\cdot\nabla\,e^{\kappa(\sigma_n - \sigma_{n-1})\nabla^2}\cdots \\
&\qquad\qquad \cdots e^{\kappa(\sigma_2 - \sigma_1)\nabla^2}\,\mathbf{v}(\sigma_1)\cdot\nabla\,e^{\kappa(\sigma_1 - s)\nabla^2}. \tag{2.8}
\end{aligned}$$

The Markov process with the transition probabilities $P_\kappa^{t,s}(\mathbf{x}, \mathbf{y}|\mathbf{v})$ may be defined for a large class of velocity fields $\mathbf{v}$, without much restriction on the regularity of $\mathbf{v}$, see Stroock & Varadhan (1979). For a regular velocity, one recovers then in the limit $\kappa \searrow 0$ the deterministic process with the transition kernels given by Eq. (2.4). The $\kappa \searrow 0$ limit of the transition probabilities $P_\kappa^{t,s}(\mathbf{x}, \mathbf{y}|\mathbf{v})$ exists, however, in many other cases (for subsequences, in general) even when the equation (1.1) has many solutions passing through a given point or has no such solutions. The limiting transition probabilities give weak solutions of the evolution equation (2.6), i.e. the ones satisfying the equation in the distributional sense. We shall say in such a situation that Eq. (1.1) admits a Markov process solution.

We conjecture that the typical turbulent 3-dimensional velocities at the infinite Reynolds number admit (diffuse) Markov process solutions of the flow equation (1.1), i.e. that they lead to stochastic Lagrangian flows. Such velocities are weak solutions of the incompressible Euler equation

$$\partial_t \mathbf{v} + \mathbf{v}\cdot\nabla\mathbf{v} = \mathbf{f} - \nabla p, \qquad \nabla\cdot\mathbf{v} = 0,$$

which dissipate energy, in accordance with the energy cascade picture of fully developed turbulence. $\mathbf{f}$ denotes the external (intensive) force inducing the flow and p is the pressure. As shown recently by Duchon & Robert (2000) by a refinement of the Kolmogorov (1941) argument, weak solutions of the Euler equation that dissipate energy with a locally finite rate have the spatial

---

[1]For later convenience, we write the formulae for the backward transition probabilities with $t \geq s$; the case $t < s$ may be treated similarly

Hölder continuity exponent (defined appropriately) equal to the Kolmogorov value $\frac{1}{3}$. The stochastic character of the Lagrangian flow for such velocities should provide the mechanism for the local energy dissipation.

This important idea which seems to be a direct consequence, in the limit of high Reynolds numbers, of the Richardson dispersion law or of the Kolmogorov scaling of velocity differences has been rarely stressed in the past. A somewhat different version of it appeared in the study of weak solutions of the Euler equation by Brenier (1989) and Shnirelman (2000). In order to make our conjecture about a stochastic character of the Lagrangian flow at $Re = \infty$ more plausible, we shall discuss in what follows a recent analytic study of the Lagrangian trajectories in velocities of a simple statistical ensemble with the spatial Hölder continuity of typical realizations built in. As we shall see, such velocities lead, indeed, to stochastic Lagrangian flows.

3   Stochastic Lagrangian flows, an example

Following R.H. Kraichnan (1968, 1994) who initiated a study of transport properties of simple multiscale velocities decorrelated in time, let us consider a Gaussian ensemble of d-dimensional velocity fields with mean zero and 2-point function given by

$$\langle v^\alpha(t,\mathbf{x})\, v^\beta(s,\mathbf{x}')\rangle \;=\; \delta(t-s)\left(D_0\delta^{\alpha\beta} - d^{\alpha\beta}(\mathbf{x}-\mathbf{x}')\right) \qquad (3.1)$$

with D_0 a constant and $d^{\alpha\beta}(\mathbf{x}) \propto r^\xi$ for small $r \equiv |\mathbf{x}|$; $0 < \xi < 2$ is the parameter of the ensemble. The constant D_0 drops out in the correlations of the velocity differences. E.g.

$$\langle (v^\alpha(t,\mathbf{x}) - v^\alpha(t,\mathbf{0}))\,(v^\beta(s,\mathbf{x}) - v^\beta(s,\mathbf{0}))\rangle \;=\; \delta(t-s)\left(d^{\alpha\beta}(\mathbf{x}) + d^{\beta\alpha}(\mathbf{x})\right).$$

One may take

$$D_0 - d^{\alpha\beta}(\mathbf{x}) \;=\; \int \frac{e^{i\mathbf{k}\cdot\mathbf{x}}}{(\mathbf{k}^2 + L^{-2})^{(d+\xi)/2}}\left(A\frac{k^\alpha k^\beta}{k^2} + B(\delta^{\alpha\beta} - \frac{k^\alpha k^\beta}{k^2})\right)\, d\mathbf{k} \quad (3.2)$$

with the infrared cutoff L. For $A = 0$ the typical velocities are incompressible, $\nabla \cdot \mathbf{v} = 0$, whereas for $B = 0$ one obtains potential flows, $\mathbf{v} = \nabla\phi$. It is convenient to characterize the resulting velocity ensemble by, besides ξ, the **compressibility degree**

$$\wp \;=\; \frac{\langle (\nabla \cdot \mathbf{v})^2\rangle}{\langle (\nabla\mathbf{v})^2\rangle} \;=\; \frac{1}{1 + (d-1)\frac{B}{A}}$$

contained between 0 and 1. The degree $\wp = 0$ corresponds to the incompressible case whereas $\wp = 1$ corresponds to the potential one.

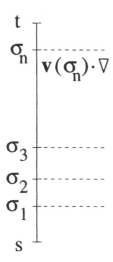

Figure 1

The above ensemble is not very realistic in its assumption of temporal velocity decorrelation. It builds in, however, the scaling behavior of the velocities in space with the Hölder continuity of typical realizations: $|\mathbf{v}(t,\mathbf{x})-\mathbf{v}(t,\mathbf{x}')| \sim |\mathbf{x}-\mathbf{x}'|^{\xi/2}$ up to logarithmic corrections.

In the velocity fields of the Kraichnan ensemble, we shall consider the Lagrangian trajectories perturbed first by a small noise, i.e. satisfying the stochastic differential equation (2.7). As noticed by Le Jan & Raimond (1998), the limit $\kappa \searrow 0$ of the transition probabilities $P_\kappa^{t,s}(\mathbf{x},\mathbf{y}|\mathbf{v})$ of the Markov process solutions of Eq. (2.7) may be easily controlled despite the poor regularity of the velocities of the Kraichnan ensemble. These authors have rewritten the expression (2.8) for $P_\kappa^{t,s}(\mathbf{v})$ in the Wick-ordered form:[2]

$$P_\kappa^{t,s}(\mathbf{v}) \;=\; :\mathcal{T}\,\mathrm{e}^{\int_s^t[-\mathbf{v}(\sigma)\cdot\nabla + (\kappa+\frac{1}{2}D_0)\nabla^2]\,d\sigma}: \;=\; \sum_{n=0}^{\infty} P_{\kappa,n}^{t,s}(\mathbf{v}),$$

where

$$P_{\kappa,n}^{t,s}(\mathbf{v}) \;=\; (-1)^n \int_s^t d\sigma_n \dots \int_s^{\sigma_2} d\sigma_1 \;:\mathrm{e}^{(\kappa+\frac{1}{2}D_0)(t-\sigma_n)\nabla^2}\,\mathbf{v}(\sigma_n)\cdot\nabla$$

$$\times \mathrm{e}^{(\kappa+\frac{1}{2}D_0)(\sigma_n-\sigma_{n-1})\nabla^2}\dots\,\mathrm{e}^{(\kappa+\frac{1}{2}D_0)(\sigma_2-\sigma_1)\nabla^2}\,\mathbf{v}(\sigma_1)\cdot\nabla\,\mathrm{e}^{(\kappa+\frac{1}{2}D_0)(\sigma_1-s)\nabla^2}:.$$

---

[2]The Wick ordering of a monomial in Gaussian random variables, denoted by colons around the monomial, replaces it by a polynomial with the same highest degree term but orthogonal to all lower degree polynomials in the L^2 scalar product defined by the Gaussian probability measure

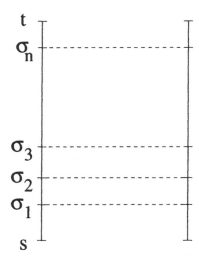

Figure 2

The Wick-ordered monomials $P_{\kappa,n}^{t,s}(\mathbf{v})$ of $\mathbf{v}$ may be represented by Figure 1. The homogeneous Wick polynomials of different degrees are orthogonal in the L^2 scalar product w.r.t. the Gaussian measure of the velocity process. By establishing the bound

$$\sum_{n=0}^{\infty}\langle |P_{\kappa,n}^{t,s}(\mathbf{v})g|^2\rangle \leq e^{\frac{1}{2}D_0(t-s)\nabla^2}|g|^2$$

where g are functions on $\mathbf{R}^d$ and the average is over the Gaussian ensemble of velocities, one shows then that the series giving $P^{t,s}(\mathbf{v})f$ converges in the space of square-integrable functionals of $\mathbf{v}$, so also for almost all (a.a.) $\mathbf{v}$, as long as the g are bounded. It defines for a.a. velocities the transition probabilities $P_{\kappa}^{t,s}(\mathbf{x},\mathbf{y}|\mathbf{v})$ of a Markov process which are continuous as functions of $\kappa \geq 0$. In particular, one has

$$\mathcal{P}_{1,\kappa}^{t,s}(\mathbf{x},\mathbf{y}) \equiv \langle P_{\kappa}^{t,s}(\mathbf{v})(\mathbf{x},\mathbf{y})\rangle = P_{\kappa,0}^{t,s}(\mathbf{v}) = e^{(\kappa+\frac{1}{2}D_0)|t-s|\nabla^2}(\mathbf{x},\mathbf{y}). \quad (3.3)$$

Indeed, the expectations of Wick-ordered monomials of positive degree vanish so that only the $\mathbf{v}$-independent $n = 0$ term contributes to the expectation.

The essential question that we wish to address is about the nature of the Markov process with the transition probabilities $P^{t,s}(\mathbf{x},\mathbf{y}|\mathbf{v})$ obtained in the limit $\kappa \to 0$. Are the limiting transition probabilities concentrated at single points $\mathbf{y}$ leading to deterministic Lagrangian trajectories in a fixed velocity realization, or, on the contrary, do they stay diffuse? A way to study this question is to examine the joint PDF (probability distribution function) of the equal-time values of two solutions of Eq. (2.7) averaged over the velocity ensemble:

$$\mathcal{P}_{2,\kappa}^{t,s}(\mathbf{x}_1, \mathbf{x}_2; \mathbf{y}_1, \mathbf{y}_2) = \langle P_\kappa^{t,s}(\mathbf{x}_1, \mathbf{y}_1|\mathbf{v})\, P_\kappa^{t,s}(\mathbf{x}_2, \mathbf{y}_2|\mathbf{v})\rangle$$
$$= \sum_{n=0}^{\infty} \langle P_{\kappa,n}^{t,s}(\mathbf{x}_1, \mathbf{y}_1|\mathbf{v})\, P_{\kappa,n}^{t,s}(\mathbf{x}_2, \mathbf{y}_2|\mathbf{v})\rangle.$$

The averages on the right hand side may be represented by Figure 2 with the broken-line propagators given by the spatial part $(D_0 - d(\cdot))$ of the velocity 2-point functions (3.1). The whole sum becomes the perturbative expansion for the heat kernel (i.e. kernel of the exponential) of the second-order differential operator:

$$\mathcal{P}_{2,\kappa}^{t,s}(\underline{\mathbf{x}}; \underline{\mathbf{y}}) = e^{-|t-s|\mathcal{M}_{2,\kappa}}(\underline{\mathbf{x}}; \underline{\mathbf{y}}), \qquad (3.4)$$

where

$$\mathcal{M}_{2,\kappa} = -(\kappa + \tfrac{1}{2}D_0)\,(\nabla_{\mathbf{x}_1}^2 + \nabla_{\mathbf{x}_2}^2) - (D_0\delta^{\alpha\beta} - d^{\alpha\beta}(\mathbf{x}_1 - \mathbf{x}_2))\nabla_{x_1^\alpha}\nabla_{x_2^\beta}$$
$$= d^{\alpha\beta}(\mathbf{x}_1 - \mathbf{x}_2)\nabla_{x_1^\alpha}\nabla_{x_2^\beta} - \kappa\,(\nabla_{\mathbf{x}_1}^2 + \nabla_{\mathbf{x}_2}^2) + \tfrac{1}{2}D_0\,(\nabla_{\mathbf{x}_1} + \nabla_{\mathbf{x}_2})^2. \quad (3.5)$$

The last term drops out in the translation-invariant sector. In other words, two solutions of Eq. (2.7) undergo, in their relative motion averaged over the velocity realizations, an effective diffusion with the diffusion coefficient dependent on their relative position.

The PDF $\mathcal{P}_{2,\kappa}^{t,s}(r; \rho)$ of the distance ρ between the time s positions of two solutions, given their time t distance r, is expressed by the heat kernel of the operator $\mathcal{M}_{2,\kappa}$ restricted to the translation- and rotation-invariant sector. The latter becomes an explicit second order differential operator in the radial variable:

$$\mathcal{M}_{2,\kappa}^{\mathrm{inv}} = -C\,r^{\xi-a}\partial_r\, r^a\partial_r - 2\,\kappa\, r^{-d+1}\partial_r\, r^{d-1}\partial_r$$

with the exponent $a = a(\xi, \wp)$ (for simplicity, we give the formula after the removal of the infrared cutoff L in Eq. (3.2)). The operator $\mathcal{M}_{2,\kappa}^{\mathrm{inv}}$ may be transformed by a change of variables and a similarity transformation to a Schrödinger operator on the half-line. In particular, in the $\kappa \to 0$ limit,

$$\mathcal{M}_2^{\mathrm{inv}} = C'\,u^c\,[-\partial_u^2 + \tfrac{b^2 - \frac{1}{4}}{u^2}]\,u^{-c}, \qquad (3.6)$$

where $u = r^{\frac{2-\xi}{2}}$ and $b = b(\xi, \wp)$, $c = c(\xi, \wp)$. The $\kappa \to 0$ limit $\mathcal{P}_2^{t,s}(r; \rho)$ of the PDF $\mathcal{P}_{2,\kappa}^{t,s}(r, \rho)$ can be explicitly controlled. As discovered in Gawędzki & Vergassola (2000), two different regimes appear in this limit, depending on the strength of the singularity at $r = 0$ in the operator $\mathcal{M}_2^{\mathrm{inv}}$.

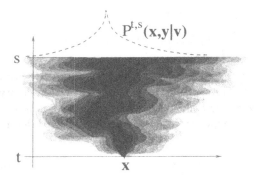

Figure 3

1. Weakly compressible regime

For weak compressibility $\wp < \frac{d}{\xi^2}$, which corresponds to $b < 1$, the distance
PDF $\mathcal{P}_2^{t,s}(r, \rho)$ is an integrable function of ρ and stays such in the limit $r \to 0$:

$$\lim_{r \to 0} \mathcal{P}_2^{t,s}(r; \rho) \, d\rho \ \propto \ \left(\frac{\rho^{2-\xi}}{|t-s|}\right)^{1-b} e^{-\frac{\rho^{2-\xi}}{4C'|t-s|}} \frac{d\rho}{\rho} . \tag{3.7}$$

This behavior excludes the concentration of the transition kernels $P^{t,s}(\mathbf{x}, \mathbf{y}|\mathbf{v})$
at single points $\mathbf{y}$. In particular, we obtain the Richardson dispersion law in
the form

$$\lim_{r \to 0} \int \rho^2 \, \mathcal{P}_2^{0,t}(r, \rho) \, d\rho \ \propto \ t^{\frac{2}{2-\xi}}$$

indicating an explosive separation of the Lagrangian trajectories and repro-
ducing for $\xi = \frac{4}{3}$ the mean growth (2.2). As we see, the trajectories in a fixed
typical realization of the velocity fields are not determined by the initial position but rather form a Markov process with diffuse transition probabilities
$P^{t,s}(\mathbf{x}, \mathbf{y}|\mathbf{v})$, as predicted above and illustrated by Figure 3.

2. Strongly compressible regime

For strong compressibility $\wp \geq \frac{d}{\xi^2}$, i.e. for $b \geq 1$, one observes a different
behavior due to the strong repulsive singularity at $u = 0$ in the operator
(3.6):

$$\mathcal{P}_2^{t,s}(r; \rho) \, d\rho \ = \ p^{t,s}(r) \, \delta(\rho) \, d\rho \ + \ \text{regular}$$

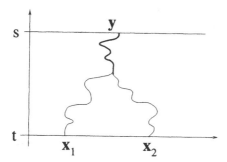

Figure 4

with the coefficient $p^{t,s}(r)$ of the delta-function converging to 1 when $r \to 0$ or $|t - s| \to \infty$. In particular,

$$\lim_{r \to 0} \mathcal{P}_2^{t,s}(r; \rho)\, d\rho \;=\; \delta(\rho)\, d\rho$$

in this regime implying the concentration of the transition kernels $P^{t,s}(\mathbf{x}, \mathbf{y}|\mathbf{v})$ at single points $\mathbf{y}$ and the existence, in a fixed typical velocity realization, of Lagrangian trajectories determined by their initial positions. The presence of the singular term in $\mathcal{P}_2^{t,s}(r; \rho)$ indicates, however, an **implosive collapse** of distinct Lagrangian trajectories with a positive probability which grows in time, see Figure 4.

One infers that in Hölder-continuous velocity fields, there is a competition of the tendencies of two Lagrangian trajectories to separate or to collapse explosively. For weak compressibility, this is the first tendency that wins. The trapping effects, however, increase with the degree of compressibility and lead, in the Kraichnan ensemble, to a sharp transition in the behavior of the trajectories at $\wp = \frac{d}{\xi^2}$.

4   Advection of scalars by Kraichnan velocities

What are the consequences on the hydrodynamics of fluids of the observed dramatic behaviors of Lagrangian trajectories, violating the Newton-Leibniz existence-and-uniqueness of solutions paradigm of the ordinary differential equations, a prerequisite of the theory of differentiable dynamical systems? The answer to this question may be the clue to a consistent theory of fully developed turbulence. Here, we shall content ourselves with discussing the transport properties in the flows induced by the velocities of the Kraichnan ensemble. Following Gawędzki & Vergassola (2000), we shall observe that these properties differ drastically for weak and strong compressibility as a result of the different behaviors of Lagrangian trajectories. For concreteness, we shall concentrate on the passive transport of a scalar quantity $\theta(t, \mathbf{x})$ (called the tracer) whose evolution is described by the advection-diffusion equation

$$\partial_t \theta + (\mathbf{v} \cdot \nabla)\theta - \kappa \nabla^2 \theta = f, \tag{4.1}$$

where $f(t, \mathbf{x})$ denotes a scalar source. In the incompressible case, θ may also describe the temperature field or the density of pollutant.

When the velocity field is sufficiently regular, it is easy to solve the above linear equation for θ. For $\kappa = 0$ and $f = 0$, the scalar is simply carried by the Lagrangian flow

$$\theta(t, \mathbf{x}) = \theta(s, \mathbf{x}_{t,\mathbf{x}}(s)),$$

where, as before, $\mathbf{x}_{t,\mathbf{x}}(s)$ denotes the Lagrangian trajectory passing at time t through point $\mathbf{x}$. Note that the forward evolution of θ corresponds to the backward Lagrangian flow. In the presence of the source f, the scalar is also created or depleted along the trajectory:

$$\theta(t, \mathbf{x}) = \theta(s, \mathbf{x}_{t,\mathbf{x}}(s)) + \int_s^t f(\sigma, \mathbf{x}_{t,\mathbf{x}}(\sigma)) \, d\sigma.$$

Finally, for $\kappa > 0$, the same formulae hold except that $\mathbf{x}_{t,\mathbf{x}}(s)$ should be taken as a solution of Eq. (2.7) for the noisy Lagrangian trajectories and the right hand side should be averaged over the noise. Recalling that $P_\kappa^{t,s}(\mathbf{x}, \mathbf{y}|\mathbf{v})$ is the PDF of the values of $\mathbf{x}_{t,\mathbf{x}}(s)$, we obtain in this case

$$\theta(t, \mathbf{x}) = \int P_\kappa^{t,s}(\mathbf{x}, \mathbf{y}|\mathbf{v}) \, \theta(s, \mathbf{y}) \, d\mathbf{y} + \int_s^t \left(\int P_\kappa^{t,\sigma}(\mathbf{x}, \mathbf{y}|\mathbf{v}) \, f(\sigma, \mathbf{y}) \, d\mathbf{y} \right) d\sigma. \tag{4.2}$$

The right hand side still makes sense for a.a. velocities of the Kraichnan ensemble. In the incompressible case, it defines in this case a weak solution[3] of the linear differential equation (4.1).

---

[3] Eq. (4.1) is an infinite-dimensional stochastic differential equation due to the white temporal dependence of $\mathbf{v}$ and it should be treated according to the Stratonovich prescription; the weakness of the solution (4.2) is, however, due to its poor spatial regularity resulting from the similar property of typical velocities leading to stochastic Lagrangian trajectories

Let us assume that we are given a random distribution of the scalar at time s, independent of the (later) velocities, and we wish to study its distribution at a later time t. In free decay, i.e. in the absence of the source f, by taking averages over the initial distribution and over the velocities, we obtain for the 1-point function of θ

$$\langle \theta(t,\mathbf{x}) \rangle = \int \mathcal{P}^{t,s}_{1,\kappa}(\mathbf{x},\mathbf{y}) \, \langle \theta(s,\mathbf{y}) \rangle \, d\mathbf{y} = \int e^{(\kappa+\frac{1}{2}D_0)|t-s|\nabla^2}(\mathbf{x},\mathbf{y}) \, \langle \theta(s,\mathbf{y}) \rangle \, d\mathbf{y},$$

see Eq. (3.3). In other words, the 1-point function decays diffusively with the initial diffusivity κ increased by the **eddy diffusivity** $\frac{1}{2}D_0$. Similarly, for the 2-point function,

$$\langle \theta(t,\mathbf{x}_1)\,\theta(t,\mathbf{x}_2) \rangle = \int \mathcal{P}^{t,s}_{2,\kappa}(\mathbf{x}_1,\mathbf{x}_2;\mathbf{y}_1,\mathbf{y}_2) \, \langle \theta(s,\mathbf{y}_1)\,\theta(s,\mathbf{y}_2) \rangle \, d\mathbf{y}_1\,d\mathbf{y}_2 , \quad (4.3)$$

and, for the N-point one,

$$\left\langle \prod_{n=1}^{N} \theta(t,\mathbf{x}_n) \right\rangle = \int \mathcal{P}^{t,s}_{N,\kappa}(\underline{\mathbf{x}};\underline{\mathbf{y}}) \left\langle \prod_{n=1}^{N} \theta(s,\mathbf{y}_n) \right\rangle d\underline{\mathbf{y}},$$

where

$$\mathcal{P}^{t,s}_{N,\kappa}(\underline{\mathbf{x}};\underline{\mathbf{y}}) = \left\langle \prod_{n=1}^{N} P^{t,s}_{\kappa}(\mathbf{x}_n,\mathbf{y}_n|\mathbf{v}) \right\rangle = \mathcal{P}^{s,t}_{N,\kappa}(\underline{\mathbf{x}};\underline{\mathbf{y}})$$

is the joint PDF of the time s positions $\underline{\mathbf{y}}$ of the Lagrangian trajectories, given their time t positions $\underline{\mathbf{x}}$, see Figure 5 (the last equality is implied by the stationarity and time-reflection invariance of the Kraichnan ensemble of velocities).

It is easy to see that the PDF's $\mathcal{P}^{t,s}_{N,\kappa}(\underline{\mathbf{x}};\underline{\mathbf{y}})$ are again given by the heat kernels of second order differential operators:

$$\mathcal{P}^{t,s}_{N,\kappa}(\underline{\mathbf{x}};\underline{\mathbf{y}}) = e^{-|t-s|\mathcal{M}_{N,\kappa}}(\underline{\mathbf{x}};\underline{\mathbf{y}}),$$

where

$$\mathcal{M}_{N,\kappa} = \sum_{n<m} d^{\alpha\beta}(\mathbf{x}_n - \mathbf{x}_m)\nabla_{x_n^\alpha}\nabla_{x_m^\beta} - \kappa \sum_{n=1}^{N} \nabla^2_{\mathbf{x}_n} + \tfrac{1}{2}D_0\left(\sum_n \nabla_{\mathbf{x}_n}\right)^2.$$

If the initial distribution of θ is homogeneous and isotropic then the PDF $\mathcal{P}^{t,s}_{2,\kappa}(\mathbf{x}_1,\mathbf{x}_2;\mathbf{y}_1,\mathbf{y}_2)$ in (4.3) may be replaced by its version $\mathcal{P}^{t,s}_{2,\kappa}(r,\rho)$ invariant under translations and rotations. In particular, taking the limit $r \to 0$, we infer that, for $\kappa = 0$ and in the weakly compressible case $\wp < \frac{d}{\xi^2}$, the mean tracer energy density

$$\bar{e}_\theta(t) \equiv \langle \tfrac{1}{2}\theta(t)^2 \rangle = \tfrac{1}{2}\int \mathcal{P}^{t,s}_2(0,\rho) \, \langle \theta(s,\mathbf{y})\,\theta(s,\mathbf{0}) \rangle \, d\rho < \bar{e}_\theta(s) \quad (4.4)$$

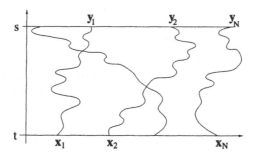

Figure 5

if the 2-point function $\langle\,\theta(s,\mathbf{y})\,\theta(s,\mathbf{0})\rangle$ decays for large $\rho = |\mathbf{y}|$. Indeed, the latter is bounded by its value at $\rho = 0$ so that the relation (4.4) follows since $\mathcal{P}_2^{t,s}(0,\rho)$ is a strictly positive PDF, see Eq. (3.7). Hence the mean energy density of the scalar decreases with time (i.e. is dissipated) even for $\kappa = 0$ due to the diffuse character of the Lagrangian flow. To the contrary, in the strongly compressible case $\wp \geq \frac{d}{\xi^2}$, the limiting PDF $\mathcal{P}_2^{t,s}(0,\rho) = \delta(\rho)$ and the mean energy density is conserved in the unforced evolution at $\kappa = 0$:

$$\bar{e}_\theta(t) \;=\; \bar{e}_\theta(s)\,.$$

In the presence of the random Gaussian source f, independent of velocities and the initial distribution of θ, with mean zero and covariance

$$\langle f(t,\mathbf{x})\,f(s,\mathbf{x}')\rangle \;=\; \delta(t-s)\,\chi(\tfrac{|\mathbf{x}-\mathbf{x}'|}{L})\,,$$

the 1-point function of the scalar diffuses as before and for the 2-point function, we obtain from Eq. (4.2)

$$\langle\,\theta(t,\mathbf{x}_1)\,\theta(t,\mathbf{x}_2)\rangle \;=\; \int \mathcal{P}_{2,\kappa}^{t,s}(\underline{\mathbf{x}};\underline{\mathbf{y}})\,\langle\,\theta(s,\mathbf{y}_1)\,\theta(s,\mathbf{y}_2)\rangle\,d\underline{\mathbf{y}}$$

$$+ \int\limits_s^t \Big(\int \mathcal{P}_{2,\kappa}^{t,\sigma}(\underline{\mathbf{x}};\underline{\mathbf{y}})\,\chi(\tfrac{|\mathbf{y}_1-\mathbf{y}_2|}{L})\,d\underline{\mathbf{y}}\Big)d\sigma. \qquad (4.5)$$

As noticed by Kraichnan (1968), this 2-point function solves the closed evolution equation

$$\partial_t \langle\,\theta(\mathbf{x}_1)\,\theta(\mathbf{x}_2)\rangle \;=\; -\mathcal{M}_{2,\kappa}\,\langle\,\theta(\mathbf{x}_1)\,\theta(\mathbf{x}_2)\rangle \;+\; \chi(\tfrac{|\mathbf{x}_1-\mathbf{x}_2|}{L})$$

$$= -d^{\alpha\beta}(\mathbf{x}_1 - \mathbf{x}_2)\,\langle\nabla_\alpha\theta(\mathbf{x}_1)\,\nabla_\beta\theta(\mathbf{x}_2)\rangle \;-\; 2\kappa\,\langle\nabla\theta(\mathbf{x}_1)\cdot\nabla\theta(\mathbf{x}_2)\rangle$$

$$+\, \chi\!\left(\tfrac{|\mathbf{x}_1-\mathbf{x}_2|}{L}\right), \qquad (4.6)$$

as a consequence of the heat kernel form of the PDF $\mathcal{P}^{t,\sigma}_{2,\kappa}(\mathbf{x};\underline{\mathbf{y}})$, see Eqs. (3.4) and (3.5). In the weakly compressible regime $\wp < \frac{d}{\xi^2}$, the scalar 2-point function reaches a stationary state. Taking the limit $\mathbf{x}_2 \to \mathbf{x}_1$ for $\kappa > 0$, one infers from Eq. (4.6) the stationary energy balance for the tracer:

$$\bar{\epsilon}_\theta \;=\; \tfrac{1}{2}\chi(0)\,,$$

where $\bar{\epsilon}_\theta$ denotes the mean dissipation rate $\bar{\epsilon}_\theta = \langle\kappa(\nabla\theta)^2\rangle$ of tracer energy. Thus $\bar{\epsilon}_\theta$ turns out to be κ-independent (the **dissipative anomaly**) and equal to the rate $\frac{1}{2}\chi(0)$ of the scalar energy injection by the random source f. The energy injected at large scales of order $\gtrsim L$ cascades in a **direct cascade** process towards short distances or large wavenumbers where it is dissipated. Eq. (4.6) may be easily solved for the stationary 2-point function of the scalar giving, for $\kappa = 0$ and $r \ll L$,

$$\langle\theta(\mathbf{x})\,\theta(\mathbf{0})\rangle \;=\; A_2(\chi)\,L^{2-\xi} - \text{const.}\,\bar{\epsilon}_\theta\, r^{2-\xi} + \mathcal{O}(L^{-2})\,,$$

where $r \equiv |\mathbf{x}|$ or, for the scalar 2-point structure function,

$$S_2(r) \equiv \langle(\theta(\mathbf{x}) - \theta(\mathbf{0}))^2\rangle \;\propto\; \bar{\epsilon}_\theta\, r^{2-\xi}\,,$$

in agreement with a dimensional analysis of the stationary version of Eq. (4.6).

In the strongly compressible regime $\wp > \frac{d}{\xi^2}$, the behavior of the scalar 2-point function (4.5) in the limit $\kappa \to 0$ is quite different. Now the 2-point function does not stabilize but has a constant contribution growing linearly in time. The dissipation rate vanishes in the $\kappa \to 0$ limit and the tracer energy, instead of cascading towards short distances, is pumped into the constant mode at a constant rate equal to the injection rate $\frac{1}{2}\chi(0)$. This signals the presence of an **inverse cascade** of tracer energy towards small wavenumbers. The 2-point structure function of the scalar stabilizes and its stationary limit satisfies the equation

$$\mathcal{M}_{2,\kappa}\,\langle(\theta(\mathbf{x}) - \theta(\mathbf{0})^2\rangle \;=\; 2(\chi(0) - \chi(\tfrac{|\mathbf{x}|}{L}))$$

from which one infers that

$$S_2(r) \;\propto\; r^{2-\xi}$$

for $r \gg L$ now.

The higher point functions of the tracer also show very different behavior in the two regimes. The important question is whether the higher structure functions $S_N(r) \equiv \langle (\theta(\mathbf{x}) - \theta(0))^N \rangle$ scale with powers $\frac{N}{2}(2 - \xi)$, as the dimensional analysis first discussed for the passive scalar problem by Obukhov (1949) and Corrsin (1951) would suggest, or whether they show scaling with anomalous powers depending non-linearly on the order of the function, signaling **intermittency** of the scalar statistics, with enhanced fluctuations at short distances.

First, it is not difficult to see that the higher equal-time correlators of the scalar advected by the Kraichnan velocities again satisfy closed evolution equations generalizing Eq. (4.6):

$$\partial_t \left\langle \prod_{n=1}^{N} \theta(\mathbf{x}_n) \right\rangle = -\mathcal{M}_{N,\kappa} \left\langle \prod_{n=1}^{N} \theta(\mathbf{x}_n) \right\rangle + \sum_{p<q} \left\langle \prod_{n\neq p,q} \theta(\mathbf{x}_n) \right\rangle \chi \left(\frac{|\mathbf{x}_p - \mathbf{x}_q|}{L} \right). \quad (4.7)$$

In the weakly compressible regime, the correlation functions stabilize at long times and, besides, we expect that the limits $t \to \infty$ and $\kappa \to 0$ commute. The stationary equal-time correlations satisfy a similar equation but with the vanishing left hand side. By inverting operators $\mathcal{M}_{N,\kappa}$, one may then compute the stationary N-point functions of the scalar recursively, a rare situation, indeed, since in most hydrodynamical problems the evolution equations for the correlation functions do not close.

It was realized in Shraiman & Siggia (1995), Gawędzki & Kupiainen (1995) and Chertkov, Falkovich, Kolokolov & Lebedev (1995) that, for large L, the $\kappa = 0$ higher point functions of the scalar solving the stationary version of Eqs. (4.7) are dominated by the contributions from the scaling zero modes of the operators $\mathcal{M}_N$. The scaling dimensions of such modes are not constrained by the dimensional analysis but are accessible perturbatively or numerically (note that the zero modes drop out in the stationary version of Eq. (4.7)). Indeed, it was shown in Gawędzki & Kupiainen (1995) and Bernard, Gawędzki & Kupiainen (1996) for the incompressible case and extended in Gawędzki & Vergassola (2000) to the weakly compressible one with $\wp < \frac{d}{\xi^2}$ that, for small ξ (i.e. for rough Kraichnan velocities),

$$\left\langle \prod_{n=1}^{N} \theta(\mathbf{x}_n) \right\rangle = A_N(\chi) \, L^{\frac{N}{2}(2-\xi) - \zeta_N} \, \varphi_N(\underline{\mathbf{x}}) + \mathcal{O}(L^{-2+\mathcal{O}(\xi)}) + \cdots,$$

where $\varphi_N(\underline{\mathbf{x}})$ are scaling zero modes of the operators $\mathcal{M}_N$, i.e.

$$\mathcal{M}_N \, \varphi_N(\underline{\mathbf{x}}) = 0, \qquad \varphi_N(\lambda \underline{\mathbf{x}}) = \lambda^{\zeta_N} \, \varphi_N(\underline{\mathbf{x}})$$

with

$$\zeta_N = \frac{N}{2}(2-\xi) - \frac{N(N-2)(1+2\wp)}{2(d+2)} \xi + \mathcal{O}(\xi^2).$$

The coefficients $A_N(\chi)$ are non-universal amplitudes and the dots denote terms that do not depend on, at least, one variable and, as such, do not

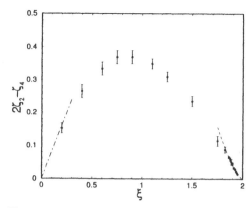

Figure 6

contribute to the correlations of scalar differences. In particular, one infers the anomalous scaling of the structure functions for small r and large L,

$$S_N(r) \propto L^{\frac{N}{2}(2-\xi)-\zeta_N} r^{\zeta_N},$$

demonstrating the short distance intermittency of the tracer.

This **zero mode dominance** of the stationary higher point functions of the scalar has been exhibited by the perturbative analysis of the Green functions of operators $\mathcal{M}_N$ around $\xi = 0$. A parallel perturbative analysis by Chertkov, Falkovich, Kolokolov & Lebedev (1995) and Chertkov & Falkovich (1996) in the powers of inverse dimension gave concordant results. A more difficult perturbative expansion around $\xi = 2$ further confirmed the zero mode dominance picture, see Shraiman & Siggia (1995, 1996), Pumir, Shraiman & Siggia (1997), Gat, L'vov & Procaccia (1997) and Balkovsky, Falkovich & Lebedev (1997). The numerical results give a similar picture for all values of ξ, see Figure 6 representing the values of the 4-point function anomalous exponent $(2\zeta_2 - \zeta_4)$ obtained by Frisch, Mazzino & Vergassola (1998) in numerical simulations of the 3-dimensional incompressible Kraichnan model.

What is the physical meaning of the zero modes of the operators $\mathcal{M}_N$ that dominate the short distance asymptotics of the N-point functions of scalar differences? They constitute **slow modes** of the effective diffusion of Lagrangian trajectories with generators $\mathcal{M}_N$. Indeed, for a generic scaling function $\psi_N(\mathbf{x})$ of scaling dimension σ_N, viewed as a function of time t positions of N Lagrangian trajectories, the effective time evolution is

$$\langle \psi_N \rangle_t \equiv \int P_N^{0,t}(\mathbf{x}, \mathbf{y}) \, \psi_N(\mathbf{y}) \, dy \ \sim \ t^{\frac{\sigma_N}{2-\xi}} \tag{4.8}$$

for large t, i.e. it exhibits a **super-diffusive** growth. But if $\psi_N = \phi_N$ is a zero mode of $\mathcal{M}_N$ then the above expectation is conserved in time (such

statistically conserved modes are accompanied by descendent ones for which
the time growth is slower than (4.8), see Bernard, Gawędzki & Kupiainen
(1997)).

In the strongly compressible phase with the inverse cascade of scalar energy,
the behavior of the higher structure functions is different. In fact, only the
lower ones stabilize, but the ones that do, scale normally on large distances.
In this regime one can find exactly the stationary form of the PDF of the
scalar difference:

$$\langle \delta(\vartheta - \tfrac{\theta(\mathbf{x}) - \theta(\mathbf{0})}{r^{(2-\xi)/2}}) \rangle \;\; \propto \;\; [\chi(0) + C'\vartheta^2]^{-b-\frac{1}{2}}$$

for $r \gg L$. Its scaling form indicates that there is no intermittency in the
inverse cascade of the tracer (the non-Gaussianity is scale-independent). The
poor decay of the PDF at infinity corresponds to the fact that only lower
order functions reach a stationary regime. For $r \ll L$, all stabilizing structure
functions scale with the same power equal to $(2 - \xi)b$ signaling an extreme
short distance intermittency.

As we see, the transport of a scalar tracer by velocities distributed according to the Kraichnan ensemble shows two different phases characterized
by different directions of the scalar energy cascades and different degrees of
intermittency. The phase transition occurs at the value $\wp = \frac{d}{\xi^2}$ of the compressibility degree, where the behavior of the Lagrangian trajectories changes
drastically from the explosive separation to the implosive collapse. These two
phases are quite reminiscent of the behavior of the 3-dimensional versus 2-
dimensional developed turbulence. That suggests that one should put more
stress on the Lagrangian methods in studying the latter, not quite a new
lesson, see e.g. Pope (1994), but with a new twist pointing to the stochastic
character of the Lagrangian flow in weak solutions of the Euler equations. Of
course, the NS and the Euler equations, unlike the scalar advection one, are
non-linear, a difference that, as many understood cases of non-linear dynamics
teach us, is far from being minor. Also, they describe velocity fields that are
tranformed when carried along their own Lagrangian trajectories and, due
to pressure, there are non-local interactions present. Some of these effects,
however, may be examined already in synthetic velocity ensembles. It seems
that the study of such ensembles has a potential to teach us important lessons
that we have to master on the way to an understanding of fully developed
turbulence.

References

Balkovsky, E., Falkovich, G., Lebedev, V. (1997) 'Three-point correlation function
    of a scalar mixed by an almost smooth random velocity field', *Phys. Rev.* **E 55**,
    R4881–R488.

Bernard, D., Gawędzki, K., Kupiainen, A. (1996) 'Anomalous scaling of the N-point functions of a passive scalar', *Phys. Rev.* **E 54**, 2564–2572.

Bernard, D., Gawędzki, K., Kupiainen, A. (1997) 'Slow modes in passive advection', *J. Stat. Phys.* **90**, 519–569.

Brenier, Y. (1989) 'The least action principle and the related concept of generalized flow for incompressible perfect fluids', *J. Amer. Math. Soc.* **2**, 225–255.

Chertkov, M., Falkovich, G. (1996) 'Anomalous scaling exponents of a white-advected passive scalar', *Phys. Rev Lett.* **76**, 2706–2709.

Chertkov, M., Falkovich, G., Kolokolov, I., Lebedev, V. (1995) 'Normal and anomalous scaling of the fourth-order correlation function of a randomly advected scalar', *Phys. Rev.* **E 52**, 4924–4941.

Corrsin, S. (1951) 'On the spectrum of isotropic temperature fluctuations in an isotropic turbulence', *J. Appl. Phys.* **22**, 469–473.

Duchon, J., Robert, R. (2000) 'Inertial energy dissipation for weak solutions of incompressible Euler and Navier–Stokes equations', *Nonlinearity* **13**, 249–255.

Frisch, U., Mazzino, A., Vergassola, M. (1998) 'Intermittency in passive scalar advection', *Phys. Rev. Lett.* **80**, 5532–5535

Gat, O., L'vov, V.S., Procaccia, I. (1997) 'Perturbative and non-perturbative analysis of the 3'rd order zero modes', *Phys. Rev.* **E 56**, 406–416.

Gawędzki, K., Kupiainen, A. (1995) 'Anomalous scaling of the passive scalar, *Phys. Rev. Lett.* **75**, 3834–3837.

Gawędzki, K., Vergassola, M. (2000) 'Phase transition in the passive scalar advection', cond-mat/9811399, to appear in *Physica D.*

Kolmogorov, A.N. (1941) 'The local structure of turbulence in incompressible viscous fluid for very large Reynolds' numbers', *C. R. Acad. Sci. URSS* **30**, 301–305.

Kraichnan, R.H. (1968) 'Small-scale structure of a scalar field convected by turbulence', *Phys. Fluids* **11**, 945–963.

Kraichnan, R.H. (1994) 'Anomalous scaling of a randomly advected passive scalar', *Phys. Rev. Lett.* **72**, 1016–1019.

Le Jan, Y., Raimond, O. (1998) 'Solutions statistiques fortes des équations différentielles stochastiques', *C. R. Acad. Sci.,* **327**, 893.

Obukhov, A.M. (1949) 'Structure of the temperature field in a turbulent flow', *Izv. Akad. Nauk SSSR, Geogr. Geofiz.* **13**, 58–69.

Pope, S.B. (1994) 'Lagrangian pdf methods for turbulent flows', *Ann. Rev. Fluid Mech.* **26**, 23–63.

Pumir, A., Shraiman, B.I., Siggia, E.D. (1997) 'Perturbation theory for the δ-correlated model of passive scalar advection near the Batchelor limit', *Phys. Rev.* **E 55**, R1263–R1266.

Richardson, L.F. (1926) 'Atmospheric diffusion shown on a distance-neighbour graph', *Proc. R. Soc. Lond.* **A110**, 709–737.

Ruelle, D. (1991) 'The turbulent fluid as a dynamical system', in: *New Perspectives in Turbulence*, L. Sirovich, ed., Springer, 123–138.

Shnirelman, A. (2000) 'Weak solutions with decreasing energy of incompressible Euler equations', *Commun. Math. Phys.* **210**, 541–603.

Shraiman, B., Siggia, E. (1995) 'Anomalous scaling of a passive scalar in turbulent flow', *C.R. Acad. Sci.* **321**, 279–284.

Shraiman, B.I., Siggia, E.D. (1996) 'Symmetry and scaling of turbulent mixing', *Phys. Rev. Lett.* **77**, 2463–2466.

Stroock, D.W., Varadhan, S.R.S. (1979) *Multidimensional Diffusion Processes*, Springer.

Growth of Magnetic Fluctuations in a Turbulent Flow

Gregory Falkovich

1 Introduction

It is believed that the magnetic fields of planets, stars and galaxies have their origin in dynamo action driven by motions of conducting fluids; see Moffatt (1978), Parker (1979), Childress and Gilbert (1995) and references therein. Inhomogeneous flow stretches magnetic lines, amplifying the field, while the field produces electric currents that dissipate energy and diffuse the field due to finite resistivity. The outcome of the competition between amplification and diffusion depends on the type of the flow. Most flows in Nature are turbulent. Kinematic dynamo theory is presented here for turbulent conductive fluids. We consider the evolution of magnetic noise on a time scale much larger than the correlation time of a fluctuating flow. A consistent description of the long-time evolution of the small-scale field taking into account diffusion remained elusive for a long time, only the second moment being found (Kraichnan and Nagarajan, 1967; Kazantsev, 1968). When diffusivity κ is small, the field is almost frozen into the fluid and is expected to grow exponentially like an infinitesimal material line element. To what extent this is offset by a transversal contraction that eventually brings diffusion into play depends on the flow statistics. We describe magnetic fluctuations generated below the viscous scale of turbulence where the velocity field can be locally approximated by a linear profile. There is no folding of magnetic lines at such small scales, just stretching and contraction. A spatial smoothness of the velocity permits a systematic analysis of the Lagrangian path dynamics and of the respective evolution of the magnetic field. We find analytically the moments and multi-point correlation functions of the magnetic field at small yet finite diffusivity κ. We show that the field is concentrated in long narrow strips and describe anomalous scalings and angular singularities of the multi-point correlation functions which are manifestations of the field's intermittency. We relate the field growth rate to the Lyapunov spectrum of the flow for both incompressible and compressible cases. We show that the magnetic flux always decays at large time.

2 Growth rate in an incompressible flow

We consider the long-standing problem of how turbulence excites inhomogeneous fluctuations of magnetic field (Batchelor, 1950; Kraichnan and Nagarajan, 1967; Kazantsev, 1968). Since the growth rate is inversely proportional to the turnover time, the fastest growth is for the fluctuations shorter than the viscous scale of turbulence, they are the first to reach saturation and strongly influence the subsequent evolution of the system (Kulsrud and Anderson, 1992). It is then important to have a systematic description of the field that has emerged from the linear dynamo phase when the only equation to solve is that for the magnetic field

$$\partial_t \mathbf{B} + (\mathbf{v} \cdot \partial)\mathbf{B} = (\mathbf{B} \cdot \partial)\mathbf{v} + \kappa \triangle \mathbf{B}, \qquad (2.1)$$

while the velocity statistics are presumed to be known. In many astrophysical applications the viscosity-to-diffusivity ratio is large and there is a wide interval of scales between viscous and diffusive cut-offs where velocity is spatially smooth while the magnetic field has a nontrivial spatial structure to be described below. Given the initial condition, the solution of (2.1) is then conveniently written in Fourier space

$$\mathbf{B}(\mathbf{k}, t) = \widehat{W}(t)\, \mathbf{B}(\mathbf{k}(0), 0) \exp\left(-\kappa \int_0^t k^2(t')\, dt'\right)$$

where the wavevectors evolve as $\mathbf{k}(t') = \widehat{W}^T(t, t')\mathbf{k}(t)$ and the final condition is $\mathbf{k}(t) = \mathbf{k}$ (Zeldovich *et al.* 1984). The evolution matrix $\widehat{W}$ satisfies $d\widehat{W}(t, t')/dt = \hat{\sigma}(t)\widehat{W}(t, t')$, with $\widehat{W}(t', t') = 1$ and $\widehat{W}(t) = \widehat{W}(t, 0)$. We adopt here the methods of Lagrangian path analysis developed recently for the related problem of passive scalar (Shraiman and Siggia, 1994; Chertkov *et al.* 1995). The moments of $\mathbf{B}$ are to be calculated by two independent averaging: first, (trivial) average over initial statistics and, second, average over velocity statistics. Without any loss of generality, we assume the initial statistics to be homogeneous, isotropic and Gaussian, with zero mean and the variance $\langle B_\alpha(\mathbf{k}, 0)B_\beta(\mathbf{k}', 0)\rangle = P_{\alpha\beta}(\mathbf{k})k^2 f(k^2)\delta(\mathbf{k} + \mathbf{k}')$. The solenoidal projector $P_{\alpha\beta} = \delta_{\alpha\beta} - k_\alpha k_\beta/k^2$ ensures that $\mathbf{B}$ is divergence-free. The initial magnetic noise is concentrated at the scale L, we use $f(k^2) = L^5 \exp(-k^2 L^2)$ whenever an explicit calculation is performed. Inhomogeneous advection produces smaller and smaller scales and balances with diffusion at the scale $r_d = \sqrt{\kappa/\lambda_1}$, here λ_1 is the growth rate of a material line element (the largest Lyapunov exponent). The magnetic Reynolds number L/r_d is assumed to be large. Inhomogeneous advection also produces larger scales, the theory below is valid until $L \exp(\lambda_1 t)$ is less than the viscous scale.

The wavectors are of order $1/L$ initially and for some period they remain much larger than $1/r_d$. This is the stage where dynamics is insensitive to diffusion so that the field is frozen into the fluid like in a perfect conductor, see Childress and Gilbert (1995) and (4.1) below. At some time $t_d \propto \log(L/r_d)$, the wavevectors reach $1/r_d$, transversal contractions bring diffusion into play and the new regime starts which is the main subject of this paper. It is supposed that t_d is much larger than the velocity correlation time τ and we can then carry over from random matrix theory the well-known Iwasawa decomposition (see e.g. Goldhirsch *et al.* 1987): $\hat{W}(t) = \hat{R}\hat{D}\hat{S}$. For any fixed time, $\hat{R}$ is an SO(3) rotation matrix, $\hat{D}$ is diagonal with $D_{ii}(t) = \exp[\rho_i(t)]$ and the shearing matrix $\hat{S}$ is upper triangular with unit elements on the diagonal. The sum of ρ_i's vanishes by incompressibility and the ratios ρ_i/t tend at $t \to \infty$ to the three Lyapunov exponents λ_i (arranged in decreasing order); see e.g. Crisanti *et al.* (1993).

Since **B** is expressed via $\hat{W}$, our aim now is to reformulate the average over $\hat{\sigma}$ into that over $\hat{W}$. The average over $\hat{R}$ is equivalent to the integration over the directions of the vectors involved. The matrix $\hat{S}$ tends for each realization to a time-independent form. Indeed, it will be shown below that the magnetic field moments grow exponentially in time and the dominant contributions come from realizations with $\rho_1 \gg \rho_2$ at $t \to \infty$. For such realizations the matrix $\hat{S}$ is frozen at large times, the eigendirections of $\widehat{W}^T\widehat{W}$ do not fluctuate in time and fix an orthogonal basis: $\mathbf{n}_1$, $\mathbf{n}_2$, $\mathbf{n}_3$, with respective stretching (contraction) rates $2\lambda_1$, $2\lambda_2$, $2\lambda_3$ (Goldhirsch *et al.* 1987). We shall also see below that the time-independent random matrix elements of $\hat{S}$ influence only constant factors, not of interest for the space-time dependencies studied here. The problem is reduced then, on one hand, to the integration over the angles of the vectors involved and, on the other hand, to the average over the statistics of ρ_1 and ρ_2.

The moments of the magnetic field can be obtained by considering $\mathbf{B}^2(t)$ averaged over initial statistics:

$$\mathbf{B}^2(t) = \int d^3q f(q) e^{-2\kappa \mathbf{q}\hat{\Lambda}\mathbf{q}} q^2 \mathrm{Tr}[\widehat{W}\hat{P}(\mathbf{q})\widehat{W}^T],$$

then taking powers and averaging over velocity. We have changed the variables $\mathbf{q} = \widehat{W}^T\mathbf{k}$ to reexpress the average in terms of the wavevectors at $t = 0$. When $t \gg |\lambda_3|^{-1}$ the main contribution to $\hat{\Lambda}(t) = \int_0^t dt' \widehat{W}^{-1}(t')\widehat{W}^{-1,T}(t')$ is given by $t' \gg |\lambda_3|^{-1}$. Changing the variables $\mathbf{q} = \hat{S}^T\mathbf{Q}$ we eliminate the constant $\hat{S}$-matrix from the diffusive exponent. The dependence on $\hat{S}$ remains only in the prefactor quadratic in $\mathbf{Q}$ and in f, so that in averaging over velocity we may replace $\hat{S}$ by the unit matrix. It follows that $\mathbf{q}\hat{\Lambda}\mathbf{q} = \int_0^t dt \sum Q_i^2 e^{-2\rho_i} \equiv U(Q, \rho)$ and in the $\mathbf{q}^2\mathrm{Tr}$ term the main contribution is $e^{2\rho_1}(Q_2^2 + Q_3^2)$. In any given realization the growth of the field is thus described by a simple formula

$$\mathbf{B}^2(t) \simeq \int d^3Q f(Q) \exp(-2\kappa U) e^{2\rho_1}(Q_2^2 + Q_3^2).$$

This shows that initially diffusion is unimportant ($\kappa U \ll 1$) and B^2 grows as $e^{2\rho_1}$, i.e. as a square of a material line element. At $t \simeq t_d = |\lambda_3|^{-1}\log(L/r_d)$, the diffusive exponent starts to decrease substantially and the growth rate is reduced. Asymptotically for $t \gg t_d$, it is clear that the realizations and the q's dominating the growth are such that the quadratic form κU at the diffusive exponent remains $O(1)$. Note that for growing functions one has with exponential accuracy $\int^t dt' \exp(-\rho_3(t')) \propto \exp(-\rho_3(t))$. The integration over Q_3 is thus restricted within the exponentially small interval: the diffusion exponent remains $O(1)$ only for initial wavevectors with such a small projection on the contraction direction $\mathbf{n}_3$ that the respective component does not reach $1/r_d$ during the time t (Zeldovich *et al.* 1984). Neglecting Q_3^2 comparatively to Q_2^2 and omitting numerical factors, we get after the integration over $\mathbf{Q}$

$$\mathbf{B}^2(t) \simeq \exp[2\rho_1(t)]\{1 + (r_d/L)^2 \exp[-2\rho_3(t)]\}^{-1/2}$$
$$\times \left\{1 + (r_d/L)^2 \int^t dt' \exp[-2\rho_2(t')]\right\}^{-3/2}. \tag{2.2}$$

The first curly bracket reduces for large times to $(L/r_d) \exp \rho_3$, the geometrical factor due to the condition of orthogonality to $\mathbf{n}_3$. In the second line the exponential term can be either comparable or larger than unity depending on the sign of ρ_2, which corresponds to the geometrical pictures (cone vs pancake in k-space) illustrated in Zeldovich *et al.* (1984).

Moments of (2.2) are to be averaged over the probability distribution $\mathcal{P}_t(\rho_1, \rho_2)$. When $t \gg \lambda_1^{-1}, \tau$, the theory of large deviations ensures that $\mathcal{P}_t(\rho_1, \rho_2) \propto \exp[-tH(\rho_1/t, \rho_2/t)]$, where the entropy $H(x, y)$ has a sharp minimum $H = 0$ at $x = \lambda_1$, $y = \lambda_2$ whose width decreases as $t^{-1/2}$ (Ellis, 1985) for the vector case; see e.g. Balkovsky and Fouxon (1999). The mean growth rate $\gamma(t) = \langle \log B^2(t) \rangle / 2t$ is then simply obtained by taking the logarithm of (2.2) and substituting there $\rho_i = \lambda_i t$ (strictly speaking, one had to average the logarithm over the initial measure as well, yet this differs by a correction decreasing as t^{-1}, the growth rate doesn't fluctuate at large time). At $t \ll t_d$ the growth rate $\gamma = \lambda_1$, as it has to be for a perfect conductor. During an intermediate stage $t \sim t_d$, γ decreases and eventually at $t \gg t_d$ it comes to an asymptotic value:

$$\gamma_\infty = \begin{cases} \lambda_1 + \lambda_3/2 = (\lambda_1 - \lambda_2)/2 & \text{for } \lambda_2 \geq 0, \\ \lambda_1 + \lambda_3/2 + 3\lambda_2/2 = (\lambda_2 - \lambda_3)/2 & \text{for } \lambda_2 \leq 0. \end{cases} \tag{2.3}$$

More compactly, $\gamma_\infty = \min\{(\lambda_1 - \lambda_2)/2, (\lambda_2 - \lambda_3)/2\}$, that is the growth rate is half the difference of the Lyapunov exponents of the same sign in the incompressible turbulence. Note that γ_∞ is independent of κ (that corresponds to a so-called fast dynamo, see Vainshtein and Zeldovich, 1972, Childress and Gilbert, 1995). Further, $\gamma_\infty \geq 0$ and $\gamma_\infty \to 0$ as $\lambda_2 \to \lambda_1$ or $\lambda_2 \to \lambda_3$, corresponding to the zero growth rate for axially symmetric cases. Both for time-reversible flow statistics and for 2D flow, $\lambda_2 = 0$ and $\gamma_\infty = \lambda_1/2$. Note that while 3D magnetic fluctuations do grow in 2D linear velocity profile (since the decaying component of the field is a source for growing components – see Zeldovich *et al.* 1984), our consideration for a completely 2D case (both flow and field are 2D) gives $\gamma_\infty = -\lambda_1/2 < 0$, that is no dynamo.

For isotropic Navier-Stokes turbulence, numerical data suggest $\lambda_2 \simeq \lambda_1/4$ (Girimaji and Pope, 1990), so that our prediction for the long-time growth rate in a typical realization is $\gamma_\infty \simeq 3\lambda_1/8$.

Let us emphasize that at $t \gg t_d$, the magnetic field grows while the magnetic flux (conserved in an ideal conductor) decreases with the rate $\gamma_\infty + \lambda_2 + \lambda_3 < 0$, independent of diffusivity.

3   Moments of the magnetic field

The moments with $n > 0$ all grow in a random incompressible flow with a nonzero Lyapunov exponent since $E_n = \log\langle B^{2n}\rangle/2t$ is a convex function of n (due to Hölder's inequality) with $E_0 = 0$ and $dE_n/dn(0) = \gamma \geq 0$. Even when $\gamma = 0$, E_n are positive for $n > 0$ if H has a finite width, that is if the flow is random (for $n = 1$ this was stated in Gruzinov *et al.* 1996). The growth of the $2n$-th moment at $t \ll t_d$ is determined by the average of $\exp(2n\rho_1)$. For $t \gg t_d$, the expression to average is either $\exp(n\rho_1 - n\rho_2)$ (with $\rho_2 > \log(r_d/L)$) or $\exp(n\rho_2 - n\rho_3)$ depending on whether the entropy function favours positive or negative ρ_2 (cf. Zeldovich *et al.* 1984). The formula $\langle \mathbf{B}^2 \rangle \propto \exp[(\lambda_1 - \lambda_2)t]$ was obtained previously for a permanent strain by Pearson (1959); its use for fluctuating strain in Zeldovich *et al* (1984) is incorrect since the moments of the field are not self-averaging unlike the growth rate. The function E_n is nonuniversal since it is determined by the saddle-point of $\mathcal{D}\rho_{1,2}$ integration which depends on the particular form of the entropy function. The saddle-point falls within the (universal) parabolic region of H around the minimum only for $n \ll (\lambda_1\tau)^{-1}$. Therefore, the temporal growth of the moments we calculate for a short-correlated velocity in Section 4 below.

Here, we continue with the general case to establish what is universal in the different-point correlation functions. We consider time-reversible statistics, where the scaling laws turn out to be universal. We start from the second moment $F_2(r) = \langle \mathbf{B}(0,t) \cdot \mathbf{B}(\mathbf{r},t) \rangle$ with $r \gg r_d$ which makes the diffusion nonessential. Averaging over the direction Ω of $\mathbf{r}$ as a substitute for the account of the R-matrix we get

$$F_2(r) \propto \left\langle \int d^3\mathbf{Q}\,d\Omega e^{2\rho_1}[Q_2^2 + Q_3^2]\exp[i\,(\mathbf{r}D\mathbf{Q})]f(Q) \right\rangle. \qquad (3.1)$$

For $\lambda_1 t > \log(L/r)$ the Q_3 dependence of $f(Q)$ can be neglected, integration over dQ_3 gives the factor $\delta(r_3)e^{-\rho-\rho_2/2}$ and after Gaussian integration over dQ_2 we arrive at

$$F_2(r) \propto \left\langle \int d\Omega L\delta(r_3)e^{\rho_1-\rho_2} \left(\frac{1}{2} - \frac{r_2^2 e^{-2\rho_2}}{4L^2} \right) \exp\left(-\frac{r_2^2 e^{-2\rho_2}}{4L^2} \right) \right\rangle. \qquad (3.2)$$

Using the parametrization $r_3 = r\cos\theta$, $r_2 = r\sin\theta\sin\varphi$ we obtain:

$$
F_2(r) = \frac{L}{r}\left\langle e^{\rho_1-\rho_2}\left[\exp\left(-\frac{r^2 e^{-2\rho_2}}{4L^2}\right)\right.\right.
$$
$$
\left.\left. +2\int_0^{\pi/2} d\varphi \sin^2\frac{\varphi}{2}\left(\frac{1}{2}-\frac{r^2\sin^2\varphi e^{-2\rho_2}}{4L^2}\right)\exp\left(-\frac{r^2\sin^2\varphi e^{-2\rho_2}}{4L^2}\right)\right]\right\rangle.
$$

Note that the domain of small φ's gives zero contribution to the integral. The role of the square bracket is that it restricts the average to the realizations with $\rho_2 > \log(r/L)$. Indeed, the realizations contributing have the advective exponent $\exp(ir\widehat{W}^{T,-1}\mathbf{q})$ of order unity. This requires $\rho_2 > \log(r/L)$ and the direction of contraction $\mathbf{n}_3$ to be almost perpendicular to $\mathbf{r}$, which gives a geometrical factor $(L/r)\exp\rho_3$ so that neglecting the order-unity factor one has a simple formula:

$$
F_2(r,t) \simeq \frac{L}{r}\int_{-\infty}^{\infty} d\rho_1 \int_{\log(r/L)}^{\infty} d\rho_2 e^{\rho_1-\rho_2-tH} \propto r^{-2-h}e^{E_2 t},
$$

where $h = \partial H/\partial y$ taken at $y = 0$ and at x given by the saddle point $\partial H/\partial x = 1$. Time reversibility means that $H(x,y) = H(x+y,-y)$ so $h = 1/2$ and $F_2 \propto r^{-5/2}\exp(E_2 t)$. At $r \ll L$ and $\lambda_1 t \ll \log(L/r)$, F_2 is r-independent. This generalizes the consideration of Kulsrud and Anderson (1992) for an arbitrary time-reversible statistics. To understand the simple geometrical picture behind this derivation, note that the integral over ρ_2 comes from $\rho_2 \simeq \log(r/L)$. That means that the field configurations in the form of strips with width r dominate $F_2(r)$. The angular integral comes from $\varphi \simeq 1$, that is the strips with the stretching direction almost parallel to $\mathbf{r}$ do not contribute (because of cancellations due to solenoidality).

Consider a fourth-order correlation function

$$
\langle(\mathbf{B}(0,t)\mathbf{B}(\mathbf{R_1},t))(\mathbf{B}(\mathbf{R_2},t)\mathbf{B}(\mathbf{R_3},t))\rangle =
$$
$$
\mathcal{A}(\mathbf{R_1},\mathbf{R_2}-\mathbf{R_3}) + \mathcal{B}(\mathbf{R_2},\mathbf{R_3}-\mathbf{R_1}) + \mathcal{C}(\mathbf{R_3},\mathbf{R_1}-\mathbf{R_2}). \tag{3.3}
$$

where three terms describe the Wick decomposition due to Gaussian integration over initial measure. Representation of (3.3) as an average over $\widehat{W}(t)$ statistics is straightforward, one gets, in particular, for the first term

$$
\mathcal{A}(\mathbf{r_1},\mathbf{r_2}) = \left\langle \int\prod_{j=1}^{2} d^3\mathbf{q}_j \left\{ \left[\mathbf{q}_j^2\mathrm{Tr}(\widehat{W}^T(t)\widehat{W}(t)) - (\mathbf{q}_j\widehat{W}^T(t)\widehat{W}(t)\mathbf{q}_j)\right]\right.\right.
$$
$$
\left.\left.\times f(q_j^2)\exp\left(i\mathbf{q}_j\widehat{W}^{-1}(t)\mathbf{r}_j\right)\exp\left[-2\kappa\int_0^t d\tau\left(\mathbf{q}_j\widehat{W}^{-1}(\tau)\widehat{W}^{-1,T}(\tau)\mathbf{q}_j\right)\right]\right\}\right\rangle. \tag{3.4}
$$

Integration over rotation R is equivalent to the substitution $\mathbf{r}_{1,2} \to \hat{\mathcal{R}}\mathbf{r}_{1,2}$ in (3.4), where $\hat{\mathcal{R}}$ is a matrix from SO(3), and then integrating over the group SO(3). Such a procedure will be denoted by $\langle\ldots\rangle_{\mathcal{R}}$. Eliminating the rotation matrix $\hat{R}$, changing to $\mathbf{Q}^{(1,2)} = \hat{T}^{-1,T}\mathbf{q}$, integrating with respect to $Q_1^{(1,2)}$ and finally neglecting all the exponentially small terms one arrives at

$$\mathcal{A} \approx \left\langle \left\langle e^{4\rho_1} \int \prod_{j=1}^2 dQ_{2,3}^{(j)} \left\{ (Q_2^{(j)})^2 \exp\left[i \left(r_2^{(j)} Q_2^{(j)} e^{-\rho_2} + r_3^{(j)} Q_3^{(j)} e^{\rho_1+\rho_2} \right) \right] \right. \right. $$
$$\left. \left. \times \exp\left[-2\kappa \int_{\tau_0}^t d\tau \left((Q_2^{(j)})^2 e^{-2\rho_2} + (Q_3^{(j)})^2 e^{2(\rho_1+\rho_2)} \right) \right] \exp\left(-L^2 (Q_2^{(j)})^2 \right) \right\} \right\rangle \right\rangle_{\mathcal{R}}.$$
(3.5)

We first consider the collinear case $\mathbf{r}_i = r_i \mathbf{n}$, $\mathbf{n} = (\sin\theta\cos\varphi, \sin\theta\sin\varphi, \cos\theta)$ and $\langle \rangle_{\mathcal{R}} = \int_0^\pi \sin\theta d\theta \int_0^{2\pi} d\varphi/4\pi$. The collinear case is special because at $\kappa = 0$ the $Q_3^{(1,2)}$ integration results in $\delta^2(\cos\theta)$, i.e. a singularity which is smoothed out by diffusion giving the factor $\sim \exp[-2\rho_1 - 2\rho_2]\delta(\cos\theta)/[r\sqrt{\kappa}]$ (consider $r_1 \sim r_2 \gg r_d$). Performing the integrals over $\theta, \varphi, dQ_2^{(1,2)}$ and noting that at large distances the $d\varphi$ integration has $\varphi = 0$ as a saddle-point, one gets

$$\mathcal{A} \propto \frac{L^3 r_1^2 r_2^2}{r_d(r_1^2 + r_2^2)^3} \left\langle \frac{e^{2\rho_1(t)-\rho_2(t)}}{[1 + (r_d/L)^2 \exp(-2\rho_2)]^{5/2}} \right\rangle \propto \frac{L^5 r_1^2 r_2^2}{r_d^3(r_1^2 + r_2^2)^3} e^{E_4 t}.$$

For general (noncollinear) $\mathbf{r}_1$ and $\mathbf{r}_2$, $dQ_3^{(1,2)}$ integrations are not singular and (at $r \gg r_d$) give a κ-independent factor, $\exp[-2(\rho_1+\rho_2)]/r^2$, while the $dQ_2^{(1,2)}$ integrations give a finite function of $\exp[\rho_2]/r$. As a result, $\mathcal{A} \propto (L/r)^5 e^{E_4 t}$ which is parametrically smaller than the collinear result.

Consider now a general $F_{2n} = \langle \prod_{k=1}^n (\mathbf{B}(\mathbf{x}_{2k-1}, t)\mathbf{B}(\mathbf{x}_{2k}, t)) \rangle$. Its calculation is reduced to averaging $(2n-1)!!$ terms arising from the Wick decomposition in the Gaussian integration over the random initial condition. Each term is a product of n integrals like (3.1). The n vectors $\mathbf{r}_j$ are differences between pairs of $\mathbf{x}_k$'s. The presence of the rotation matrix $\hat{R}$ in the exponential factor requires an explicit integration over the angles. The $\mathbf{q}$ integrations proceed along the same lines: in each of the n integrals we change variables $\mathbf{q} = \hat{S}^T \mathbf{Q}$ and the dependence on $\hat{S}$ is entirely moved into the prefactors. Replacing $\hat{S}$ by the unit matrix and performing the $\mathbf{Q}$ integrations, we obtain the long-time asymptotics for any of the $(2n-1)!!$ contributions to F_{2n}:

$$\left\langle \frac{(L/r_d)^n \exp[n(\rho_1-\rho_2)]}{[1+e^{-2\rho_2} r_d^2/L^2]^{5n/2}} \int_{-1}^1 d\cos\theta \int_0^{2\pi} d\varphi \int_0^{2\pi} d\phi \right.$$
$$\left. \times \prod_{j=1}^n \left[2 - \frac{R_{2j}^2 e^{-2\rho_2}}{L^2} \right] \exp\left[-\frac{R_{2j}^2}{4[L^2 e^{2\rho_2} + 2r_d^2]} - \frac{R_{3j}^2}{8r_d^2} \right] \right\rangle,$$
(3.6)

where $\mathbf{R}_j = \hat{R}_3[\varphi]\hat{R}_2[\theta]\hat{R}_3[\phi]\mathbf{r}_j$, and $\hat{R}_{2,3}$ stand for rotation around Y and Z axis respectively.

As we already saw from the consideration of the four-point correlation function, the geometry of the vectors $\mathbf{r}_j$ becomes important for $n \geq 2$. Let us first consider the case where all the vectors are in the same plane (their length being $\simeq r$). The vectors can be either collinear or not. Almost orthogonality of $\mathbf{r}_j$ to the direction of contraction $\mathbf{n}_3$ involves therefore either one angle or two, giving the geometric factor $(L/r) \exp \rho_3$ or its square, respectively. The other difference concerns the behaviour along the intermediate direction $\mathbf{n}_2$. For non-collinear geometry, not all the vectors can be simultaneously orthogonal to $\mathbf{n}_2$ so that ρ_2 should then be constrained as $\rho_2 > \log(r/L)$. Technically, this means that the integration over φ is not saddle-point. Conversely, for collinear geometry all the vectors can be orthogonal to both $\mathbf{n}_2$ and $\mathbf{n}_3$, giving an additional angular factor $(L/r) \exp \rho_2$: the saddle-point integrations over θ and φ pick $\theta = \pi/2$, $\varphi \simeq \exp[\rho_2] L/r$. In the rest of the integrals (either $n-1$ or $n-2$, respectively) the above angular constraints ensure that the advective exponents are $O(1)$ so the diffusive exponents $\exp(-2\kappa \mathbf{q} \Lambda \mathbf{q})$ become important. The calculation of these integrals is essentially the same as for the moments and this is where diffusion comes into play. The wavevectors should be quasi-orthogonal to $\mathbf{n}_3$, giving either $n-1$ or $n-2$ factors $(L/r_d) \exp \rho_3$. The growth along $\mathbf{n}_2$ for a generic planar geometry is automatically controlled by the previous constraint $\rho_2 > \log(r/L)$; for collinear geometry it provides the bound $\rho_2 > \log(r_d/L)$. Simply speaking, the strips with the width r_d stretched along $\mathbf{r}$ contribute in the collinear case, while the width is r in the generic case. The resulting integrations over ρ_1 are saddle-point and those over ρ_2 are dominated by the lower bounds. Finally,

$$F_{2n} \simeq e^{E_n t} \left(\frac{L}{r_d}\right)^{\frac{5n}{2}} \left(\frac{r_d}{r}\right)^2 \times \left\{ \begin{array}{ll} 1 & \text{collinear,} \\ (r_d/r)^{\frac{3n}{2}} & \text{planar.} \end{array} \right. \tag{3.7}$$

Here, $E_n = x_n - H(x_n, 0)$ with $\partial H/\partial x(x_n, 0) = n$. That the integrals over ρ_2 are all dominated by the lower bounds indicates the geometric nature of the scaling universality found: the field configurations that contribute are narrow strips (not ropes and layers suggested in Zeldovich *et al.* 1984) with one direction of stretching, one of contraction and a neutral one. The factor $(r_d/r)^2$ is the probability for two points at distance r to lie within the same strip of width r_d. The peculiar nature of strips has another dramatic consequence for $n \geq 3$: the correlation functions are strongly suppressed in a generic situation when at least three vectors $\mathbf{r}_j$ do not lie in parallel planes. Indeed, they cannot then be on parallel strips and nonzero correlation appears only because the strips have exponential diffusive tails. As a result, the factor $(r_d/r)^2$ in the planar formula is replaced by $\exp(-ar^2 \sin^2 \Theta/r_d^2)$, where $a \simeq 1$ and Θ is the minimal angle between a vector and the plane formed by another two. That can be derived from (3.6) where no angular integrations are saddle-point now. For general irreversible velocity statistics, the r dependencies are different yet the qualitative conclusions (that the correlation functions are anomalously large for collinear geometry, suppressed by a power law for planar geometry and exponentially suppressed for a general geometry) are generally valid. The most remarkable manifestation of intermittency is that for collinear geometry $F_{2n}/F_2^n \propto (r/r_d)^{-2+5n/2} \exp(E_{2n}t - nE_2t) \gg 1$ both due to the nonlinear relation between E_n and n and due to anomalous scaling. Angular anomalies are peculiar to the viscous interval where advection by a smooth velocity preserves collinearity. Similar collinear anomalies have been described before for a passive scalar advected by a smooth velocity (Balkovsky *et al.* 1995; Shraiman and Siggia, 1995; Balkovsky *et al.* 1999). Note that the cliff-and-ramp structures observed in passive scalar experiments (see Sreenivasan, 1991 for review) are probably related to the strips too.

4   Short-correlated velocity field

One can calculate the whole set of E_n for the standard Kazantsev-Kraichnan model of an isotropic short-correlated Gaussian strain with $\langle \sigma_{\alpha\beta}(t)\sigma_{\alpha\beta}(0) \rangle = 10\lambda\delta(t)$. Here the average is equivalent to the path integral

$$\int \mathcal{D}\hat{\sigma} \exp\left[-(10\lambda)^{-1} \int_0^t I[\hat{\sigma}(t')]dt'\right]$$

with $I = [4\text{Tr}(\hat{\sigma}\hat{\sigma}^T) + \text{Tr}\hat{\sigma}^2]$. Passing in the path integral to the variables ρ_i (elements of $\hat{D}$) and χ_i (elements of $\hat{S}$) we get $I(\rho, \chi) = 2(\dot{\rho}_1^2 + \dot{\rho}_2^2 + \dot{\rho}_1\dot{\rho}_2) + [\dot{\chi}_{12}^2 \exp(2\rho_1 - 2\rho_2) + \dot{\chi}_{23}^2 \exp(2\rho_1 + 4\rho_2) + (\dot{\chi}_{13} - \dot{\chi}_{12}\chi_{13})^2 \exp(4\rho_1 + 2\rho_2)]/2$. While calculating the long-time asymptotic of the magnetic field averaged with $I(\rho, \chi)$ one indeed sees that the temporal fluctuations of χ are exponentially suppressed since their time derivatives are multiplied in $I(\rho, \chi)$ by large exponential factors (Chertkov *et al.* 1998). We thus integrate over $\hat{S}$ already in the measure (that is independently of the moment we are to average). Straightforward derivation gives the Gaussian $\mathcal{P}_t(\rho_1, \rho_2)$ with

$$H = (\rho_1 + \rho_2/2 - \lambda t)^2/\lambda t + 3\rho_2^2/4\lambda t.$$

Note that $\lambda_1 = -\lambda_3 = \lambda$ and $\lambda_2 = 0$. Now we integrate the moments of (2.2) with such $\mathcal{P}_t$. First, we integrate $\exp(2n\rho_1)$ and get the answer for the perfect conductor (see Childress and Gilbert, 1995, Zeldovich *et al.* 1984)

$$\langle \mathbf{B}^{2n} \rangle \simeq \exp\left[2\lambda n(2n + 3)t/3\right]. \tag{4.1}$$

The main contribution to the nth moment comes from $\rho_1 = \lambda t(4n+3)/3, \rho_2 = -2n\lambda t/3$ so (4.1) is valid until $L \exp \rho_3 > r_d$ that is for $t < 3t_d/(2n + 3)$ with $t_d \equiv \lambda^{-1} \log(L/r_d)$. Then there is a logarithmically wide cross-over interval when the growth is non-exponential. The asymptotic regime starts at $t > 3t_d/(n + 2)$ when unity in the first braces of (2.2) may be neglected. The integral over ρ_1 now comes from $\rho_1 + \rho_2/2 = (n + 2)\lambda_1 t/2$ while that over ρ_2 is dominated by the lower bound $\rho_2 \simeq \log(r_d/L)$:

$$\langle \mathbf{B}^{2n} \rangle \simeq (L/r_d)^{5n/2} \exp(E_n t), \quad E_n = \lambda n(n + 4)t/4. \tag{4.2}$$

For $n = 1$, this was obtained by Kazantsev (1968). The difference between (4.1) and (4.2) formally means that the two limits $t \to \infty$ and $\kappa \to 0$ do not commute (what is called the dissipative anomaly). The physical reason is quite clear: realizations with continuing contraction along two directions contribute most in a perfect conductor, while with diffusion present one direction is neutral. A magnetic field initially concentrated in the ball with the radius L will have the fastest growth rate γ if the ball turns into a strip with the dimensions $L(L/r_d)^{1/2} \exp(3Dt)$, r_d, $L(r_d/L)^{1/2} \exp(-3Dt)$. The main contribution to the n-th moment will be given by strips with the dimensions $L(L/r_d)^{1/2} \exp[3(n+2)Dt/2]$, r_d and $L(r_d/L)^{1/2} \exp[-3(n+2)Dt/2]$.

To conclude the part devoted to incompressible flow, we related the growth rate of the small-scale dynamo to the Lyapunov exponents of the flow and described analytically the strip structure of magnetic field.

5   Growth and decay in a compressible turbulence

In the compressible case, the extra term appears in the equations

$$\partial_t \mathbf{B} + (\mathbf{v} \cdot \partial)\mathbf{B} = (\mathbf{B} \cdot \partial)\mathbf{v} - \mathbf{B}(\partial \cdot \mathbf{v}) + \kappa \triangle \mathbf{B}. \tag{5.1}$$

It is B/ρ which is now frozen in for a perfect conductor. Compressible flow doesn't preserve Lagrangian measure so that one has to be careful in distinguishing forward-in-time statistics (usually used to introduce Lyapunov exponents) and backward-in-time ones necessary to describe the statistics of passive quantities (scalar or vector). Euler and Lagrangian means differ by the Jacobian $|\det W|$ which is not unity now. That makes the PDF $\mathcal{P}_t(\rho_1, \rho_2, \rho_3)$ equal to

$$\mathcal{P}_t(\rho_1, \rho_2, \rho_3) \propto \exp\left[\sum \rho_i - tH(\rho_1/t, \rho_2/t, \rho_3/t)\right]. \tag{5.2}$$

As a result, the effective Lyapunov exponents (that determine the growth rate of the magnetic field) are $\tilde{\lambda}_i = \lambda_i + x_i$ where λ_i are standard Lyapunov exponents determined by Lagrangian dynamics and x_i are given by the condition $\partial H/\partial x_i = 1$. We see that there is no universal relation between the field growth rate and Lyapunov exponent for a compressible flow, the relation depends on the strain statistics. This is because every trajectory comes now with its own weight $|detW| = \exp \sum \rho_i$. Note also that there is no condition now that the largest Lyapunov exponent is nonnegative while the smallest one is nonpositive. In particular, all the exponents may be of the same sign. When both the field and the flow are 2D, in a perfect conductor (at $t < t_d$) $\gamma = -\tilde{\lambda}_2$, while at $t > t_d$ the magnetic field always decays as long as there exists a nonzero Lyapunov exponent:

$$\gamma_\infty = \min\{-2\tilde{\lambda}_2, \tilde{\lambda}_2\}. \tag{5.3}$$

In 3D, we get a general formula valid for both compressible and incompressible cases:

$$\gamma_\infty = -\tilde{\lambda}_2 - \tilde{\lambda}_3 + \tilde{\lambda}_1\theta(-\tilde{\lambda}_1)/2 + 3\tilde{\lambda}_2\theta(-\tilde{\lambda}_2)/2 + \tilde{\lambda}_3\theta(-\tilde{\lambda}_3)/2, \qquad (5.4)$$

where the step function $\theta(x) = 1$ for $x > 0$ and $\theta(x) = 0$ for $x < 0$. Even though there are cases when the field grows (say, when $-\tilde{\lambda}_3/2 > \tilde{\lambda}_2 > 0$) the magnetic flux always decays since $\gamma_\infty + \lambda_2 + \lambda_3 < 0$.

I briefly present here the simple limit of a fast-fluctuating velocity field. For a compressible short-correlated velocity characterized by the degree of compressibility $\Gamma = \langle(\nabla \cdot \mathbf{v})^2\rangle/\langle|\nabla\mathbf{v}|^2\rangle$,

$$\tilde{\lambda}_1 = \lambda(1 + 2\Gamma), \quad \tilde{\lambda}_2 = 5\lambda\Gamma/3, \quad \tilde{\lambda}_3 = \lambda\left(\frac{4\Gamma}{3} - 1\right). \qquad (5.5)$$

Note that $0 \le \Gamma \le 1$. For $\Gamma < 3/4$ one has

$$\gamma = \frac{\lambda}{2}\left(1 - \frac{14\Gamma}{3}\right). \qquad (5.6)$$

For a short-correlated velocity, compressibility suppresses dynamo activity, so for $\Gamma > 3/14$ the magnetic field decays at large time. Using the convexity of $E(n)$ and $E(0) = 0$ one can show that for $\Gamma < 3/14$ all the moments of the magnetic field grow exponentially (the details will be published in Balkovsky *et al.* 2000).

This article is based on two papers (Chertkov *et al.* 1999; Balkovsky *et al.* 2000). I am grateful to my co-authors E. Balkovsky, A. Fouxon, M. Chertkov, I. Kolokolov and M. Vergassola for pleasure of fruitful collaboration. The work was supported by the grants from the Minerva Foundation and the Edward and Anna Mitchell Research Fund at the Weizmann Institute. Very useful discussions with A. Gruzinov, V. Lebedev, H.K. Moffatt, A. Newell and B. Shraiman are gratefully acknowledged. The article was partially prepared during the turbulence programme at the Isaac Newton Institute. I am grateful to J.C. Vassilicos and the Institute staff for their hospitality.

References

Balkovsky, E., Chertkov, M., Kolokolov, I., Lebedev, V. (1995) 'Fourth-order correlation function of randomly advected passive scalar', *JETP Lett.* **61**, 1049–1054.

Balkovsky, E., Falkovich, G., Lebedev, V., Lysiansky, M. (1999) 'Large-scale properties of passive scalar advection', *Phys. Fluids* **11**, 2269–2279.

Balkovsky, E., Falkovich, G., Fouxon, A. (2000) 'Lagrangian dynamics of a compressible turbulent flow', in preparation.

Balkovsky, E. and Fouxon, A. (1999) 'Universal long-time properties of Lagrangian statistics in the Batchelor regime and their application to the passive scalar problem', preprint chao-dyn/9905020, *Phys. Rev. E* (in press).

Batchelor, G.K. (1950) 'On the spontaneous magnetic field in a conducting fluid in turbulent motion', *Proc. Roy. Soc. Lond.* **A261**, 405–416.

Chertkov, M., Falkovich, G., Kolokolov, I., Lebedev, V. (1995) 'Statistics of a passive scalar advected by a large-scale two-dimensional velocity field: analytic solution', *Phys. Rev. E* **51**, 5609–5627.

Chertkov, M., Falkovich, G., Kolokolov, I. (1998) 'Intermittent dissipation of a passive scalar in turbulence', *Phys. Rev. Lett.* **79**, 2121–2124.

Chertkov, M., Falkovich, G., Kolokolov, I., Vergassola, M. (1999) 'Small-scale magnetic dynamo', preprint chao-dyn/9906030, to appear in *Phys. Rev. Lett.*

Childress, S. and Gilbert, A. (1995) *Stretch, Twist, Fold: The Fast Dynamo*, Springer-Verlag.

Crisanti, A., Paladin, G., Vulpiani, A. (1993) *Products of Random Matrices*, Springer-Verlag.

Ellis, R. (1985) *Entropy, Large Deviations and Statistical Mechanics*, Springer-Verlag.

Girimaji, S. and Pope, S. (1990) 'Material-element deformation in isotropic turbulence', *J. Fluid Mech.* **220**, 427–458.

Goldhirsch, I., Sulem, P., Orszag, S. (1987) 'Stability and Lyapunov stability of dynamical systems', *Physica D* **27**, 311.

Gruzinov, A., Cowley, S., Sudan, R. (1996) 'Small-scale-field dynamo', *Phys. Rev. Lett.* **77**, 4342–4345.

Kazantsev, A.P. (1968) 'Enhancement of a magnetic field by a conducting fluid', *Sov. Phys. JETP* **26**, 1031–1034.

Kraichnan, R.H. and Nagarajan, S. (1967) 'Growth of turbulent magnetic fields', *Phys. Fluids* **10**, 859–870.

Kulsrud, R. and Anderson, S. (1992) 'The spectrum of random magnetic fields in the mean field dynamo theory of the galactic magnetic field', *Astrophys. J.* **396**, 606–630.

Moffatt, H.K. (1978) *Magnetic Field Generation in Electrically Conducting Fluids*, Cambridge University Press.

Parker, E.N. (1979) *Cosmic Magnetic Fields, their Origin and Activity*, Clarendon Press.

Pearson, J. (1959) 'The effect of uniform distortion on weak homogeneous turbulence', *J. Fluid Mech.* **5**, 274–288.

Shraiman, B., and Siggia, E. (1994) 'Lagrangian path integrals and fluctuations in random flow', *Phys. Rev. E* **49**, 2912–2927.

Shraiman, B. and Siggia, E. (1995) 'Anomalous scaling of a passive scalar in a turbulent flow', *C.R. Acad. Sci.* **321**, Ser. II, 279–284.

Sreenivasan, K. (1991) 'On local isotropy of passive scalars in turbulent shear flows', *Proc. Roy. Soc. Lond.* **A434**, 165–182.

Vainshtein, S. and Zeldovich, Ya. (1972) 'Origin of magnetic fields in astrophysics', *Sov. Phys. Usp.* **15**, 159–172.

Zeldovich, Ya., Ruzmaikin, A., Molchanov S., Sokolov, V. (1984) 'Kinematic dynamo problem in a linear velocity field', *J. Fluid Mech.* **144**, 1–11.

Non-Homogeneous Scalings in Boundary Layer Turbulence

Sergio Ciliberto, Emmanuel Lévêque and Gerardo Ruiz Chavarria

1 Introduction

In order to characterize the statistical properties of fully developed turbulence, one usually studies the scaling properties of velocity structure functions $S_n(\vec{x}, \vec{r})$, defined as the moments of longitudinal velocity increments across a separation $\vec{r}$ at the location $\vec{x}$:

$$S_n(\vec{x}, \vec{r}) = \langle |\delta V(\vec{x}, \vec{r})|^n \rangle, \qquad (1.1)$$

where angular brackets $\langle , \rangle$ stand for average and the longitudinal increment $\delta V(\vec{x}, \vec{r})$ is given by

$$\delta V(\vec{x}, \vec{r}) \equiv (\vec{V}(\vec{x} + \vec{r}) - \vec{V}(\vec{x})) \cdot \frac{\vec{r}}{r}. \qquad (1.2)$$

In homogeneous and isotropic turbulence, $S_n(\vec{x}, \vec{r})$ only depends on the distance (or scale) r. At very high Reynolds number Re, the $S_n(r)$ exhibit a power-law dependence on r:

$$S_n(r) \propto r^{\zeta(n)} \qquad (1.3)$$

for $L > r >> \eta$, where L is the integral scale of the flow, $\eta = (\nu^3/\epsilon)^{1/4}$ is the Kolmogorov dissipation scale, ϵ is the mean energy dissipation rate and ν is the kinematic viscosity. According to Kolmogorov theory (Kolmogorov 1941), one usually assumes that for $r << L$ the statistical properties of turbulence are locally isotropic. Furthermore Kolmogorov theory (K41) predicts $\zeta(n) = n/3$ (Kolmogorov 1941). Experiments (Anselmet *et al.* 1984, Benzi *et al.* 1995, Arneodo *et al.* 1996) and numerical simulations (Vincent & Meneguzzi 1991, Briscolini *et al.* 1994) show that $\zeta(n)$ deviates substantially from this linear law. This phenomenon is believed to be related to the intermittent behavior of the energy dissipation, and it can be taken into account by rewriting (1.3) in the following way:

$$S_n(r) \propto \langle \epsilon_r^{n/3} \rangle r^{n/3} \propto r^{\tau(n/3)+n/3} \qquad (1.4)$$

where ϵ_r is the average of the local energy dissipation on a volume of size r (Kolmogorov 1962, Monin & Yaglom 1975). A comparison of (1.3) and (1.4)

118

leads to the conclusion that the scaling exponents $\tau(n/3)$ of the energy dissipation are related to those of S_n by $\zeta(n) = \tau(n/3) + n/3$. In the isotropic case the well-known Kolmogorov equation can be derived (Monin 1975, Frisch 1995) from the Navier–Stokes equations:

$$\langle \delta V^3 \rangle = -\frac{4}{5} \, \epsilon \, r \, + 6 \, \nu \, \frac{\partial S_2(r)}{\partial r}. \tag{1.5}$$

When viscosity effects are negligible, that is for r in the inertial range, then $\langle \delta V^3 \rangle \simeq -\frac{4}{5} \, \epsilon \, r$. As a consequence $\zeta(3) = 1$ and $\tau(1) = 0$.

Experiments in homogeneous and isotropic turbulence at very high Re support the scaling laws (1.3) and (1.4). Furthermore it turns out that the $\zeta(n)$ are independent of the stirring process, that is wake turbulence, grid turbulence, jet turbulence, ..., exhibit the same scaling laws (Benzi *et al.* 1995, Arneodo *et al.* 1996). However, this universality may not persist in regions where turbulent fluctuations are non homogeneous or anisotropic. Specifically, the $\zeta(n)$ may have different values if measurements are done either far away from boundaries, where turbulence is almost homogeneous and isotropic, or in locations of the flow where a strong mean shear is present (Benzi *et al.* 1993b, Stolovitzky & Sreenivasan 1993, Mordant *et al.* 1997, Gaudin *et al.* 1998).

The purpose of this paper is to make a direct comparison between the scaling laws in the isotropic case and in boundary layer turbulence. Specifically we analyze how the properties of Extended Self Similarity (ESS) (Benzi *et al.* 1993a) and its generalized form (GESS), (Benzi *et al.* 1996) are modified in wall-bounded turbulence. We have observed that the statistics depend on the distance from the wall. However, the relative scaling obtained from ESS remains constant in the logarithmic sublayer. A new form of GESS, recently proposed for boundary layer turbulence, is found in good agreement with our experimental data. This result, which may be useful for practical applications, has been indirectly checked by using the hierarchy of S_n derived from the She–Lévêque model of intermittency (She & Lévêque 1994). This hierarchy has been quite well verified in the isotropic case and it turns out to be a quite good indicator that (1.4) fails close to the wall.

Our experimental observations show that the study of scaling laws may provide new insight into the mechanisms by which energy is actually dissipated at different scales. This is clearly an important issue in order to safely use large eddy simulations in real applications.

The paper is organized as follows. In the next section we summarize the general statistical properties of isotropic turbulence measured in experiments. In section 3 we will discuss how these properties, observed in isotropic flows, are modified in boundary layer turbulence. Finally, concluding remarks are given in section 4.

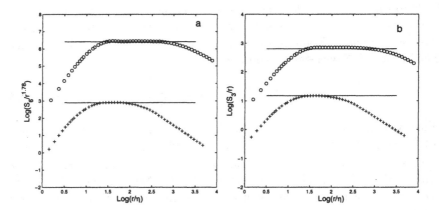

Figure 1: Structure functions $S_n(r)/r^{\zeta(n)}$ as a function of r. Data taken from an experiment on a jet at $R_\lambda = 800$ (o). Data taken from the wake behind a cylinder at $R_\lambda = 140$ (+): (a) $n = 6$ and $\zeta(6) = 1.78$; (b) $n = 3$ and $\zeta(3) = 1$. Logarithms are base 10 in all of the figures if not otherwise indicated.

2 Isotropic Turbulence

2.1 Extended Self Similarity

Extended Self Similarity (ESS) is a property of velocity structure functions of homogeneous and isotropic turbulence (Benzi *et al.* 1993a, Benzi *et al.* 1995). It has been shown using experimental and numerical data (Benzi *et al.* 1994) that the structure functions present an extended scaling range when one plots one structure function against another, namely

$$S_n(r) \propto S_m(r)^{\gamma_{n,m}} \tag{2.1}$$

for $r > 5\eta$ and

$$S_n(r) \propto S_m(r)^{n/m} \tag{2.2}$$

for $r < 5\eta$. It turns out that $\gamma_{n,m} = \zeta(n)/\zeta(m)$. As $\zeta(3) = 1$ because of (1.5) then $\gamma_{n,3} = \zeta(n)$. The details of ESS have been reported elsewhere (Benzi *et al.* 1995). In the following, we will only describe its main features.

As an example we consider two experimental data sets at different R_λ, which is the Reynolds number based on the Taylor scale ($R_\lambda \simeq 1.4\,Re^{1/2}$) (Monin & Yaglom 1975). The two experiments are a jet at $R_\lambda = 800$ and the wake behind a cylinder at $R_\lambda = 140$. In both cases data have been recorded at about 25 integral scales downstream (Benzi *et al.* 1995). In Figure 1a $S_6/r^{\zeta(6)}$, computed for the two experiments, is plotted as a function of r. In Figure 1b we show $S_3/r^{\zeta(3)}$ as a function of r. In both figures a scaling

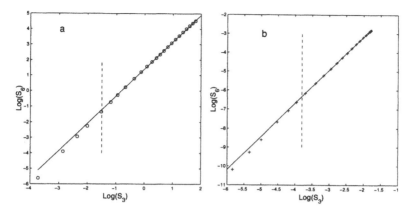

Figure 2: Structure functions S_6 as a function of S_3 at $R_\lambda = 800$ (a) and $R_\lambda = 140$ (b) computed from the same data set as Figure 1. Vertical dashed lines indicate the value of S_3 at 5η.

region is observed only for the highest R_λ. In contrast if the relative scaling (2.1) is used, see Figure 2, a clear scaling range is present for both R_λ with $\gamma_{6,3} \simeq 1.78$. The vertical dashed lines in Figure 2 correspond to $S_3(5\eta)$ and they roughly indicate the extension of the scaling (2.1), that is $5\eta < r < L$. The ESS scaling has been checked both on numerical data and in experiments, in the range $30 < R_\lambda < 2000$ (Benzi *et al.* 1995, Belin *et al.* 1996, Arneodo *et al.* 1996).

ESS has been also checked for the temperature and velocity fields in Rayleigh–Bénard convection (Benzi *et al.* 1994) and in the case of a passive scalar (Ruiz *et al.* 1996). It turns out that ESS is a very useful tool in order to distinguish between Kolmogorov and Bolgiano scaling (Benzi *et al.* 1994, Cioni *et al.* 1995).

Another interesting observation concerns the behavior of $\gamma_{n,3}$ with respect to R_λ (Benzi *et al.* 1996, Belin *et al.* 1996). The values of $\gamma_{n,3}$ have been measured in the range $30 < R_\lambda < 10^4$, without noticing, within error bars, any change or trend of $\gamma_{n,3}$ as a function of R_λ (Benzi *et al.* 1996, Belin *et al.* 1996). This means that far away from boundaries the $\gamma_{n,3}$ are constants which do not depend on Re and on the integral scale stirring process.

2.2 Generalized Kolmogorov Similarity Hypothesis

A final point regarding ESS concerns the generalization of the Refined Kolmogorov Similarity Hypothesis (RKS) defined by (1.4). The RKS hypothesis, as stated in (1.4), supports the idea that ϵ_r has the same scaling as $\delta V^3/r$. We can generalize the RKS hypothesis by introducing an effective

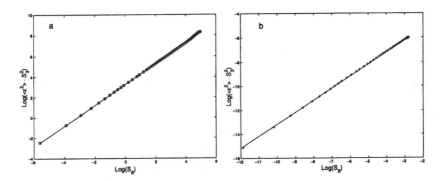

Figure 3: Log-log plot of $< \epsilon_r^2 > S_3(r)^2$ against $S_6(r)$ at $R_\lambda = 800$ (a) and $R_\lambda = 140$ (b). The straight lines refer to best fits with slope 1 ± 0.005.

scale $L(r) = S_3(r)/\epsilon$, as suggested by ESS, and we obtain the following relation: $\epsilon_r = \delta V^3 \, \epsilon/S_3$. Generalization of RKS simply states that:

$$S_n(r) = \frac{\langle \epsilon_r^{n/3} \rangle}{\epsilon^{n/3}} \, S_3(r)^{n/3}. \tag{2.3}$$

Equation (2.3) has been proposed in (Ruiz 1994, Dubrulle 1994) and carefully checked in (Ruiz 1994). A typical experimental result is shown in Figure 3(a) and Figure 3(b) where $\langle \epsilon_r^2 \rangle \, S_3^2$ is plotted as a function of $S_6(r)$. The energy dissipation has been computed using the one-dimensional surrogate that is

$$\epsilon_r = \frac{15\nu}{r} \int_x^{x+r} \left(\frac{\partial V(x')}{\partial x} \right)^2 dx'. \tag{2.4}$$

In Figure 3 one can see a clear scaling extending over almost ten orders of magnitude from the integral scale to η. The slope of the straight line is 1.005 showing that (2.3) is compatible with experimental data.

One can argue that (2.3) is trivial because for $r < \eta$, ϵ_r is constant and $S_n \propto r^n$, thus the scaling $S_n \propto S_3^{n/3}$ is obviously satisfied. Furthermore for r in the inertial range (2.3) is verified because $(S_3/\epsilon) \propto r$. However, in principle the proportionality constants of (2.3) in the inertial and in the dissipative range could be different. The fact that experimentally they are found equal has several important consequences discussed by Benzi *et al.* (1996). In that paper the extended scaling implied in (2.3) has been named generalized ESS (GESS).

2.3 The hierarchy of structure functions

She and Lévêque (1994) proposed an interesting theory to explain the anomalous scaling exponents of velocity structure functions. The theory yields a prediction

$$\zeta(n) = n/9 + 2(1 - (2/3)^{n/3})$$

which is in very good agreement with available experimental data (Benzi *et al.* 1995, Arneodo *et al.* 1996). The She–Lévêque model is based upon a fundamental assumption on the hierarchy of the moments, $\langle \epsilon_r^n \rangle$, of the coarse grained energy dissipation. Specifically they consider that

$$\frac{\langle \epsilon_r^{n+1} \rangle}{\langle \epsilon_r^n \rangle} = A_n \left(\frac{\langle \epsilon_r^n \rangle}{\langle \epsilon_r^{n-1} \rangle} \right)^{\beta} (\epsilon_r^{(\infty)})^{(1-\beta)} \qquad (2.5)$$

where A_n are geometrical constants and $\epsilon_r^{(\infty)} = \lim_{n \to \infty} (\frac{\langle \epsilon_r^{n+1} \rangle}{\langle \epsilon_r^n \rangle})$ is associated in (She & Lévêque 1994) with filamentary structures of the flow. On the basis of simple arguments it is assumed that $\epsilon_r^{(\infty)} \propto r^{-2/3}$. The value of β predicted by She & Lévêque (1994) is 2/3. Notice that in (2.5) for $n = 1$, taking into account that $\langle \epsilon_r \rangle = \epsilon$ is constant in r, one immediately finds that

$$(\epsilon_r^{(\infty)})^{(1-\beta)} \propto \langle \epsilon_r^2 \rangle = \frac{S_6}{S_3^2} \qquad (2.6)$$

where (2.3) has been used.

Equation (2.5), which has been experimentally tested in (Ruiz *et al.* 1995a), can be extended to the velocity structure functions (Ruiz *et al.* 1995b). Taking in (2.5) the value $n = p/3$ and using equations (2.3) and (2.6), after some algebra one finds the following relation for the velocity structure functions:

$$F_{p+1}(r) = C_p \, (F_p(r))^{\beta'} \cdot G_\infty(r) \qquad (2.7)$$

where

$$F_{p+1}(r) = \frac{S_{p+1}(r)}{S_p(r)}$$

and

$$G_\infty(r) = \left(\frac{S_6}{S_3^{(1+\beta)}} \right)^{\frac{(1-\beta')}{3(1-\beta)}},$$

C_p are geometry-dependent constants and $\beta' = \beta^{1/3}$.

Notice that (2.7) is certainly valid for any β in the dissipative range where $S_n \propto r^n$. Equation (2.7) has been experimentally tested in (Ruiz *et al.* 1995b).

This can be seen in Figure 4(a) and Figure 4(b), where the scaling obtained for various n using (2.7) is reported for two different *Re*. As we have already

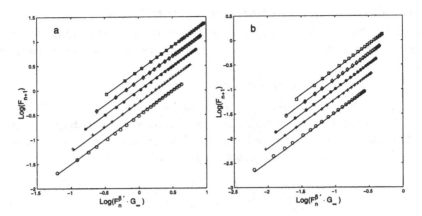

Figure 4: The function $F_{n+1}(r)$ defined in (2.7) is plotted, for several values of n, as a function of $[(F_n(r))^{\beta'} \cdot G_\infty(r)]$ with $\beta = 2/3$, $\beta' = \beta^\delta$ and $\delta = 1/3$. $R_\lambda = 800$ in (a) and $R_\lambda = 140$ in (b). The five curves in (a) and (b) correspond to $n = 1, 2, 3, 4, 5$ starting from the bottom lines. They have been vertically shifted by $-0.4, -0.2, 0, 0.2, 0.4$ in order to separate them. The solid lines have slope 1.

observed in the case of (2.3) (Figure 3), the scaling extends from large to small scales even for values of r where ESS is no longer satisfied. It is important to point out that the validity of (2.7) for all scales is not necessarily true for any turbulent field, but it is an important feature of the velocity field. Indeed it has been shown that the passive scalar field does not present the same property (Ruiz *et al.*1996).

3 Boundary Layer Turbulence

To check how the properties of isotropic turbulence, summarized in section 2, are modified in the case of shear we have performed an experiment on boundary layer turbulence. We have measured the statistical properties of the streamwise component of the velocity field near a wall. We summarize here the main results, all the details are given elsewhere (Ruiz *et al.* 1999).

3.1 The experimental set-up

The experiment is performed in a recirculating wind tunnel with a 3m long working section and a 25cm×50cm cross section. The boundary layer develops on a smooth 2m long horizontal plate. The velocity probe is located at a distance of 1.5 m downstream from the plate edge. In the present study, the focus is only on the vertical dependence of the statistics. The horizontal distance from the plate edge is therefore kept constant. The measurements are made with a constant temperature anenometer. The distance d between the probe and the plate varies from 1.5mm to 12.5cm. The maximal distance corresponds to the middle between the plate and the upper wall of the wind tunnel. At distances smaller than 1.5mm, the probe would greatly disturb the flow and therefore influence the measurements. The mean velocity at $d = 12.5$cm is 6.9m/s.

The mean streamwise velocity profile expressed in terms of the standard dimensionless distance $\xi = V_* d/\nu$ is in good agreement with previously reported mean velocity profiles (Schlichting 1968, Landau & Lifschitz 1959), with $V_* = 0.34$m/s.

The statistical features have been investigated at only 12 different distances from the plate, in the range $32 < \xi < 2670$. The logarithmic sublayer is for $60 < \xi < 200$. In order to ensure a reasonable convergence of the statistics, 10 million points have been sampled at each position. Velocity structure functions have been estimated up to the sixth order. Energy spectra and probability density functions of velocity increments have also been evaluated (Ruiz *et al.* 1999). The standard deviation of velocity fluctuations is about 15% of the mean velocity. This has allowed us to use the Taylor hypothesis for recasting temporal measurements into the space domain.

3.2   The structure functions as a function of ξ

In order to characterize the dependence on ξ and r of the velocity difference statistics, we study the scaling behavior of the velocity structure functions $S_n(\xi, r)$ for various distances from the wall.

We start our analysis with the third-order structure function. In Figure 5(a), $S_3(\xi, r)/r$ is plotted as a function of r for various ξ. One notices that $S_3(\xi, r)$ exhibits a power-law behavior, with a scaling exponent $\zeta(\xi, 3)$ depending on the distance from the plate. The extension of the scaling range also differs with ξ.

The scaling exponent $\zeta(\xi, 3)$ is an increasing function of ξ. It seems to tend to its homogeneous and isotropic value $\zeta(3) = 1$ as one moves off the plate. This is in agreement with the picture that anisotropy effects vanish with the distance from the boundary. However, the isotropic value $\zeta(3) = 1$ is not reached in our configuration. Indeed, the maximum attainable distance $\xi = 2670$ already sits at the boundary between the turbulent logarithmic sublayer and the laminar zone. At $\xi = 2670$, the velocity signal exhibits periods of calm, indicating that the probe has reached the (oscillating) upper boundary of the turbulent layer.

The compensated sixth-order structure function $S_6(\xi, r)/r^{1.78}$ is plotted in Figure 5(b), for the same three characteristic distances. An evolution

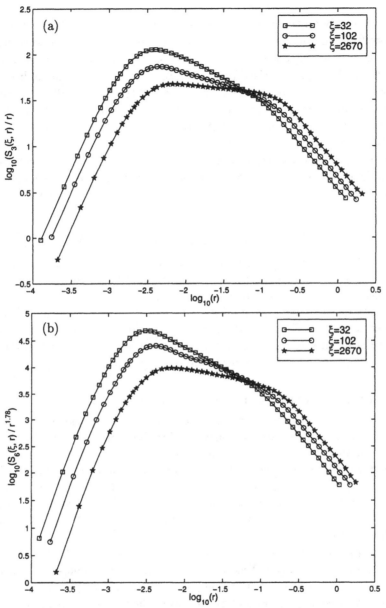

Figure 5: (a) The compensated third-order structure function $S_3(\xi, r)/r$, for three characteristic distances from the plate. A power-law dependence on r is observed for each distance. The scaling exponent $\zeta(\xi, 3)$ is an increasing function of ξ. (b) The sixth-order structure function, compensated by the isotropic scaling law $S_6(r) \sim r^{1.78}$, for three characteristic distances from the plate.

ξ	$\zeta(\xi,1)$	$\zeta(\xi,2)$	$\zeta(\xi,3)$	$\zeta(\xi,4)$	$\zeta(\xi,5)$	$\zeta(\xi,6)$
32	0.20	0.39	0.56	0.71	0.85	0.98
37	0.21	0.40	0.57	0.72	0.86	0.99
59	0.23	0.44	0.62	0.79	0.93	1.05
80	0.25	0.48	0.68	0.86	1.01	1.15
102	0.28	0.52	0.74	0.92	1.09	1.23
124	0.29	0.55	0.78	0.98	1.15	1.31
146	0.31	0.57	0.81	1.00	1.18	1.34
168	0.32	0.60	0.84	1.06	1.24	1.41
189	0.33	0.61	0.86	1.07	1.26	1.42
211	0.34	0.63	0.88	1.09	1.28	1.44
233	0.34	0.64	0.89	1.11	1.30	1.47
2670	0.36	0.66	0.92	1.14	1.35	1.53

Table 1: Scaling exponents of the velocity structure functions. The results show a clear dependence on the distance ξ from the plate. Relative error in the estimate of scaling exponents is about 3%.

towards the isotropic scaling law $S_6(r) \sim r^{1.78}$ (see Figure 1 and Figure 2 for comparison) is also observed. The scaling exponents for $n = 1 \ldots 6$ are finally summarized in Table 1.

3.3 Relative scaling behavior in the turbulent logarithmic sublayer

Figure 5 shows that the scaling properties of the velocity fluctuations depend on the distance from the plate. However, other quantities exhibit a behavior independent of ξ; these are relative scaling exponents, computed using the ESS scaling (2.1.

The structure function $S_6(\xi, r)$ is displayed versus $S_3(\xi, r)$ in Figure 6, for three distances ξ lying in the logarithmic sublayer. All points collapse into a single curve, indicating that relative scaling is independent of ξ, for ξ in the logarithmic sublayer.

As for the isotropic turbulence (see Figure 2), for the boundary layer turbulence we observe in Figure 6 two (relative) scaling ranges. At very small scales, structure functions display relative dissipative scaling, $S_n(\xi, r) \sim S_3(\xi, r)^{n/3}$. At larger scales an extended range of anomalous scaling is exhibited,

$$S_n(\xi, r) \sim S_3(\xi, r)^{\gamma_{n,3}} \tag{3.1}$$

with $\gamma_{n,3}$ nonlinear in n.

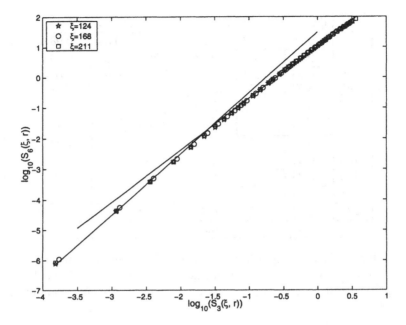

Figure 6: $S_6(\xi, r)$ versus $S_3(\xi, r)$ for various distances in the logarithmic sublayer. At very small scales, structure functions display relative dissipative scalings; $S_6(\xi, r) \sim S_3(\xi, r)^2$. At larger scales, data points align on the same straight line of slope $\zeta(\xi, 6)/\zeta(\xi, 3) \simeq 1.68$. The property of Extended Self-Similarity is well satisfied. The relative scaling is clearly different from the isotropic value $\zeta(6)/\zeta(3) = 1.78$.

It is found that the relative exponents $\gamma_{n,3} = \zeta(\xi, n)/\zeta(\xi, 3)$ remain constant in the logarithmic sublayer. The values are presented in Table 2. Deviations from constant values only occur near the wall as one approaches the viscous sublayer ($\xi = 32, \ 37$). Relative exponents then gets closer to the non-intermittent scaling $\gamma_{n,3} = n/3$. The values of $\gamma_{n,3}$ in the logarithmic sublayer are compared in Figure 7 with those of homogeneous and isotropic turbulence. We observe that intermittent corrections to the linear scaling $\gamma_{n,3} = n/3$ are larger than those usually observed in the homogeneous and isotropic case.

3.4   A new form of GESS for wall bounded turbulence

In section 2.2 we discussed a generalized form (2.3) of Kolmogorov's refined similarity hypothesis (1.4). We have also seen that this property, known as

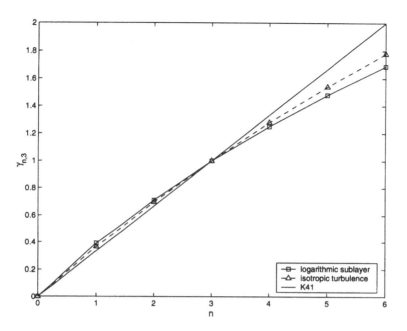

Figure 7: Mean (in the logarithmic sublayer) relative scaling exponents $\gamma_{n,3}$ are plotted as a function of n. These values are different from those usually reported in homogeneous and isotropic turbulence.

Generalized Extended Self-Similarity, is observed in isotropic turbulence over the whole range of scales $\eta \leq r \leq L$ at different Re (see Figure 3).

However in a recent numerical simulation (Benzi et $al.$ 1999), it has been observed that (2.3) was not satisfied in a turbulent boundary layer flow. Instead

$$S_n(\xi, r) = \frac{\langle \epsilon_r(\xi)^{n/2} \rangle}{\bar{\epsilon}(\xi)^{n/2}} S_2(\xi, r)^{n/2} \qquad (3.2)$$

should be considered. Two key arguments are given in order to justify (3.2). The first is that in the presence of shear (1.5) is no longer valid but an extra term appears (Hinze 1959),

$$\langle \delta V^3 \rangle = -\frac{4}{5} \epsilon r + 6 \nu \frac{\partial \langle \delta V^2 \rangle}{\partial r}$$
$$+ \frac{1}{r^4} \int_0^r \langle \delta V(r') \, \delta W(r') \rangle \frac{d\langle V(z) \rangle}{dz} r'^4 \, dr' \qquad (3.3)$$

where z denotes the direction of the shear, $\langle V(z) \rangle$ is the mean velocity component in the flow direction and W is the velocity component normal to the

ξ	$\gamma_{1,3}$	$\gamma_{2,3}$	$\gamma_{4,3}$	$\gamma_{5,3}$	$\gamma_{6,3}$
32	0.36	0.70	1.27	1.52	1.75
37	0.37	0.70	1.27	1.52	1.74
59	0.37	0.70	1.26	1.49	1.68
80	0.37	0.71	1.26	1.48	1.68
102	0.38	0.71	1.25	1.48	1.68
124	0.38	0.71	1.25	1.48	1.68
146	0.38	0.71	1.25	1.47	1.67
168	0.38	0.71	1.25	1.48	1.68
189	0.38	0.71	1.25	1.47	1.66
211	0.39	0.71	1.25	1.46	1.64
233	0.38	0.71	1.25	1.47	1.65
2670	0.39	0.72	1.25	1.47	1.67

Table 2: Relative scaling exponents of velocity structure functions, $\gamma_{n,3}$ remain constant in the logarithmic sublayer. Intermittent corrections to linear scalings are larger than in the isotropic case. Relative error in the estimate of relative scaling exponents is about 2%.

wall. This equation, which is derived under strong physical assumptions, is very useful to analyze the scaling properties in shear turbulence. First of all, one notices that the integral term may, in presence of a strong shear, give the largest contribution in the right hand side of (3.3). This could explain why the scaling exponent $\zeta(3)$ is different from 1 in the logarithmic sublayer. Furthermore, because of (3.3) and by dimensional arguments, one may guess that an extra term proportional to

$$\frac{\partial \langle V(z) \rangle}{\partial z} \delta V(z, r)^2, \tag{3.4}$$

enters into the estimate of the energy flux on the scale r. Because of (3.3) and (3.4), it can be speculated that in shear turbulence, δV^2 plays the same role as $\delta V(r)^3$ in isotropic turbulence (compare with section 2.2).

Experimentally (3.2) cannot be tested easily, the major difficulty arising from the estimation of the dissipation rate. Indeed, the one-dimensional surrogate of ϵ, usually used in the isotropic case, is no more valid. All components of the velocity must be considered (Tsinober 1992). We propose here to test the validity of (3.2) by using the symmetry relation between velocity structure functions discussed in section 2.3 for the isotropic case.

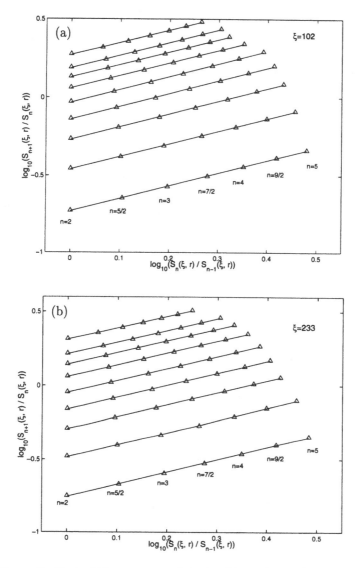

Figure 8: Test of the structure function hierarchy with measurements carried out in the logarithmic sublayer at the distance $\xi = 102$ (a) and $\xi = 233$ (b). $S_n(\xi, r)$ are computed at fixed scale r and varying order n, the data points are aligned in agreement with the symmetry relation (3.5). This symmetry relation remains satisfied over the whole range of scales $\eta \leq r \leq L$; the ratio $r/\eta = 3, 5, 10, 15, 25, 50, 100, 200$ and 500 from the bottom line to the top. These lines are parallel, which indicates that $\beta(\xi, r)$ is independent of r for $\eta \leq r \leq L$.

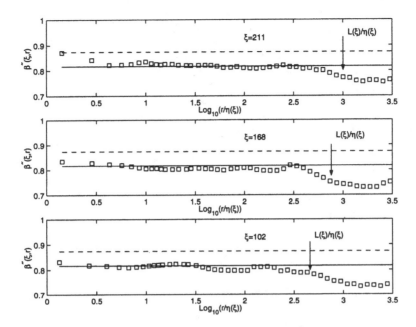

Figure 9: The dependence on r of the modulation factor $\beta''(\xi, r)$ is displayed for different ξ in the logarithmic sublayer; $\beta''(\xi, r)$ is estimated by least square fitting. For each ξ, we find that $\beta''(\xi, r)$ remains almost constant over a wide range of scales extending from the dissipation to the integral scale. At larger scales, the symmetry remains satisfied but the modulation factor exhibits a dependence on r. The inertial-range value of β'' is found to be independent of ξ in the logarithmic sublayer. This value is in full agreement with the prediction given by Benzi *et al.*, i.e. $\beta''(\xi) = (2/3)^{1/2}$ (plain line). The dashed line indicates the value $\beta' = (2/3)^{1/3}$ usually observed in homogeneous and isotropic turbulence.

3.5 Hierarchy of velocity structure functions in the logarithmic sublayer

One may expect that in the presence of shear (2.7) persists with a modulation factor dependent on the distance ξ:

$$\frac{S_{n+1}(\xi, r)}{S_n(\xi, r)} = \left(\frac{S_n(\xi, r)}{S_{n-1}(\xi, r)} \right)^{\beta''(\xi, r)} G_\infty(\xi, r). \qquad (3.5)$$

We have tested the symmetry relation (3.5) for various ξ in the logarithmic sublayer. For fixed scale r and varying order n, we observe in Figure 8(a)

($\xi = 102$) and Figure 8(b) ($\xi = 233$) that data points are aligned (plain lines) in agreement with (3.5). The modulation factor $\beta''(\xi, r)$ is given by the slope of the lines. The different lines (corresponding to different scales r) are parallel, which gives that $\beta''(\xi, r)$ is independent of r over the whole range of scales $\eta(\xi) \leq r \leq L(\xi)$ (here $L(\xi)$ denotes the integral scale which of course is a function of ξ).

In Figure 9, $\beta''(\xi, r)$ is displayed for various ξ in the logarithmic sublayer and r ranging from dissipative to very large scales; $\beta''(\xi, r)$ is estimated by least square fitting. We clearly observe that $\beta''(\xi, r)$ remains constant for $\eta(\xi) \leq r \leq L(\xi)$. Moreover, $\beta''(\xi, r)$ does not depend on ξ either. The symmetry relation (3.5) is therefore uniformly satisfied in the logarithmic sublayer. The value of β'' is found very close to the theoretical value $(2/3)^{1/2}$ (see below) and obviously different from the isotropic value $(2/3)^{1/3}$, used in sec.2.3 to verify (2.7) in the isotropic case.

It is shown numerically in (Benzi *et al.* 1999) that the moments of $\epsilon_r(\xi)$ in wall bounded turbulence satisfy the same symmetry relation as those of ϵ_r in homogeneous and isotropic turbulence (She & Lévêque 1994, Ruiz *et al.* 1995a,b). In concrete terms,

$$\frac{\langle \epsilon_r(\xi)^{n+1} \rangle}{\langle \epsilon_r(\xi)^n \rangle} = \left(\frac{\langle \epsilon_r(\xi)^n \rangle}{\langle \epsilon_r(\xi)^{n-1} \rangle} \right)^{\beta(\xi, r)} \epsilon_\infty(\xi), \tag{3.6}$$

with $\beta(\xi, r) = 2/3$ in the inertial range and $G_\infty(\xi, r)$ independent of n. The new form of GESS (3.2) then suggests that the velocity structure function $S_n(\xi, r)$ should satisfy (3.5) with $\beta''(\xi, r) = (2/3)^{1/2}$. This prediction is found in full agreement with our experimental results.

4   Concluding remarks

In this paper we have directly compared the statistical features of isotropic turbulence with those observed in an experiment on turbulence near a wall. The particular focus has been on the dependence of structure function scaling properties on the distance from the wall. The following features have been observed:

(1) The scaling exponents of structure functions depend on the distance from the wall;

(2) The relative exponents in the logarithmic layer are constant;

(3) In the logarithmic sublayer, a new form of GESS, making use of the second-order structure function, is supported by our experimental data;

(4) The structure function scalings change near the wall because GESS changes form. Instead the intermittent properties of ϵ are, within error bars, the same in boundary layer and in homogeneous turbulence.

These results are in agreement with numerical simulation by Benzi *et al.* (1999). The existence of a new form of GESS near a wall is quite important because it can be useful to improve large eddy simulations near walls. Indeed in these simulations RKS is indirectly used to close the equations. If this relation fails, new forms of closure, taking into account our results (3) and (4), should be introduced.

The point (4) merits special comment. The independence of the intermittent properties of ϵ has been indirectly proved by showing that the β of the She–Lévêque model is constant within error bars. Indeed ϵ cannot be correctly measured experimentally in strong anisotropic cases. It has to be pointed out that the She–Lévêque hierarchy has been used here as a sensible tool to distinguish between the two situations, namely isotropic and anisotropic. We do not want to discuss here the kind of statistics which is more appropriate to describe a turbulent field. Our data show that β, on a first approximation, depends neither on r nor on n, which is consistent with log-Poisson statistics. However, a small dependence on r and n, which would imply different statistics, such as for example log-normal (see Arneodo *et al.* 1998), cannot be a priori excluded. This topic is outside the interest of this paper.

References

F. Anselmet, Y. Gagne, E.J. Hopfinger, R.A. Antonia, (1984), *J. Fluid Mech.* **140** 63.

A. Arneodo *et al.*, (1996), *Europhysics Lett.* **34** 411.

A. Arneodo, S. Manneville, J.F. Muzy, (1998), *Eur. Phys. J* **B1** 129.

F. Belin, P.Tabeling, H. Willaime, (1996), submitted to *Physica D*.

R. Benzi, S. Ciliberto, R. Tripiccione, C. Baudet, S. Succi, (1993a), *Phys. Rev. E* **48** R29.

R. Benzi, S. Ciliberto, C. Baudet, G. Ruiz Chavarria, R. Tripiccione, (1993b), *Europhys. Lett.* **24** 275.

R. Benzi, R. Tripiccione, F. Massaioli, S. Succi, S. Ciliberto, (1994), *Europhys. Lett.* **25** 331.

R. Benzi, S. Ciliberto, C. Baudet, G. Ruiz Chavarria, (1995), *Physica D* **80** 385.

R. Benzi, L. Biferale, S. Ciliberto, M. V. Struglia, R. Tripiccione, (1996), *Physica D* **96** 162.

R. Benzi, G. Amati, C.M. Casciola, F. Toschi, R. Piva, (1999), 'Intermittency and scaling laws for wall bounded turbulence', to be published in *Physics of Fluids*.

M. Briscolini, P. Santangelo, S. Succi, R. Benzi, (1994), *Phys. Rev. E* **50** 1745.

B. Castaing, Y. Gagne, E. Hopfinger, (1990), *Physica D* **46** 177.

S. Cioni, S. Ciliberto, J. Sommeria, (1995), Europhysics. Lett. 32, 413.

B. Dubrulle, (1994), *Phys. Rev. Lett.* **73** 959.

U. Frisch, (1995), *Turbulence: the Legacy of A.N. Kolmogorov*, Cambridge University Press.

Y. Gagne, B. Castaing, (1991), *C.R. Acad. Sci. Paris* **312** 441.

E. Gaudin, B. Protas, S. Goujon-Durand, J. Wojciechowski, J.E. Wesfreid, (1998), *Phys. Rev. E* **57** R9.

J.O. Hinze, (1959), *Turbulence*, McGraw-Hill.

A.N. Kolmogorov, (1941), *C.R. Acad. Sci. USSR* **30** 299.

A.N. Kolmogorov, (1962), *J. Fluid Mech.* **13** 83.

L. Landau, E. Lifshitz., (1959) *Fluid Mechanics*, Pergamon Press.

A.S. Monin, A.M. Yaglom, (1975), *Statistical Fluid Mechanics* MIT Press.

N. Mordant, J.F. Pinton, F. Chillá, (1997), *J. Physique* **7** 1729.

G. Ruiz Chavarria, (1994), *J. Physique* **4** 1083.

G. Ruiz Chavarria, C. Baudet, S. Ciliberto, (1995a), *Phys. Rev. Lett.* **74** 1986.

G. Ruiz Chavarria, C. Baudet, R. Benzi, S. Ciliberto, (1995b), *J. Physique* **5** 485.

G. Ruiz Chavarria, C. Baudet, S. Ciliberto, (1996), *Physica D* **99** 369.

G. Ruiz Chavarria, S.Ciliberto, C. Baudet, E. Lévêque, (1999), submitted to *Physica D*.

Z S. She, E. Lévêque, (1994), *Phys. Rev. Lett.* **72** 336.

H. Schlichting, (1968), *Boundary Layer Theory*, McGraw-Hill.

G. Stolovitzky, K.R. Sreenivasan, (1993), *Phys Rev. E* **48** 32.

A. Tsinober, E. Kit, T. Dracos, (1992), *J. Fluid Mech.* **242** 169.

A. Vincent, M. Meneguzzi, (1991), *J. Fluid Mech.* **225** 1.

Turbulent Wakes of 3D Fractal Grids

D. Queiros-Conde and J.C. Vassilicos

1 Introduction

The work reported in this article is motivated by a series of developments in turbulence research to do with the multiple scale structure of realisations of the turbulence. This particular vein of turbulence research was initiated with the works of Novikov (1970), Mandelbrot (1974), Frisch *et al.* (1978) and Parisi & Frisch (1985) who introduced in the study of turbulence the concept that scaling exponents of structure functions may depend on the fractal structure of the turbulence. More recently, a lot of effort has been expended in deriving the scaling properties of the Kraichnan model of scalar turbulence (Kraichnan 1994) where the velocity field is assumed to be a fractional Brownian motion in space. In other words, this model assumes that the iso-surfaces of realisations of the velocity field are fractal in the Hausdorff sense, and that the second order velocity structure function $\langle \delta u^2 \rangle$ has a power-law dependence on the spatial separation distance r,

$$\langle \delta u^2 \rangle \sim r^H,$$

with a scaling exponent H that is a simple function of the Hausdorff dimension D_H of these iso-surfaces. The scalar structure functions $\langle \delta \theta^n \rangle$ in the Kraichnan model turn out (Gawędzki 2000) to also be power laws in r,

$$\langle \delta \theta^n \rangle \sim r^{\xi_n},$$

and the exponents ξ_n are functions of H (Gawędzki 2000, Gawędzki & Kupiainen 1995, Chertkov & Falkovich 1996, Balkovsky & Lebedev 1998) and thereby, indirectly, of the Hausdorff dimension D_H.

Another recent model of scalar turbulence in two dimensions (Khan & Vassilicos 2000) consists of a spatial distribution of 2D vortices which generate spiral scalar interfaces of Kolmogorov capacity (box-counting dimension) D_K. The scalings of the scalar structure functions $\langle \delta \theta^n \rangle$ are determined by this spiral geometry and, specifically,

$$\langle \delta \theta^n \rangle \sim r^{2-D_K}$$

where $1 \leq D_K < 2$. Furthermore, the scalar dissipation rate χ has a power-law dependence on the scalar diffusivity κ in the limit where $\kappa \to 0$, which reads (see Flohr & Vassilicos 1997)

$$\chi \sim \kappa^{1-D_K/2}.$$

136

The dependence of the kinetic energy dissipation rate ϵ of the turbulence on the kinematic viscosity ν may also involve a fractal dimension of the turbulence if the multiple scale structure of realisations of the turbulence is of a fractal or spiral nature or both. Consequently, the Reynolds number dependence of drag may also be determined in certain regimes by the quantitative properties of the geometry of turbulent wakes.

The question that arises from this line of research is whether it is possible, in the laboratory, to vary the dimensions D and D_H and, more generally, the multiple scale geometry and spatio-temporal structure of realisations of the turbulence in such a way as to change the scalings of $\langle \delta u^n \rangle$, $\langle \delta \theta^n \rangle$ and the dependences of χ on κ and ϵ on ν at least for a while. Is it possible to modify the turbulence directly in its inner scaling structure? This question raises two issues: (i) the issue of universality, for if turbulence generated by a large-scale forcing is universal then it should not to be possible to modify the inner scalings of a turbulence forced at the large scales; and (ii) the issue of departures from the presumed range of applicability of Kolmogorov's 1941 theory (Kolmogorov 1941a, 1941b, Frisch 1995), for example by explicitly forcing not the large scales only but a wide range of scales of the turbulence.

(i) Universality arguments such as those applied to structure functions following Kolmogorov (1941a, 1941b) lead to power laws

$$\langle \delta u^n \rangle \sim r^{\zeta_n}.$$

However, experimental measurements even at very high Reynolds numbers do not support such power laws unequivocally (Kholmyansky & Tsinober 2000), and it is safer to assume that the exponents ζ_n are functions of r, i.e. $\zeta_n = \zeta_n(r)$. The fact that the exponents ζ_n are not independent of r does not necessarily imply that the turbulence is not universal. But the finding by Praskovsky *et al.* (1993) that small-scale turbulence fluctuations are statistically correlated with large-scale velocities where these velocities are large *is* evidence that the turbulence is not universal since the large-scale velocities depend on boundary and initial conditions and on the type of forcing (see also Hunt *et al.* 1988, Hunt & Vassilicos 1991, Sreenivasan & Stolovitzky 1996, Kholmyansky & Tsinober 2000). More to the point, a recent series of evidence obtained from a numerical simulation of channel flow turbulence (Toschi *et al.* 1999), an experimental study of a turbulent boundary layer flow near a solid plate (Ciliberto *et al.* 2000) and an experimental study of a cylinder's turbulent wake (Gaudin *et al.* 1998) suggest that the exponent functions $\zeta_n(r)$ are not universal but depend on the local mean shear (see Benzi *et al.* 1999) and are therefore different at different locations in a turbulent shear or wake flow. This is evidence that the scalings of $\langle \delta u^n \rangle$ and $\langle \delta \theta^n \rangle$ vary with the location in inhomogeneous flows in response to local mean shear.

(ii) However, our goal is different. We want to find a systematic way to tamper directly with the inner multiple scale geometry and spatio-temporal structure of realisations of the turbulence and we believe that by doing so we may be able to also alter the dependences of χ on κ and ϵ on ν as well as the scalings of structure functions. The way we propose to achieve this goal is by placing a 3D fractal grid in a wind tunnel and studying its turbulent wake as a function of the fractal dimension D_f of the 3D fractal grid. This amounts to forcing over a wide range of scales with some control over the scaling of the forcing. This procedure is different from all the turbulence experiments mentioned above (and in the references cited) where the forcing is applied only at the large scales.

In this paper, we present laboratory evidence in support of the idea that the scaling properties of the near turbulent wake are different for 3D fractal grids with different fractal dimensions D_f. In other words this paper's purpose is to establish the feasibility of modifying the multiple scale structure of the turbulence in some controlled way by using different 3D fractal grid objects.

In the next section, we describe the experimental setup and, in Section 3, we present our results. We conclude in Section 4.

2 Experimental setup

The experiments were performed in the wind tunnel located in the basement of the Department of Applied Mathematics and Theoretical Physics of the University of Cambridge. A detailed description of this wind tunnel can be found in Ramsey (1988) and Ushijima (1998). The wind tunnel is a non-recirculating suction-type tunnel with a horizontal test section that is 176cm long and 45.7cm tall and wide. The air screw fan located downstream of the test section can generate air mean velocities up to 12ms^{-1} in the test section. Some classical square grids are available with mesh sizes $M = 2.54\text{cm}$, 5.08cm, 11.4cm and 15.2cm, their blockage ratio being equal to 4. We have used these grids to verify the classical characteristics of our wind tunnel. During our experiment on turbulent wakes generated by the fractal grids, the mean velocity of the laminar input flow was kept at 10ms^{-1} with a residual turbulence of 0.6% in the test section upstream of the 3D fractal grid in all the runs involving a fractal grid that we report in this paper. Measurements of velocities $u(t)$ at different times t were made with a Dantec constant temperature anemometer. The hot wire's thickness is 5μm and its length 1.25mm. The velocities $u(t)$, which, strictly speaking, are amplitudes of velocity vectors, are derived from voltage measurements using a slightly modified King's law the validity of which we have established and checked. The data acquisition and analysis were carried out with Labview 5 from National Instrument with a sampling frequency of 100kHz.

A schematic diagram of the design of our 3D fractal objects is shown in Figs. 1 and 2 (see also Table 1). The generating elementary structure of the fractal objects consists of two horizontal X crosses held together by a vertical

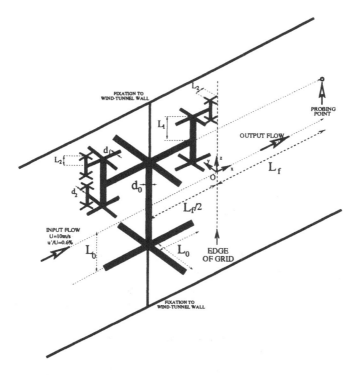

Figure 1: Construction of a 3D fractal grid and experimental configuration.

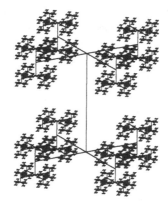

Figure 2: Sketch of a 3D fractal grid. For the sake of clarity, the object is represented with only three iterations and without thickness.

Iteration	Lengths	Horiz. thick.	Vertic. thick.
generator	L_0	d_0	d_0
iteration 1	$L_1 = RL_0$	$d_1 = Rd_0$	d_0
iteration 2	$L_2 = RL_1$	$d_2 = Rd_1$	$d_1 = Rd_0$
iteration 3	$L_3 = RL_2$	$d_3 = Rd_2$	$d_2 = Rd_1$
iteration 4	$L_4 = RL_3$	$d_4 = Rd_3$	$d_3 = Rd_2$
.	.	.	.
iteration N_{it}	$L_{N_{it}} = RL_{N_{it}-1}$	$d_{N_{it}} = Rd_{N_{it}-1}$	$d_{N_{it}-1} = Rd_{N_{it}-2}$

Table 1. Dimensional characteristics of the 3D fractal grids. The smallest scales are fixed at $L_{min} = L_{N_{it}} = 2$mm and $d_{min} = d_{N_{it}} = 1$mm which is the minimum cross section.

rod. The generation of the fractal structure is achieved by sticking at the four ends of each X cross the same pattern again but downsized. The lengths of the X crosses and of the connecting rod in the largest such elementary structure are $2L_0$. This rod and the rods making the X crosses are slender bars with a square cross section of side length d_0. At the four ends of each one of these X crosses are smaller versions of the generating elementary structure connected from the centre of their horizontal rod as in Fig. 1. The smaller version's rods have length $2L_1 = R2L_0$ and cross section side lengths, horizontal and vertical respectively, $d_1 = Rd_0$ and d_0 (with $R < 1$) and the rods' aspect ratio L_1/d_1 is conserved, i.e. $L_1/d_1 = L_0/d_0 = A$. This generating pattern is repeated over and over again with ever smaller-sized elementary structures of lengths $2L_{j+1} = R2L_j$ if $RL_j \geq L_{min}$ but $L_{j+1} = L_{min}$ if $RL_j \leq L_{min}$; of horizontal cross section side lengths $d_{j+1} = Rd_j$ if $Rd_j \geq d_{min}$ but $d_{j+1} = d_{min}$ if $Rd_j \leq d_{min}$; and of vertical cross section side length d_j. L_{min} is the minimum length and d_{min} the minimum cross section size and they are reached after a number N_{it} of iterations (i.e. $0 \leq j \leq N_{it}$) (see Table 1).

Four such objects (see Fig. 2) with four different fractal dimensions D_f have been manufactured for us by Amalgamated Research Inc. in Idaho, USA (see Kearney 1997). These objects are the same in all respects but two. They all have the same aspect ratio $A = 6.66$ and the same minimum length $L_{min} = 2$mm and minimum cross section size $d_{min} = 1$mm (this is the smallest cross section size that can be achieved by the technique for manufacturing 3D fractal grids used by Amalgamated Research Inc.), but they have different iteration ratios R and in one case a slightly different number of iteration N_{it}. The iteration ratio determines the value of the fractal dimension D_f and the number of iterations N_{it} relates to the range of scales over which D_f is well-defined (see Appendix). The numbers N_{it} have been chosen as large as possible under the constraint however that the fractal objects' overall

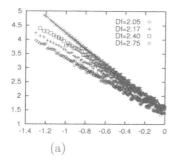
(a)

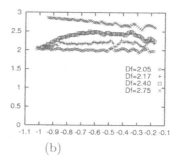
(b)

Figure 3: (a) Fractal analysis of the 3D fractal grids: Log(N) vs log(r/L_0) where N is the number of cubes of size r needed to cover the grid. (b) Local fractal dimension: D_f vs log(r/L_0) where $D_f = -d(\log(N))/d(\log(r/L_0))$.

sizes $L_f = 2L_0 + 2L_1 + \cdots + 2L_{N_{it}}$ should be significantly smaller than the width and height of the wind tunnel's test section (which is 45.7cm). Our four fractal objects have been constructed with the following four pairs of specification: $R = 0.36$ and $N_{it} = 4$; $R = 0.4$ and $N_{it} = 4$; $R = 0.45$ and $N_{it} = 4$; $R = 0.5$ and $N_{it} = 5$. The values of L_0 and L_f which follow from these specifications and from the value of L_{min} (which is the same for all our fractals) are, in corresponding order: $L_0 = 119.1$mm, $L_f = 37$cm; $L_0 = 78.1$mm, $L_f = 25.8$cm; $L_0 = 48$mm, $L_f = 17.4$cm; $L_0 = 64$mm, $L_f = 25.2$cm.

A detailed analytical and numerical discussion of D_f and of its relation to R and N_{it} is given in the Appendix. Suffice it to say here that the fractal dimensions D_f of our 3D fractal grids can be thought of as being the box-counting dimensions obtained from minimal coverings of the 3D fractal objects, which means that $D_f \leq 3$. As can be observed in Figs. 3a and 3b where numerical results obtained from the fractal analysis of the grids are plotted, all four objects have sufficiently well-defined values of D_f over about one decade of length scales within the broader range defined by L_0 from above and d_{min} from below. These values are $D_f = 2.05$ for $R = 0.36$, $D_f = 2.17$ for $R = 0.4$, $D_f = 2.40$ for $R = 0.45$, $D_f = 2.75$ for $R = 0.5$.

For every set of measurements involving a 3D fractal grid, one of the fractal objects was placed in the test section at equal distances from the test section's top and bottom and such that one of the arms of each X cross was aligned with the incoming flow downstream (Fig. 1). We denote by x the streamwise coordinate (parallel to the incoming flow and to one of the arms of the X crosses) with origin at the edge of the fractal object (see Fig. 1), z the vertical coordinate and y the transversal coordinate both with the same origin.

3 Results

3.1 Mean profiles

We started with a visualisation of the flow using a smoke point source placed at various points upstream of the fractal objects. Invariably, the smoke did not travel around the objects but through them, except when the point source was placed significantly above or below the vertical extent of the fractals. For a basic description of the flow in the wake of 3D fractal grids we have measured velocity and turbulence intensity profiles at a distance $x = L_f$ behind two of our 3D fractal grids ($D_f = 2.17$ and $D_f = 2.40$).

We have measured such profiles at $x = L_f$ as functions of z and y (see Figs. 4 to 12). In some cases (Figs. 6 to 8), only negative values of y have been investigated. However, because of the symmetry properties of the present velocity profiles (as can be expected from the symmetry properties of the experimental setup), we extrapolate our results to positive values of y. In the case of the fractal with $D_f = 2.4$ we find that the mean velocity U at $x = L_f$ and $y = 0$ remains about constant between $z = -L_f/2$ and $z = L_f/2$ with about one tenth of the value it takes upstream and at the edges of the object (which are at $z = \pm L_f/2$) (see Fig. 4a). Similar behaviour can be observed for the r.m.s. u' of velocity fluctuations i.e. $u' = \langle (u(t) - U)^2 \rangle^{1/2}$ (where the averages are taken by integrating over t) (Fig. 4b) and the turbulent intensity u'/U (Fig. 5).

We also studied how the profiles along the vertical coordinate of U, u' and u'/U vary with the transversal coordinate y. Two series of curves are given here: the first (Figs. 6a, 7a and 8a) shows the variations of these profiles for small transversal displacements from the centre of the grid ($y/L_f \leq 0.17$) and the second (Figs. 6b, 7b and 8b) shows these profiles for larger transversal displacements ($0.23 \leq y/L_f \leq 0.46$). It can be observed that the profile of U as a function of z varies only very slightly in the transversal direction from approximately $y = -0.2L_f$ to $y = 0.2L_f$ (Fig. 6a) but then changes quite significantly beyond $|y| \approx 0.2L_f$ with what appear to be two jets on either side of the origin along the transversal direction (see Fig. 6b). Similarly, but with more scatter, the vertical profiles of u' and u'/U do not present a significant change along the transversal coordinate from $y = -0.2L_f$ to $y = 0.2L_f$ (Figs. 7a, 8a). However, they change as soon as $y/L_f \geq 0.2$ (Figs. 7b, 8b).

The same measurements have been made at a distance $x = L_f$ downstream from the fractal grid $D_f = 2.17$ (Figs. 9 to 10). Behind this object, the mean velocity profile cannot be considered uniform along the y and z axes (Fig. 9a) and a direct comparison with the fractal grid $D_f = 2.40$ (Figs. 11a, 11b) shows that the variations in U along the vertical direction are much more important for the fractal $D_f = 2.17$ than for the fractal $D_f = 2.40$. Concerning u', it can be observed in Figs. 9b and 12a that it also displays

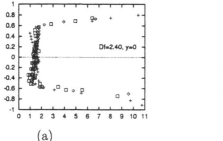 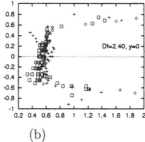

(a) (b)

Figure 4: (a) Mean velocity U (ms^{-1}) (on the abscissa) as a function of the vertical coordinate z/L_f for $x/L_f = 1$, $y/L_f = 0$ and $D_f = 2.40$. The input velocity is 10ms^{-1}. (b) Profile of r.m.s. fluctuations u' (ms^{-1}) (on the abscissa) along the vertical axis z/L_f at $x/L_f = 1$, $y/L_f = 0$ and for $D_f = 2.40$. The different symbols correspond to different realisations.

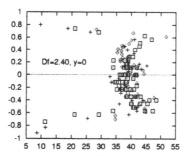

Figure 5: Turbulent intensity u'/U (in percentage points on the abscissa) as a function of the vertical coordinate z/L_f at $x/L_f = 1$, $y/L_f = 0$ and for $D_f = 2.40$. The input velocity is 10ms^{-1} and the input residual turbulent intensity is 0.6%. The different symbols correspond to different realisations.

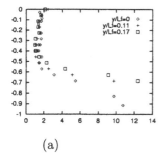 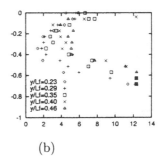

(a) (b)

Figure 6: (a) Mean velocity U (ms^{-1}) (on the abscissa) as a function of the vertical coordinate z/L_f at three different transversal positions all such that $y/L_f \leq 0.17$, at $x/L_f = 1$ and for $D_f = 2.40$. (b) Profile of mean velocity U (on the abscissa) as a function of the vertical coordinate z/L_f at five different transversal positions all such that $0.23 \leq y/L_f \leq 0.46$, at $x/L_f = 1$ and for $D_f = 2.40$.

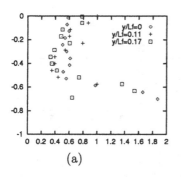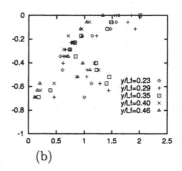

Figure 7: (a) R.m.s. u' (ms^{-1}) (on the abscissa) as a function of the vertical coordinate z/L_f at $x/L_f = 1$ for $y/L_f \leq 0.17$ and $D_f = 2.40$. (b) Profile of r.m.s u' as a function of the vertical coordinate z/L_f for $0.23 \leq y/L_f \leq 0.46$, $x/L_f = 1$ and $D_f = 2.40$.

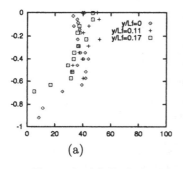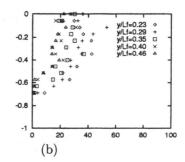

Figure 8: (a) Turbulent intensity u'/U (in percentage points on the abscissa) as a function of the vertical coordinate z/L_f at $x/L_f = 1$, $y/L_f \leq 0.17$ and for $D_f = 2.40$. (b) Turbulent intensity u'/U (in percentage points on the abscissa) as a function of the vertical coordinate z/L_f at $x/L_f = 1$, $0.23 \leq y/L_f \leq 0.46$ and for $D_f = 2.40$.

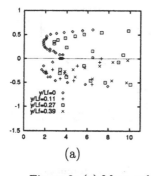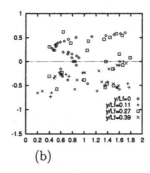

Figure 9: (a) Mean velocity U (ms^{-1}) (on the abscissa) as a function of the vertical coordinate z/L_f at $x/L_f = 1$ and $0 \leq y/L_f \leq 0.39$ and for $D_f = 2.17$. (b) R.m.s u' (ms^{-1})(on the abscissa) as a function of the vertical coordinate z/L_f at $x/L_f = 1$ and $0 \leq y/L_f \leq 0.39$ and for $D_f = 2.17$.

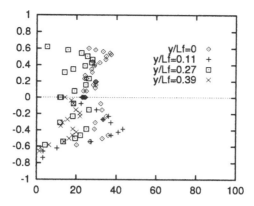

Figure 10: Turbulent intensity u'/U (in percentage points on the abscissa) as a function of the vertical coordinate z/L_f at $x/L_f = 1$ and $0 \leq y/L_f \leq 0.39$ and for $D_f = 2.17$.

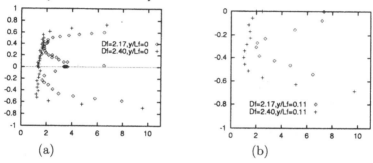

Figure 11: (a) Comparison of mean velocity U (ms^{-1}) vertical profiles (on the abscissa) for the two fractal grids $D_f = 2.17$ and $D_f = 2.40$ at $y/L_f = 0$, $x/L_f = 1$. (b) Comparison of mean velocity vertical profiles (on the abscissa) for the two fractal grids $D_f = 2.17$ and $D_f = 2.40$ at the transversal position $y/L_f = 0.11$, $x/L_f = 1$.

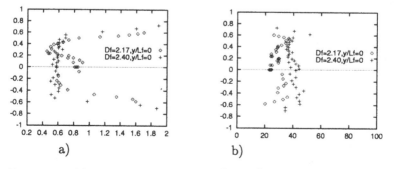

Figure 12: (a) Comparison of r.m.s. u' (ms^{-1}) vertical profiles (on the abscissa) for fractal grids $D_f = 2.17$ and $D_f = 2.40$ at $y/L_f = 0$, $x/L_f = 1$. (b) Comparison of turbulent intensity u'/U profiles (in percentage points on the abscissa) for fractal grids $D_f = 2.17$ and $D_f = 2.40$ at $y/L_f = 0$, $x/L_f = 1$.

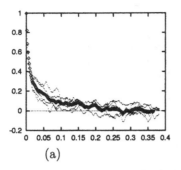 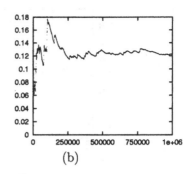

(a)   (b)

Figure 13: (a) Autocorrelation function $R(\tau)$ versus τ (in seconds) at a distance $x = L_f$ behind the fractal $D_f = 2.40$. Each dashed line represents one realisation. The diamonds are obtained by an average over five realisations. (b) Statistical convergence of $\langle|\delta u(\tau)|^3\rangle$ for $\tau/T_E = 0.48$ with $x = L_f$ and $D_f = 2.40$. The abscissa represents the number of data points in one realisation over which the statistical average is calculated. Data points are taken every $10\mu s$.

fluctuations that are more important for $D_f = 2.17$ than for $D_f = 2.40$. However, the turbulent intensity u'/U seems to be approximately uniform at $y = 0$ along the vertical direction for both fractals (Figs. 10 and 12b). In some sense, the variations in U and u' for the grid $D_f = 2.17$ seem to compensate each other.

We stress the conclusion that the mean velocity profile behind the fractal with $D_f = 2.17$ is not as uniform in the plane $x = L_f$ as it is behind the $D_f = 2.40$ fractal in the same plane (see Figs. 11a and 11b for a direct comparison). If we can extrapolate from these measurements, it would seem that the mean transversal and vertical shear rates $\frac{\partial U}{\partial y}$ and $\frac{\partial U}{\partial z}$ at a distance $x = L_f$ behind the 3D fractal grid tend towards being uniformly zero in a region around $y = 0$ and $-L_f/2 \leq z \leq L_f/2$ as the fractal dimension D_f is increased.

3.2   Two-time statistics

We have measured the velocity autocorrelation function and structure functions on the axis $y = 0, z = L_0$ at various distances behind the fractal objects on this axis, $x/L_f = 1, 2, 3, 4, 5, 5.9$, and we report in this subsection results from $x/L_f = 1$.

At the probing point, in the turbulent wake, we can measure two velocity scales: the mean velocity U and the r.m.s. u' of velocity fluctuations. From the autocorrelation function $R(\tau) = \langle(u(t) - U)(u(t + \tau) - U)\rangle/u'^2$ (with averages taken by integrating over t and sometimes an additional average over realisations) we can derive two time scales, a Taylor time microscale τ_λ and an integral time scale T_E (see Fig. 13a). We can also define the length scales

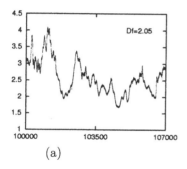
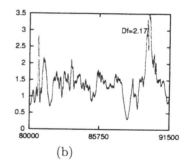

Figure 14: Velocity signals (in ms^{-1}) as a function of time (on the abscissa) for 10 integral time scales ($10T_E$) at $x = L_f, y = 0, z = L_0$. Each time unit on the plots represents $10\mu s$. (a) D_f=2.05. (b) D_f=2.17.

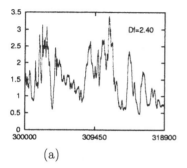
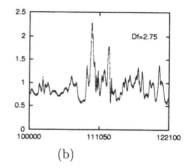

Figure 15: Velocity signals (in ms^{-1}) as a function of time (on the abscissa) for 10 integral time scales ($10T_E$) at $x = L_f, y = 0, z = L_0$. Each time unit on the plots represents $10\mu s$. (a) D_f=2.40. (b) D_f=2.75.

$L = UT_E$ and $\lambda = U\tau_\lambda$ but without guarantee that these length scales are an integral length scale and a Taylor length scale respectively. This is because, with the possible exception of the $D_f = 2.05$ fractal, turbulent intensities are way above 15 to 20% and it is therefore not justified in general to translate our time-domain results into space-domain conclusions by invoking the Taylor hypothesis. Indeed, at $x/L_f = 1$, $u'/U = 19\%$ behind the $D_f = 2.05$ fractal, $u'/U = 35\%$ behind the $D_f = 2.17$ fractal, $u'/U = 44\%$ behind the $D_f = 2.40$ fractal, and $u'/U = 47\%$ behind the $D_f = 2.75$ fractal.

Reynolds numbers may be defined strictly on the basis of the time scales measured, i.e. T_E and τ_λ, but also on the basis of the surrogate length scales L and λ. These Reynolds numbers are: $\frac{T_E}{\nu/U^2}, \frac{T_E}{\nu/u'^2}, \frac{\tau_\lambda}{\nu/u'^2}$; and $\frac{UL}{\nu}, \frac{u'L}{\nu}, \frac{u'\lambda}{\nu}$.

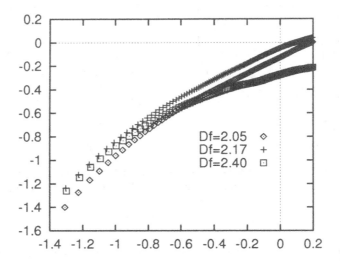

Figure 16: $\log\langle|\delta u(\tau)|^2\rangle$ vs $\log(\tau/T_E)$ at $x = L_f, y = 0, z = L_0$ for the three fractals $D_f = 2.05, 2.17, 2.40$. The smallest time plotted is $\tau = \tau_\lambda$.

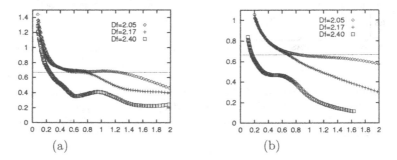

Figure 17: (a) Exponent function ζ_2 versus τ/T_E for one realisation. The smallest time plotted is $\tau = \tau_\lambda$. (b) The same as (a) but with an average of the local exponent function $\zeta_2(\tau/T_E)$ over five realisations. The horizontal dotted line in both (a) and (b) corresponds to a Kolmogorov-like prediction $\zeta_2 = 2/3$.

Note that $\frac{T_E}{\nu/U^2} = \frac{UL}{\nu}$ so that there are in fact five different Reynolds numbers. In this paper we give the values of $\Re_\lambda \equiv \frac{u'\lambda}{\nu}$ and $\Re \equiv \frac{u'L}{\nu}$. The other three Reynolds numbers can be derived from these two and from u'/U. In Table 2 we give values for $\Re_\lambda$, $\Re$, u'/U, but also for U, λ and L at $x = L_f$, $y = 0$, $z = L_0$ for the four fractal objects. Note that in all cases $\lambda \approx L_{\min}$. Also, L is much smaller than L_f and $T_E \ll x/U$ at $x = L_f$, sometimes even by a factor

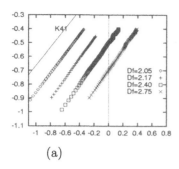 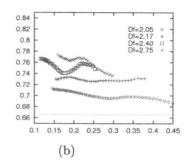

(a) (b)

Figure 18: (a) Extended self-similarity: $\log\langle|\delta u(\tau)|^2\rangle$ vs $\log\langle|\delta u(\tau)|^3\rangle$. With the exception of the case $D_f = 2.05$, the curves have been shifted in order to separate them. The smallest time plotted is $\tau = \tau_\lambda$ and the largest is the integral time scale $\tau = T_E$. (b) Relative exponent $\frac{\zeta_2}{\zeta_3}$ vs $\langle|\delta u(\tau)|^2\rangle$ between τ_λ and T_E. Average over five realisations. The horizontal dotted line corresponds to a Kolmogorov-like prediction $\zeta_2/\zeta_3 = 2/3$.

of 10, which suggests that at distances $x = O(L_f)$ and larger the turbulence generated by the fractal objects has had the time to fully develop. In fact, in the case of the $D_f = 2.4$ fractal, we find that $\frac{x}{UT_E}$=11,18,24 at $x/L_f = 1, 2, 3$ (according to the data reported in Subsection 3.3) which means in particular that T_E does not scale with x/U.

	$D_f = 2.05$	$D_f = 2.17$	$D_f = 2.40$	$D_f = 2.75$
$U =$	$2.9\mathrm{ms}^{-1}$	$1.57\mathrm{ms}^{-1}$	$1.48\mathrm{ms}^{-1}$	$1.11\mathrm{ms}^{-1}$
$u'/U =$	19%	35.6%	44.6%	46.8%
$Re_\lambda =$	75	75	88	69
$Re =$	750	672	1232	850
$\lambda =$	2mm	2mm	2mm	2mm
$L =$	2cm	1.8cm	2.8cm	2.5cm

Table 2. Characteristics of the flow at $x/L_f = 1$, $y = 0$, $z = L_0$ behind the four 3D fractal grids. The surrogate length scales λ and L are defined as $\lambda \equiv U\tau_\lambda$ and $L \equiv UT_E$ where τ_λ and T_E are, respectively, the Taylor time scale and the integral time scale of the turbulence calculated from the measured autocorrelation function. The Reynolds numbers are defined as follows: $\Re_\lambda \equiv \frac{u'\lambda}{\nu}$ and $\Re \equiv \frac{u'L}{\nu}$.

Sample signals of fluid velocities measured at $x = L_f$, $y = 0$, $z = L_0$ behind each fractal object are presented in Figs. 14 and 15. It does not seem easy to infer anything about turbulence fluctuations in the wake of fractal objects

by just looking at these sample signals. We therefore calculate the structure
functions $\langle|\delta u(\tau)|^2\rangle$ and $\langle|\delta u(\tau)|^3\rangle$ where $\delta u(\tau) \equiv u(t + \tau) - u(t)$. Strictly
speaking these structure functions are not the ones defined by Kolmogorov in
terms of longitudinal velocity differences but they have already been measured
in other turbulent flows with high turbulence intensities, for example by Labbé
et al. (1996) and Gaudin *et al.* (1998). The statistical convergence properties
of these statistics have been thoroughly tested (see example in Fig. 13b)
and the number of sample velocity measurements used to evaluate structure
functions and the global quantities U, u' and $R(\tau)$ is always $4 \cdot 10^5$ per signal
realisation, for which number the second and third order structure functions
discussed in this paper are always well converged. Except where indicated,
averages are calculated over one realisation but the experimental repeatability
of our results thus obtained has always been tested. In Fig. 16, we plot
$\log\langle|\delta u(\tau)|^2\rangle$ against $\log(\tau/T_E)$ and, in Fig. 18a, we plot $\log\langle|\delta u(\tau)|^2\rangle$ versus
$\log\langle|\delta u(\tau)|^3\rangle$ (extended self-similarity, ESS, see Benzi *et al.* 1993) for the
four different fractal objects. Defining the exponent function $\zeta_n(\tau/T_E)$ by
$\langle|\delta u(\tau)|^n\rangle = \langle|\delta u(T_E)|^n\rangle(\tau/T_E)^{\zeta_n(\tau/T_E)}$, we plot ζ_2 against τ/T_E in Figs. 17a
and 17b and ζ_2/ζ_3 against $\langle\delta u^2\rangle$ in Fig. 18b which is obtained from the
local slopes of the curves in Fig, 18a. What can be concluded from these
results with some confidence is that ζ_2/ζ_3 is an increasing function of D_f,
and perhaps with a lesser degree of confidence that ζ_2 is a decreasing function
of D_f. In other words, as D_f increases from 2.05 to 2.75, both $\langle|\delta u|^3\rangle$ and
$\langle|\delta u|^2\rangle$ become slower varying functions of τ/T_E. And as D_f decreases, $\zeta_2(r)$
and $\zeta_3(r)$ approach a classical Kolmogorov-like behaviour.

These conclusions are radically different from those of Toschi *et al.* (1999),
Ciliberto *et al.* (2000) and Gaudin *et al.* (1998) who find that the exponent
functions ζ_2 and ζ_3 decrease and that the ratio ζ_2/ζ_3 either remains constant
or increases with increasing mean shear. One of the points in our study of
mean profiles in Subsection 3.1 has been to show that the transversal and
vertical mean shear rates $\partial U/\partial y$ and $\partial U/\partial z$ decrease with increasing D_f in a
region around $y = 0$ and $-L_f/2 \leq z \leq L_f/2$ at a distance $x = L_f$ downstream
from the fractal objects. The measurements of ζ_2, ζ_3 and ζ_2/ζ_3 reported in
this subsection are made in this region, specifically at $x = L_f$, $y = 0$, $z = L_0$.
Hence, our observation that ζ_2 and ζ_3 decrease and that ζ_2/ζ_3 increase with
increasing D_f corresponds to decreasing mean shear which is exactly opposite
to the mean shear dependence of ζ_2, ζ_3 and ζ_2/ζ_3 observed by Toschi *et al.*
(1999), Ciliberto *et al.* (2000) and Gaudin *et al.* (1998) in channel flow turbulence, the turbulent boundary layer and the turbulent wake flow of a cylinder.

3.3   Comparison with turbulent wakes of bluff bodies
We have compared some of our results with similar measurements taken behind bluff bodies with the shape of cubic boxes of size L_f and volume L_f^3. This
cubic box shape of size L_f is the same as the overall cubic envelope delimiting

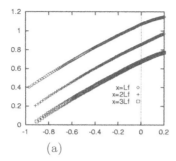 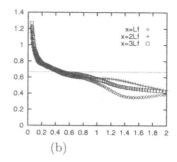

(a)   (b)

Figure 19: Downstream from the bluff body at different distances x/L_f on the axis $y = 0$, $z = L_0$. The smallest time plotted is $\tau = \tau_\lambda$. (a) $\log\langle|\delta u(\tau)|^2\rangle$ vs. $\log(\tau/T_E)$. (b) Local slope $\zeta_2(\tau/T_E)$ vs. τ/T_E. The horizontal dotted line corresponds to a Kolmogorov-like prediction $\zeta_2 = 2/3$.

the 3D fractal grid of overall size L_f. Here we report a comparative study of aspects of the turbulent wakes behind the fractal object of fractal dimension $D_f = 2.4$ on the one hand and behind a cubic box of the same overall size L_f on the other. The mean velocity in the wind tunnel upstream of either object was kept at 10ms^{-1} with a residual turbulence of 0.6%. Measurements were taken on the axis ($y = 0$, $z = L_0$) at three different distances downstream from the square bluff body and the 3D fractal object: $x/L_f = 1, 2, 3$. The first important observation is that whereas u'/U is found to be an decreasing function of x/L_f behind the bluff body in the range $1 \le x/L_f \le 3$, it is a increasing function of x/L_f behind the fractal grid in that same range (Table 3). Specifically, behind the fractal object, $U = 1.79\text{ms}^{-1}$ and $u'/U = 35.1\%$ at $x/L_f = 1$; $U = 1.85ms^{-1}$ and $u'/U = 38.3\%$ at $x/L_f = 2$; $U = 2.32\text{ms}^{-1}$ and $u'/U = 49.6\%$ at $x/L_f = 3$. Behind the bluff body, $U = 6.54\text{ms}^{-1}$ and $u'/U = 46\%$ at $x/L_f = 1$; $U = 9.7\text{ms}^{-1}$ and $u'/U = 25.2\%$ at $x/L_f = 2$; $U = 10.35\text{ms}^{-1}$ and $u'/U = 19\%$ at $x/L_f = 3$. Note also the rather dramatic drop in U behind the fractal object compared with U behind the bluff body (Table 3).

We have also made structure function measurements at these three stations behind both objects. We have calculated $\langle|\delta u(\tau)|^2\rangle$ as a function of τ/T_E (see Fig. 19a) but also ζ_2 as a function of τ/T_E (Fig. 19b). In all these plots $\tau \ge \tau_\lambda$. The time scales T_E and τ_λ were derived from the autocorrelation function (see Fig. 20) and it was found that they are both larger behind the fractal than behind the bluff body. To be specific, $T_E = 4.710^{-3}\text{s}$, 3.1310^{-3}s, 3.0410^{-3}s and $T_E/\tau_\lambda = 9.4, 8.2, 8.2$ at $x/L_f = 1, 2, 3$ behind the bluff body; $T_E = 8.7710^{-3}\text{s}$, 10^{-2}s, 9.2210^{-3}s and $T_E/\tau_\lambda = 7.5, 8.3, 8.6$ at $x/L_f = 1, 2, 3$ behind the fractal (Table 4). Note that the ratio of time scales T_E/τ_λ is about the same behind the fractal and the bluff body at downstream distances $x/L_f = 2$

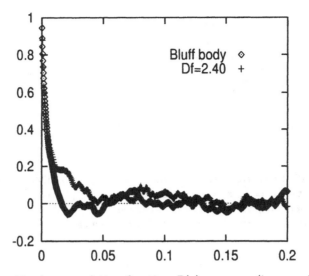

Figure 20: Autocorrelation function $R(\tau)$ versus τ (in seconds) for
bluff body and fractal grid $(D_f = 2.40)$ at the station $x = L_f$, $y = 0$,
$z = L_0$.

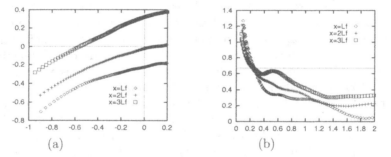

Figure 21: (a) Downstream from the fractal grid $D_f = 2.40$ at different distances x/L_f on the axis $y = 0$, $z = L_0$. The smallest time
plotted is $\tau = \tau_\lambda$. (a) $\log\langle|\delta u(\tau)|^2\rangle$ vs. $\log(\tau/T_E)$. (b) Local slope ζ_2
vs τ/T_E. The horizontal dotted line corresponds to a Kolmogorov-like
prediction $\zeta_2 = 2/3$.

and $x/L_f = 3$ but is slightly larger behind the bluff body at downstream
distance $x/L_f = 1$. Note also that this ratio is a slightly increasing function
of x/L_f behind the fractal but not behind the bluff body. In fact we also
measured T_E/τ_λ at $x/L_f = 5.9$ behind this fractal and found $T_E/\tau_\lambda \approx 10$
(see following subsection).

   A straightforward inspection of Figs. 19a and 19b shows that $\zeta_2(\tau/T_E)$

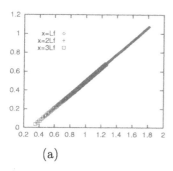 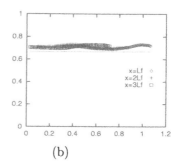

(a)   (b)

Figure 22: Extended self-similarity downstream from the bluff body at different distances x/L_f on the axis $y = 0$, $z = L_0$. The points plotted are in the range $\tau_\lambda \leq \tau \leq T_E$. (a) $\log\langle|\delta u(\tau)|^2\rangle$ vs $\log\langle|\delta u(\tau)|^3\rangle$. (b) Relative exponent $\frac{\zeta_2}{\zeta_3}$ vs. $\log\langle|\delta u(\tau)|^2\rangle$. The horizontal dotted line corresponds to a Kolmogorov-like prediction $\zeta_2/\zeta_3 = 2/3$.

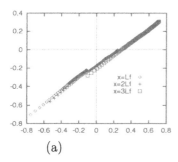 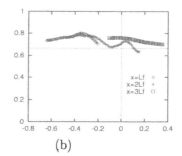

(a)   (b)

Figure 23: (a) Extended self-similarity downstream from the fractal grid $D_f = 2.40$ at different distances x/L_f on the axis $y = 0$, $z = L_0$. The points plotted are in the range $\tau_\lambda \leq \tau \leq T_E$. (a) $\log\langle|\delta u(\tau)|^2\rangle$ vs. $\log\langle|\delta u(\tau)|^3\rangle$. (b) Relative exponent $\frac{\zeta_2}{\zeta_3}$ vs $\log\langle|\delta u(\tau)|^2\rangle$. The horizontal dotted line corresponds to a Kolmogorov-like prediction $\zeta_2/\zeta_3 = 2/3$.

is about the same at the three downstream distances $x/L_f = 1, 2, 3$ from the bluff body but is definitely different at the three downstream distances $x/L_f = 1, 2, 3$ from the fractal (Fig. 21a and 21b): the values taken by the exponent function $\zeta_2(\tau/T_E)$ seem to increase with x/L_f. The same comparative observation holds for $\zeta_3(\tau/T_E)$ behind the fractal and the bluff body. We also look at this comparison from the point of view of ESS which requires plotting $\log\langle|\delta u(\tau)|^2\rangle$ against $\log\langle|\delta u(\tau)|^3\rangle$ (see Figs. 22a, 23a). ¿From these plots we obtain plots of $\frac{\zeta_2}{\zeta_3}$ versus $\log\langle|\delta u(\tau)|^2\rangle$ (Fig. 22b, 23b) which make the difference between the bluff body and the fractal evident in yet another way. The ratio of exponent functions $\frac{\zeta_2}{\zeta_3}$ is a well-defined constant over the range of

$x/L_f =$	1	2	3	5.9
U (bluff body)	6.54	9.7	10.35	
U (fractal, $D_f = 2.40$)	1.79	1.85	2.32	5.6
u'/U (bluff body)	46%	25.2%	19%	
u'/U (fractal, $D_f = 2.40$)	35.1%	38.3%	49.6%	24%

Table 3. Comparison of the mean velocity U (ms^{-1}) and the turbulence intensity u'/U at various distances behind the bluff body and behind the 3D fractal grid of dimension $D_f = 2.40$.

$x/L_f =$	1	2	3	5.9
T_E/τ_λ (bluff body)	9.4	8.2	8.2	
T_E/τ_λ (fractal, $D_f = 2.4$)	7.5	8.3	8.6	10

Table 4. Comparison of the ratio of time scales T_E/τ_λ at various distances behind the bluff body and behind the 3D fractal grid of dimension $D_f = 2.40$.

scales examined (τ_λ to T_E here) and the same for all three downstream distances $x/L_f = 1, 2, 3$ behind the bluff body. In fact, this constant is the one predicted by a theoretical approach such as Kolmogorov's with the small now experimentally well established intermittency correction, i.e. $\zeta_2/\zeta_3 \approx 0.69$ (Arneodo *et al.* 1996). However, behind the fractal the ratio ζ_2/ζ_3 is not such a well-defined constant at downstream distances $x/L_f = 1, 2, 3$. But it does become a well-defined constant further downstream as we show in the following subsection.

3.4 Comparison between a grid generated turbulence and the turbulence far downstream in the fractal's wake

We also compared the turbulence far downstream of the 3D fractal grid of fractal dimension $D_f = 2.40$ with the turbulence generated by a classical square grid in the wind tunnel. We expect the turbulence to recover the state it acquires when forced only at the large scales (as with a bluff body or a classical grid) far downstream from the fractal object, and since our previous measurements indicate that this is not the case for $x/L_f = 1, 2, 3$ (see Figs. 21a and 21b) we now measure turbulence fluctuations at a distance $x/L_f = 5.9$ from the fractal object. We compare the turbulence at the same absolute distance from the fractal and from the classical grid, that is at $x = 102$cm from both. The classical grid that we used has mesh size $M = 5.08$cm

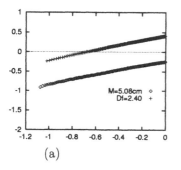 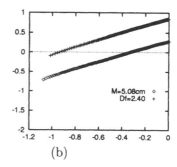

(a)　　　　　　　　　　　　　　(b)

Figure 24: (a) Comparison between classical square grid turbulence ($M = 5.08$cm, $M/b = 4$) and fractal grid $D_f = 2.40$ at the same distance $x = 20M = 5.9L_f = 102$cm. The points plotted are in the range $\tau_\lambda \leq \tau \leq T_E$. (a) $\log\langle|\delta u(\tau)|^2$ vs. $\log(\tau/T_E)$; (b) $\log\langle|\delta u(\tau)|^3\rangle$ vs. $\log(\tau/T_E)$.

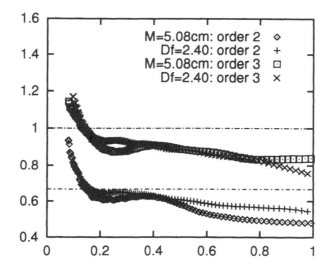

Figure 25: The local exponents $\zeta_2(\tau/T_E)$ and $\zeta_3(\tau/T_E)$ for τ ranging from τ_λ to T_E. The horizontal dot-dashed lines correspond to Kolmogorov-like predictions $\zeta_2 = 2/3$ and $\zeta_3 = 1$ respectively.

and blockage ratio $M/b = 4$. Hence, $x = 20M$ which is far enough from the classical grid to expect the turbulence to be isotropic there and with a spectrum not too dissimilar from Kolmogorov's $-5/3$ (modulo intermittency corrections) within an appropriate range of scales if the Reynolds number is large enough (Batchelor & Townsend 1947, Comte-Bellot & Corrsin 1966).

The comparison was made at equal fan powers such that the input flow

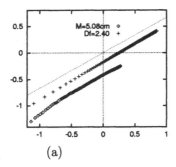
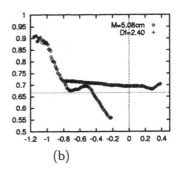

(a)   (b)

Figure 26: (a) Extended self-similarity: $\log\langle|\delta u(\tau)|^2\rangle$ vs $\log\langle|\delta u(\tau)|^3\rangle$ for τ ranging from τ_λ to T_E for the classical square grid and the 3D fractal grid ($D_f = 2.40$) at the same distance $x = 102$cm. (b) The corresponding local relative exponents ζ_2/ζ_3 vs $\log\langle|\delta u(\tau)|^2\rangle$. The horizontal dotted line corresponds to a Kolmogorov-like prediction $\zeta_2/\zeta_3 = 2/3$.

in the test section upstream of the fractal object has a mean velocity equal to 10ms^{-1} with a residual turbulence of 0.6%. With this fan power and at a streamwise distance $x = 20M = 102$cm from the grid, $U = 10\text{ms}^{-1}$ and $u'/U = 5\%$. At the same distance $x = 102\text{cm} = 5.9L_f$ from the fractal object on the axis $y = 0, z = L_0$ and with the same fan power, $U = 5.6\text{ms}^{-1}$ and $u'/U = 24\%$.

Measurements of autocorrelation functions at a distance $x = 102$cm from the grid and from the fractal with the same fan power have shown that $T_E/\tau_\lambda \approx 10$ in both cases. We are therefore comparing measurements at a point in two different turbulent flows where the ratio of scales T_E/τ_λ is the same for both flows but where the turbulence intensity u'/U is very different, small in the case of the classical grid ($u'/U = 5\%$) and large in the case of the fractal object ($u'/U = 24\%$).

Figs. 24a and 24b are plots of $\log\langle|\delta u(\tau)|^2\rangle$ and $\log\langle|\delta u(\tau)|^3\rangle$ vs. $\log(\tau/T_E)$ for both the classical grid turbulence and the fractal wake turbulence. The different curves in Fig. 25 are effectively derived from these plots and show the exponent functions $\zeta_2(\tau/T_E)$ and $\zeta_3(\tau/T_E)$ versus $\log(\tau/T_E)$. It is hard to detect much difference from these results between the grid turbulence and the turbulence in the far wake of the fractal. A difference is however detected when $\log\langle|\delta u(\tau)|^3\rangle$ is plotted against $\log\langle|\delta u(\tau)|^2\rangle$ (Fig. 26a) and this difference is made even more manifest when $\zeta_2(\tau/T_E)/(\zeta_3(\tau/T_E))$ is plotted against $\log\langle|\delta u(\tau)|^2\rangle$ (Fig. 26b). The ratio of exponent functions ζ_2/ζ_3 is very close to a constant very far in the turbulent wake of the fractal whereas ζ_2/ζ_3 is not so well behaved in grid turbulence. This constant value of ζ_2/ζ_3 in the very far turbulent wake of the fractal is about the same as the constant value of

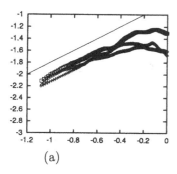
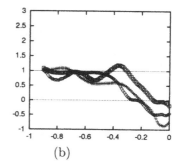

(a)   (b)

Figure 27: (a) $\log(|\langle \delta u(\tau)^3 \rangle|)$ vs. $\log(\tau/T_E)$ in the range $\tau_\lambda \leq \tau \leq T_E$ for the classical square grid ($M = 5.08$cm, $M/b = 4$) at a distance $x = 20M$. The straight line has a slope of 1 and the three differents curves represent three different realisations. (b) The corresponding local exponents $\zeta_3(\tau/T_E)$ vs. $\log(\tau/T_E)$.

ζ_2/ζ_3 at distances $x/L_f = 1, 2, 3$ behind the cubic bluff body. The behaviour of the turbulence in the wake of the bluff body is therefore recovered behind the fractal but only in its very far wake. Both in the turbulent wake of the bluff body and in the *very far* turbulent wake of the fractal the ratio ζ_2/ζ_3 is a sufficiently well-defined constant when T_E/τ_λ is close to 10 but it is not a constant in a grid turbulence with the same $T_E/\tau_\lambda = 10$. A very long stretch of the fractal's turbulent wake is very different from the turbulent wake of the bluff body, at least up to $x/L_f = 3$.

3.5   The signed third order structure function $\langle \delta u(\tau)^3 \rangle$

The signed third order structure function of the longitudinal velocity difference $\delta u_{//}$ is directly related to the dynamics of the turbulence, in particular the dynamics of energy transfer across scales (see Frisch 1995). Under the assumption that the dissipation rate of turbulent kinetic energy per unit mass ϵ is finite and independent of Reynolds number in the limit of very large Reynolds numbers and for a forcing of the turbulence that is confined to the large scales, Kolmogorov (1941b) showed from first principles that $\langle \delta u_{//}^3 \rangle \approx -4/5\epsilon r$ for sufficiently small spatial separations r and sufficiently large Reynolds numbers. Kolmogorov also assumed that the small-scale turbulence is homogeneous and isotropic but recently Nie & Tanveer (1999) rederived the Kolmogorov four-fifths law without assuming local isotropy.

We have confirmed that $\langle \delta u^3 \rangle \sim \tau$ for $\tau_\lambda \leq \tau \leq T_E$ in our grid turbulence (where we may expect that $\delta u_{//} \approx \delta u$ because the turbulence intensity is low enough) with the same classical square grid at a distance $x/M = 20$ and under the same conditions as detailed in subsection 3.4 (see Figs. 27a and 27b). But we have also measured $\langle \delta u^3 \rangle$ as a function of τ at a distance $x/L_f = 1$ behind

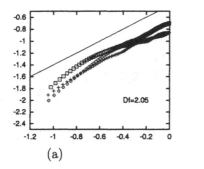 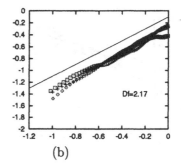

(a)   (b)

Figure 28: (a) $\log(|\langle \delta u(\tau)^3 \rangle|)$ vs. $\log(\tau/T_E)$ in the range $\tau_\lambda \leq \tau \leq T_E$ for the fractal grid $D_f = 2.05$ at $x = L_f$, $y = 0$, $z = L_0$. (b) $\log(|\langle \delta u(\tau)^3 \rangle|)$ vs. $\log(\tau/T_E)$ in the range $\tau_\lambda \leq \tau \leq T_E$ for the fractal grid $D_f = 2.17$ at $x = L_f$, $y = 0$, $z = L_0$. In both (a) and (b), the straight line has a slope of 1 and the three different curves represent three different realisations.

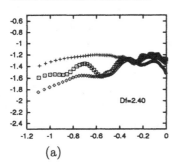 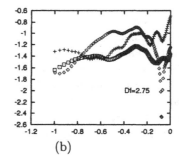

(a)   (b)

Figure 29: (a) $\log(|\langle \delta u(\tau)^3 \rangle|)$ vs. $\log(\tau/T_E)$ in the range $\tau_\lambda \leq \tau \leq T_E$ for the fractal grid $D_f = 2.40$ at $x = L_f$, $y = 0$, $z = L_0$. (b) $\log(|\langle \delta u(\tau)^3 \rangle|)$ vs. $\log(\tau/T_E)$ in the range $\tau_\lambda \leq \tau \leq T_E$ for the fractal grid $D_f = 2.75$ at $x = L_f$, $y = 0$, $z = L_0$. In both (a) and (b), the three different curves represent three different realisations.

each one of our four fractal objects (see Figs. 28 to 29). The interesting result is that whereas $\langle \delta u^3 \rangle$ seems to vary with τ almost linearly behind the fractal objects of comparatively low fractal dimension ($D_f = 2.05, 2.17$) without much difference from classical square grid turbulence (Figs. 28a and 28b), $\langle \delta u^3 \rangle$ is not too far from being effectively independent of τ (Figs. 29a and 29b) in the range $\tau_\lambda \leq \tau \leq T_E$ behind the fractal objects of high fractal dimension ($D_f = 2.4, 2.75$). The situation is very different at a distance $x/L_f = 1$ behind a cubic box bluff body. The second and third order structure functions vary very widely there from realisation to realisation, much more than at the same distance behind the fractal objects, and it is not possible to extract a scaling behaviour for these structure functions from only a few realisations as it is in the case of the fractal objects.

It is possible to work through the derivation of Kolmogorov's four-fifths law (Kolmogorov 1941b) as it is presented in Frisch (1995) without the assumption that ϵ is asymptotically independent of Reynolds number and without confining the forcing to the large scales but with the alternative assumption that, in the limit where $\Re \rightarrow \infty$, the cumulative energy injection (by the force) $\langle \underline{f}_K^< \cdot \underline{u}_K^< \rangle \sim K^p$ for $K \gg 2\pi/L$ (in the notation of Frisch (1995)) where the vector $\underline{f}_K^<(\underline{x})$ is the low-pass filtered force field, $\underline{u}_K^<(\underline{x})$ the low-pass filtered turbulent velocity field, the wave-number K is the inverse length scale of the filtering and L is an outer large length scale. The result obtained with these new assumptions is $\langle \delta u_{//}(r)^3 \rangle \sim r^{1-p}$ for $\lambda_* \ll r \ll L$ in the limit of very large Reynolds numbers and with $\lambda_* \sim \Re^{-\frac{1}{2-p}}$. If we assume the validity of the Taylor or Tennekes (1975) hypothesis linking linearly r to τ and if we also assume that $\langle \delta u(r)^3 \rangle \sim \langle \delta u_{//}(r)^3 \rangle$, then we have $\langle \delta u(\tau)^3 \rangle \sim \tau^{1-p}$ for $\tau \ll T_E$; and if p is an increasing function of the fractal dimension D_f such that $p \rightarrow 1$ as $D_f \rightarrow 3$ then we recover our experimental result that $\langle \delta u(\tau)^3 \rangle$ is close to being independent of τ in the near-turbulent wake of the fractal objects of dimensions $D_f = 2.40$ and $D_f = 2.75$.

4   Conclusions

At a distance $x/L_f = O(1)$ behind a 3D fractal grid where the turbulence is well developed because $T_E \ll x/U$ and $\frac{x}{U T_E}$ increases with x, as the fractal dimension D_f of the fractal increases, turbulent velocity fluctuations tend to become more equidistributed among scales, i.e. $\langle \delta u^2 \rangle$, $\langle |\delta u|^3 \rangle$ and $\langle \delta u^3 \rangle$ tend towards slower varying functions of the time difference τ. This effect cannot be accounted for by the mean shear rate dependence of structure functions in boundary layers, channel flows and bluff body wakes (Ciliberto et al. 2000, Toschi et al. 1999, Gaudin et al. 1998) because the mean shear rates decrease as D_f increases at the points behind the fractals where measurements were taken. Hence, the tendency of structure functions to become slower varying functions of τ is accompanied by decreasing mean shear rates which is exactly opposite to the effect in the more classical shear flows just mentioned.

We also find that the fractal's turbulent wake is very different from the bluff body's turbulent wake. In some sense the fractal's turbulent wake is longer. This is because the turbulence intensity u'/U increases behind the fractal up to at least $x/L_f = 3$, whereas u'/U decreases as x is increased beyond L_f behind the bluff body; because U increases with x to reach the upstream value of the streamwise mean velocity U_∞ at about $x \approx 3L_f$ behind the bluff body whereas U increases over a much longer distance behind the fractal and is only about $U_\infty/2$ at $x \approx 6L_f$; and because the dependences on τ of $\langle \delta u^2 \rangle$ and $\langle |\delta u|^3 \rangle$ approach the Kolmogorov forms modulo well-known intermittency and finite Reynolds number corrections only at $x \gg L_f$ behind the fractal but at least as close as $x = L_f$ behind the bluff body. Furthermore the ratio of the integral time scale to the Taylor time microscale, T_E/τ_λ, increases as x/L_f increases behind the fractal (between $x/L_f = 1$ to 6 at the very least) whereas this same ratio decreases behind the bluff body as x is increased beyond L_f.

We also mention the extra conclusions reached in this work which we have not stressed in the paper's main text: firstly, that the ESS power laws are significantly better defined in the turbulent wakes behind the bluff body and behind the fractals than in grid turbulence with similar time scale ratio T_E/τ_λ. And secondly that the dependences of $\langle \delta u^2 \rangle$ and $\langle \delta u^3 \rangle$ on τ change quite abruptly from $D_f = 2.17$ to $D_f = 2.40$, but are broadly similar for $D_f = 2.05$ and $D_f = 2.17$ on the one hand and $D_f = 2.40$ and $D_f = 2.75$ on the other.

Acknowledgements We are grateful for financial support from EPSRC, the EC TMR Research Network on Intermittency FMRX–CT98–0175 and the Royal Society. We would like to thank Mike Kearney for the fruitful discussions on questions that this project has raised.

5   Appendix: calculation of the fractal dimensions of the 3D fractal grids

The geometry described in Section 2 has been used because at the time when this project was started it was the one in agreement with the technology developed by the company Amalgamated Research Inc. (see Kearney 1997). Nevertheless, we found the resemblance of this structure with the geometry of real trees very interesting. The different branches have a length L_j with $R = L_{j+1}/L_j$, R being constant. The initial largest branch length is L_0. So we have $L_j = R^j L_0$. The number of iterations is N_{it}. To determine the covering fractal dimension D_f we calculate the number of objects (cubes) of size L_n that are needed to cover the entire 3D fractal grid.

By analysing the structure and the construction of the grid, an approximate formula giving the number $N(L_n)$ of objects of size L_n necessary to cover the grid can be given:

$$N(L_n) = 5 \sum_{j=0}^{n-1} 8^j \frac{L_j}{L_n} + 8^n \left(1 + 4\frac{\Lambda_n}{L_n} + 8\frac{\Lambda_{n+1}}{L_n} \right) \qquad (5.1)$$

where Λ_n is a cumulated length such that $\Lambda_n = \sum_{j=n}^{N_{it}} L_j$. The expression (5.1) applies only for $n \geq 1$. For $n = 0$, we have $N(L_0) = 1 + 4\frac{\Lambda_0}{L_0} + 8\frac{\Lambda_1}{L_0}$. The structure is fractal if a satisfactory scale range exists where $N(L_n) \sim L_n^{-D_f}$.

The fractal dimension D_f gives a volume-fraction f_n corresponding to the ratio of the volume occupied by the 3D fractal grid at a scale L_n to the

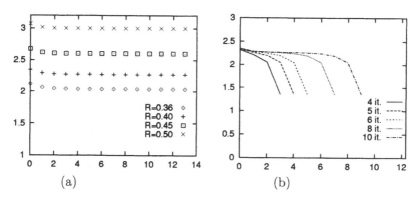

Figure 30: (a) Rods infinitely thin and infinite number of itera-
tions:calculation of $D_f(n)$ vs. n for $0 \leq n \leq 13$ using (5.2) and
for different values of R. The constant fractal dimension is given by
$\log(8)/\log(1/R)$. (b) Rods infinitely thin and finite number of iter-
ations:calculation of $D_f(n)$ vs. n for different total numbers N_{it} of
iterations in the case $R = 0.40$.

total volume given by L_f^3 where L_f is the overall size of the grid given by
$L_f = 2\sum_{j=0}^{N_{it}} L_j$. We thus have $f_n = (L_n^3 N(L_n))(L_f^3 N(L_f)) = (\frac{L_f}{L_n})^{D_f - 3}$
because $N(L_n) = N(L_f)(\frac{L_n}{L_f})^{-D_f}$. Note that $N(L_f) = 1$.

First case: rods infinitely thin and infinite number of iterations

In this case we have $\Lambda_n = \sum_{j=n}^{\infty} L_j$. It is directly obtained from (5.1) that

$$N(L_n) = \frac{5}{R^n}\left(\frac{(8R)^n - 1}{8R - 1}\right) + 8^n\left(1 + \frac{12}{1 - R}\right) \qquad (5.2)$$

and $N(L_0) = 1 + \frac{12}{1-R}$.

The fractal dimension is then calculated to be $D_f(n) = \frac{\log(N(L_{n+1})/N(L_n))}{\log(1/R)}$.
We can observe that when $n \to \infty$ the main term in $N(L_n)$ is 8^n which implies
a fractal dimension $D_f = \frac{\log(8)}{\log(1/R)}$. This is verified by a direct calculation of
$D_f(n)$ using (5.2). For four values of the scaling ratio R, we represent in
Fig. 30a the evolution of $D_f(n)$ with n. It can be observed that with an
infinite number of iterations, the grids are effectively fractal and their fractal
dimension is $D_f = \frac{\log(8)}{\log(1/R)}$. However, our real grids have a finite number of
iterations so what happens to their fractal properties in this case?

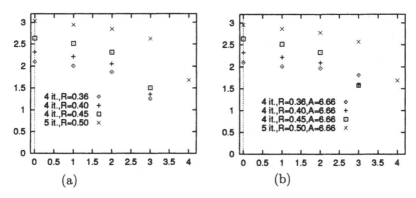

(a)   (b)

Figure 31: (a) Rods infinitely thin and finite number of itera-
tions:calculation of $D_f(n)$ vs. n for different configurations. (b) Rods
with thickness ($A = L_0/d_0 = 6.66$) and finite number of iterations
for different configurations: calculation of $D_f(n)$ vs. n for different
configurations.

Second case: Rods infinitely thin and finite number of iterations

The formula (5.1) giving $N(L_n)$ remains the same but with the cumulated
lengths Λ_n such that $\Lambda_n = \sum_{j=n}^{N_{it}} L_j$. We now obtain:

$$
\begin{aligned}
N(L_n) &= \frac{5}{R^n}\left(\frac{(8R)^n - 1}{8R - 1}\right) \\
&\quad + 8^n\left\{1 + \frac{4}{1-R}(1 - R^{N_{it}+1-n}) + \frac{8}{1-R}(1 - R^{N_{it}-n})\right\}, \quad (5.3) \\
N(L_0) &= 1 + \frac{4}{1-R}(1 - R^{N_{it}+1}) + \frac{8}{1-R}(1 - R^{N_{it}}). \quad (5.4)
\end{aligned}
$$

Using expressions (5.3) and (5.4), it is possible to calculate $D_f(n)$ in order
to observe the influence of the number of iterations on the fractal dimension.
Of course, when N_{it} is increasing, $D_f(n)$ tends to the value $\log(8)/\log(1/R)$
obtained in the first case. However, it appears that the last iteration (corresponding to the interval $[L_{N_{it}-1}; L_{N_{it}}]$) undergoes an important drop and the
object is not fractal at these smallest scales anymore. This behavior is to be
expected because as we approach the smallest scales we also reach the lower
cut-off of the fractal.

We present in Fig. 31a the calculation of $D_f(n)$ using (5.3) and (5.4) for
different configurations given by the scaling ratio R and the total number of
iterations N_{it}. Similarly to Fig. 30b, it is observed that the fractal dimension
presents a slight decreasing drift towards the small scales with a more important fall at the last iteration. To test the validity of expressions (5.3) and

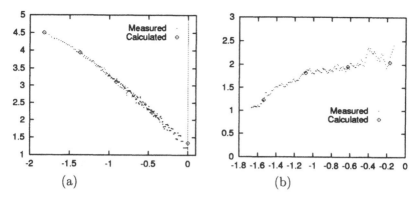

Figure 32: Rods infinitely thin and finite number of iterations: (a) $\log(N(L_n))$ vs $\log(L_n/L_0)$ calculated with (5.2) and measured by fractal analysis: grid with $N_{\mathrm{it}} = 4$, $R = 0.35$. (b) The corresponding local fractal dimension $D_f(L_n/L_0)$ measured and calculated with (5.2).

(5.4), we can compare them with the results given by a direct fractal analysis realised on the grids. For this, we have generated these grids numerically, the main parameters being the scale ratio R, the initial length L_0 and the number N_{it} of iterations. Then a fractal analysis was done. As an illustration in Fig. 32a we compare the points calculated using expression (5.3) for $N(L_n)$ (only for n integer but we could take n real) with the measurements realised by a classical fractal analysis. Good agreement is observed. This good agreement is confirmed on comparing the local fractal dimensions $D_f(L_n/L_0)$ as shown in Fig. 32b.

Third case: Rods with thickness and finite number of iterations

The previous cases with infinitely thin rods are only idealisations because in practice the rods have a thickness. We now analyse the influence that this thickness has on the fractal dimensions of the grids. The rods have a square cross section. Let us take for the initial branch a thickness d_0 and let d_j be the thickness of a branch at the level j which has a length L_j. In order to respect the fractality of the object, we keep the same scaling ratio for thicknesses: $d_j = R^j d_0$.

The number $N(L_n)$ calculated at the scale L_n depends now on the ratio d_j/L_n. If $d_j \leq L_n$, formula (5.3) remains unchanged but if $d_j > L_n$ then the number of objects of size L_n needed to cover a branch of length L_j and thickness d_j is not $\frac{L_j}{L_n}$ but $\frac{L_j}{L_n}(\frac{d_j}{L_n})^2$. The condition to have $d_j > L_n$ is given

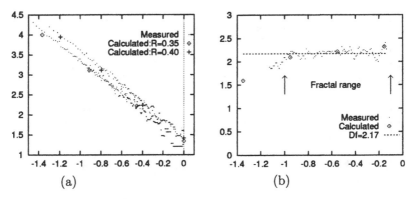

Figure 33: Rods with thickness and finite number of iterations: (a) $\log(N(L_n))$ vs $\log(L_n/L_0)$ calculated with (5.6) and (5.7) and measured by fractal analysis: the top curve corresponds to a grid with $N_{it} = 4, R = 0.35, L_0/d_0 = 6.66$, the bottom one to $N_{it} = 4, R = 0.4, L_0/d_0 = 6.66$. (b) The corresponding local fractal dimension $D_f(L_n/L_0)$ measured and calculated using (5.6) and (5.7). An approximate fractal range is evidenced and a mean fractal dimension $D_f = 2.17$ is defined.

by $j < c$ with $c = n - \frac{\log(L_0/d_0)}{\log(1/R)}$. Then,

$$N(L_n) = 5\left\{\sum_{j=0}^{c-1} 8^j \frac{L_j}{L_n}\left(\frac{d_j}{L_n}\right)^2 + \sum_{j=c}^{n-1} 8^j \frac{L_j}{L_n}\right\} + 8^n\left(1 + 4\frac{\Lambda_n}{L_n} + 8\frac{\Lambda_{n+1}}{L_n}\right). \quad (5.5)$$

After a few steps of algebra this breaks down to

$$N(L_n) = 5A_n + 8^n\left\{1 + \frac{4}{1-R}(1 - R^{N_{it}+1-n}) + \frac{8}{1-R}(1 - R^{N_{it}-n})\right\} \quad (5.6)$$

with $A_n = \frac{d_0^2}{L_0^2(8R^3-1)}\frac{(8R^3)^c-1}{R^{3n}} + \frac{1}{8R-1}\frac{(8R)^n-(8R)^c}{R^n}$ and

$$N(L_0) = 1 + \frac{4}{1-R}(1 - R^{N_{it}+1}) + \frac{8}{1-R}(1 - R^{N_{it}}). \quad (5.7)$$

Using (5.6) and (5.7), we then calculated the fractal dimension $D_f(n)$ for different configurations characterized by the scaling ratio, the number of iterations and the aspect ratio $A = L_0/d_0$. The results of this calculation are shown in Fig. 31b. The main remark to be made in comparison with the second case without thickness is that, in a similar way, $D_f(n)$ decreases with n and also presents a drop at the last iteration. It is possible to test the expressions (5.6) and (5.7) on direct measurements realised on numerical objects having now rods with thickness. An example is presented in Fig. 33a for two values of R. The comparison of the direct fractal analysis with the calculated points (using (5.6) and (5.7)) displays quite good agreement which can be seen on directly computing the local fractal dimension $D_f(L_n/L_0)$ (Fig. 33b). We can observe that even if our grids are not perfectly fractal because of the limited number of iterations, they exhibit a satisfactory fractal behaviour in a sufficiently broad scale range as evidenced in Fig. 33b.

The final grids have a slight difference from the previous case. At an iteration j the vertical thickness of vertical rods is different from that of horizontal rods. More precisely, the four horizontal rods follow the previous case with $d_j = R^j d_0$ but the vertical rod follows $d_j = R^{j-1} d_0$. We verified numerically and analytically that this modification is negligible in the estimation of fractal dimension.

The procedure described in this third case has been applied to conceive four objects displaying different mean fractal dimensions. The resulting objects are described in Section 2 and their fractal characteristics are given in Fig. 3.

As a conclusion, it can be said that our 3D fractal grids display clearly different space-filling properties. Of course, our grids are a compromise between the available techniques and fractality which is only an infinite limit. We can imagine many improvements and modifications of the grids such as the introduction of more iterations but also random characteristics, isotropic features and different porosity aspects.

References

Arneodo, A. (and 23 others) (1996) 'Structure functions in turbulence, in various flow configurations, at Reynolds number between 30 and 5000, using extended self-similarity', *Europhys. Lett.* **34**, 411–416.

Balkovsky, E. and Lebedev, V. (1998) 'Anomalous scaling exponents of a white-advected passive scalar', *Phys. Rev. E* **58**, 5776–5795.

Batchelor, G.K. and Townsend, A.A. (1947) 'Decay of vorticity in isotropic turbulence', *Proc. R. Soc. Lond. A* **190**, 534–560.

Benzi, F., Ciliberto, S., Tripiccione, C., Baudet, C., Massaioli, F. and Succi, S. (1993) 'Extended self-similarity in the dissipation range of fully developed turbulence', *Phys. Rev. E* **48**, R29–R32.

Benzi, F., Amati, G., Casciola, C.M., Toschi, F. and Piva, R. *et al.* (1999) 'Intermittency and scaling laws for wall bounded turbulence', *Phys. Fluids* **11**(6), 1284–1286.

Chertkov, M. and Falkovich, G. (1996) 'Anomalous scaling exponents of a white-advected passive scalar', *Phys. Rev. Lett.* **75**, 3834–3837.

Ciliberto, S., Lévêque, E. and Ruiz Chavarria, G. (2000) 'Non-homogeneous scalings in boundary layer turbulence', this volume.

Comte-Bellot, G. and Corrsin, S. (1966) 'The use of a contraction to improve the isotropy of grid-generated turbulence', *J. Fluid Mech.* **25**, 657–682.

Flohr, P. and Vassilicos, J.C. (1997) 'Accelerated scalar dissipation in a vortex', *J. Fluid Mech.* **348**, 295–317.

Frisch, U. (1995) *Turbulence: The Legacy of A.N. Kolmogorov*, Cambridge University Press.

Frisch, U., Sulem, P.L. and Nelkin, M. (1978) 'A simple dynamical model of intermittent fully developed turbulence', *J. Fluid Mech.* **87**, 719–736.

Gaudin, E., Protas, B., Goujon-Durand, S., Wojciechowski, J. and Wesfreid, J.E. (1998) 'Spatial properties of velocity structure functions in turbulent wake flows', *Phys. Rev. E* **57**, R9–R12.

Gawędzki, K. (2000) 'Turbulent advection and breakdown of the Lagrangian flow', this volume.

Gawędski, K. and Kupiainen, A. (1995) 'Anomalous scaling of the passive scalar', *Phys. Rev. Lett.* **75**, 3834–3837.

Hunt, J.C.R., Kaimal, J.C. and Gaynor, J.E. (1988) 'Eddy structure in the convective boundary layer – new measurements and new concepts', *Q. J. R. Met. Soc.* **114**, 827.

Hunt, J.C.R. and Vassilicos, J.C. (1991) 'Kolmogorov's contributions to the physical and geometrical understanding of small-scale turbulence and recent developments', *Proc. R. Soc. Lond.* **A434**, 183–210.

Kearney, M. (1997) 'Engineered fractal cascades for fluid control applications', *Fractals in Engineering, Institut National de Recherche en Informatique et en Automatique, Arcachon, France.*

Khan, M.A.I. and Vassilicos, J.C. (2000) 'The scalings of scalar structure functions in a velocity field with a coherent vortical structure', Submitted to *Phys. Rev. E.*

Kholmyansky, M. and Tsinober, A. (2000) 'On the origins of intermittency in real turbulent flows', this volume.

Kolmogorov, A.N. (1941a) 'The local structure of turbulence in incompressible viscous fluid for very large Reynolds number', *Dokl. Akad. Nauk SSSR* **30**, 9–13. Reprinted in *Proc. R. Soc. Lond.* **A434**, 9–13 (1991).

Kolmogorov, A.N. (1941b) 'Dissipation of energy in locally isotropic turbulence', *Dokl. Akad. Nauk SSSR* **32**, 16–18. Reprinted in *Proc. R. Soc. Lond. A* (1991), **434**, 15-17.

Kraichnan, R.H.(1994) 'Anomalous scaling of a randomly advected passive scalar', *Phys. Rev. Lett.* **72**, 1016–1019.

Labbé, R., Pinton, J.F. and Fauve, S. (1996) 'Study of the von Kármán flow between coaxial rotating disks', *Phys. Fluids* **8**(4), 914–922.

Mandelbrot, B. (1974) 'Intermittent turbulence in self-similar cascades: divergence of high moments and dimension of the carrier', *J. Fluid Mech.* **62**, 331–358.

Nie, Q. and Tanveer, S. (1999) 'A note on third-order structure functions in turbulence', *Proc. R. Soc. Lond.* **A455**, 1615–1635.

Novikov, E.A. (1970) 'Intermittency and scale similarity of the structure of turbulent flow', *Prikl. Mat. Mekh.* **35**, 266–277.

Parisi, G. and Frisch, U. (1985) 'On the singularity structure of fully developed turbulence', *Turbulence and Predictability in Geophysical Fluid Dynamics, Proc. Int. School of Physics 'E. Fermi', 1983, Varenna, Italy*, M. Ghil, R. Benzi and G. Parisi, eds., North-Holland, 4–87.

Praskovsky, A.A., Gledzer, E.B., Karyakin, M.Yu. and Zhou, Ye (1993) 'The sweeping decorrelation hypothesis and energy-inertial scale interaction in high Reynolds number flows', *J. Fluid. Mech.* **248**, 493–511.

Ramsey, S.R. (1988) *The interaction of a 2D turbulent wake with a bluff body*, PhD Thesis, University of Cambridge.

Sreenivasan, B. and Stolovitzky, G. (1996) 'Statistical dependence of inertial range properties on large scales in a high Reynolds number shear flow', *Phys. Rev. Lett.* **77**, 2218–2221.

Tennekes, H. (1975) 'Eulerian and Lagrangian time microscales in isotropic turbulence', *J. Fluid Mech.* **67**, 561.

Toschi, F., Amati, G., Succi, S., Benzi, R. and Piva, R. (1999) 'Intermittency and structure functions in channel flow turbulence', *Phys. Rev. Lett.* **82**, 5044–5047.

Ushijima, T. (1998) *The motion of heavy particles in turbulent flows*, PhD Thesis, University of Cambridge.

Vassilicos, J.C. and Hunt. J.C.R. (1991) 'Fractal dimensions and spectra of interfaces with application to turbulence', *Proc. R. Soc. Lond.* **A435**, 505–534.

Vorticity Statistics in the 2D Enstrophy Cascade and Tracer Dispersion in the Batchelor Regime

Marie-Caroline Jullien, Patrizia Castiglione, Jérôme Paret and Patrick Tabeling

Abstract

We describe a series of experiments on the 2D enstrophy cascade and on the dispersion of a passive tracer in the enstrophy cascade. A detailed characterization of the statistics of the vorticity and concentration fields is achieved. The experiments provide a support to recent theoretical studies.

1 Introduction

The enstrophy cascade is one of the most important processes in two-dimensional turbulence, and its investigation, at fundamental level, provides cornerstones for the analysis of atmosphere dynamics. The existence of this cascade was first conjectured by Kraichnan [1], and later by Batchelor [2]. Both of them proposed that in 2D turbulence, enstrophy injected at a prescribed scale is dissipated at smaller scales, undergoing a cascading process at constant enstrophy transfer rate η; this led to predicting a k^{-3} spectrum for the energy, in a range of scales extending from the injection to the dissipative scale. Later, logarithmic corrections were incorporated in the analysis to ensure constancy of the enstrophy transfer rate [3]. The advent of large computers revealed surprising deviations from the classical expectation, especially in decaying systems [4, 5, 6, 7]. It was soon realized that in 2D systems, long lived coherent structures inhibit the cascade locally and therefore the self-similarity of the process, assumed to fully apply in the classical approach, is broken. Expressions like 'laminar drops in a turbulent background' were coined to illustrate the role of coherent structures in the problem [4]. Along with the observations of unexpected exponents, models, emphasizing the role of particular vortical structures [12, 11], or based on conformal theory [13], suggested non-classical values. In the recent period however, high resolution simulations [8, 21, 10, 19] underlined that, provided long lived coherent structures are disrupted, classical behaviour holds; furthermore, theoretical studies [14, 15] suggested the absence of small scale intermittency, placing the direct enstrophy cascade in a position strikingly different from the 3D energy cascade. The recent soap film experiments, developing single point measurements of the velocity field [16, 18, 17, 20], obtained spectral exponents consistent with these views.

Nonetheless, investigating small scale intermittency in this problem requires measuring the statistics of quantities such as the vorticity increments, which has not been done yet, either in physical or in numerical experiments. Efforts in this direction were made in the numerical study of Borue [21], but difficulties arose in obtaining convergent results. An analysis of the enstrophy fluxes in the numerical experiment of Babiano *et al.* [9] led the authors to underlining the presence of weak intermittency in the enstrophy cascade; thus, although the theory on the problem is at a well advanced stage (at least compared to the 3D situation), it is not yet known, even in situations where self-similarity fully holds, to what extent classical theory, based on mean field arguments 'à la Kolmogorov', applies for the enstrophy cascade. The first objective of the experiments we describe here is to address this issue.

Concerning turbulent dispersion of passive tracers, it is involved in an extremely large number of situations, pertaining to industrial, environmental, astrophysical contexts [29, 30]. Formulated in general terms, the problem of turbulent dispersion is broad: it covers different physical situations, depending on the kinematic and molecular diffusivities of the fluids at hand, and the characteristics of the flow motion. Further to Batchelor's and Kraichnan's pioneering work [27, 31, 32], considerable theoretical progress has been made in recent years [33, 34] on the problem. At the present time however, the problem of turbulent dispersion has been solved in only one case (the Batchelor regime), a situation in which it is assumed the Péclet number (which we will define below) is high and the scale of motion is large compared to those involved in the concentration field [33]. Among the turbulence problems, the Batchelor regime turns out to be the only situation for which the two-point statistics have been calculated analytically, in a consistent way, without a closure scheme. The theory predicts a scalar variance spectrum $E_\theta(k)$ decreasing as k^{-1}, exponential tails for the concentration [34] and concentration increment distributions [35], and logarithmic behaviour for the structure functions [33]. There have been several investigations of the problem, experimental and numerical. The k^{-1} spectrum has been observed in a number of experiments [29, 30, 36], but its universality has been questionned [37, 38, 39]. Recent direct numerical simulations have confirmed the relevance of the k^{-1} spectral law in 3D isotropic turbulence [40], and others, based on synthetic velocity fields, have yielded valuable information on the distributions of the concentration and concentration increments [41]. However, none has been dedicated to investigating whether the predictions in [33] may have physical support. This is the second objective of the present work.

2   The experimental set-up

The experimental set-up has been described in a series of papers [25, 23, 24]. It appears to be formidable tool for investigating fundamental issues of 2D turbulence. It provides reliable data on quantities reputed hard to measure. We believe this is an interesting situation, since it would be unpleasant to elaborate a rationale for 2D turbulence, solely on virtual inputs. Briefly speaking, the flow is generated in a square PVC cell, 15cm × 15cm. The bottom of the cell is made of a thin (1mm thick) glass plate, below which permanent magnets, 5 × 8 × 4mm in size, and delivering a magnetic field, of maximum strength 0.3T, are placed. In order to ensure 2Dity [22], the cell is filled with two layers of NaCl solutions, each 2.5mm thick, with different densities, placed in a stable configuration, i.e. the heavier underlying the lighter. Under typical operating conditions, the stratification remains unaltered for periods of time extending up to 10 min. The interaction of an electrical current driven across the cell with the magnetic field produces local stirring forces. The current density vector applies horizontally, parallel to the side walls, and owing to the electrical conductivities at hand, the resulting electromagnetic forces are larger in the lower layer. The flow is visualized by using clusters of latex particles 2μm in size, placed at the free surface, and the velocity fields $\mathbf{v}(\mathbf{x}, t)$ are determined using particle image velocimetry technique, implemented on 64 × 64 grids, each interrogation cell incorporating 8 × 8 pixels; in physical units, the spatial resolution is thus 2.5mm. The temporal resolution is excellent (4×10^{-2}s), in comparison with the typical flow time scales. We estimate the accuracy on the velocity on the order of a few percent and that on the vorticity on the order of 10%. In such experiments, the dissipative scale for the enstrophy cascade – defined as $l_d - \eta^{1/6}\nu^{1/2}$ (where ν is the kinematic viscosity and η is the enstrophy pumping rate) – is on the order of 1mm; it is thus unresolved. Moreover, l_d lying below the layer thickness, it is reasonable to consider that the way which enstrophy is dissipated in our system is not purely 2D. Concerning measurement accuracy, we estimate, from the measurement of local divergence, that the accuracy on the velocity is a few per cent and that on the vorticity is 10%.

In the experiments we describe here, magnets are arranged into four triangular aggregates, each including roughly one hundred units, with the same magnetic orientation, as shown schematically in Fig. 1. By doing so, the electromagnetic forcing is defined on a large scale, and its spatial structure does not favour any particular permanent pattern.

The electrical current is unsteady: it is a non-periodic, zero mean, square waveform, of amplitude equal to 0.75A (see Fig. 1). The corresponding Reynolds number – defined as the square of the ratio of the forcing to the dissipative scale – is on the order of 10^3; this estimate is one order of magnitude above the largest simulation performed on the subject, using normal viscosity (see [21]). In the statistically steady state, the instantaneous flow pattern consists of transient recirculations of sizes comparable to one fourth of the box size. The formation of permanent large scale structures, which might tend to break the self-similarity of the process, seems disrupted by our particular forcing.

The passive scalar is a mixture of fluorescein and water, of density $\rho = 1030\text{gl}^{-1}$, and diffusivity $\kappa = 10^{-6}\text{cm}^2\text{s}^{-1}$; in most cases, the colorant matches the upper layer density, and is vertically homogeneously spread across it,

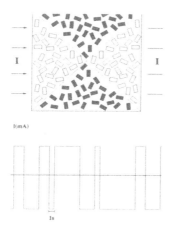

I(mA)

Is

Figure 1: A sketch of the arrangement of the magnets (as seen from above) and the time dependence of the electrical current crossing the cell. Black units have the same magnetic orientation, grey ones the opposite. The average lapse of time between two successive current switches is 2.5s.

throughout the experiment. In several cases, we have used a lighter dye, of density $\rho = 1002\mathrm{gl}^{-1}$ i.e. 3% lighter than the upper layer, so as to operate with thin colorant sheets. The concentration field illuminated by ultraviolet light, is visualized using a 512×512 CCD camera; we have checked the intensity is proportional to the scalar concentration. The images are stored and further processed. The overall spatial resolution for the concentration field is 0.2mm. The 'molecular' Péclet number UL/κ (where U and L are typical velocity and length scales) is typically on the order of 10^8. To inject the colorant, we first enclose a blob of dye within a cylinder, 5cm in diameter, in the upper layer, then switch the electrical current on, wait for the transient state to vanish, and eventually delicately remove the cylinder. In these experiments, the flow is statistically stationary, while the concentration field is in a freely decaying regime. Owing to the experimental conditions at hand (large scale forcing for the flow, high Péclet numbers), we may expect the Batchelor regime to take place for the dispersion of the tracer.

Additionally, in order to facilitate the analysis of the experimental observations, we have simulated the concentration field, using a method similar to that in [42]: the trajectories of 6×10^5 particles, located initially in a disk, 5cm in diameter, are calculated by integrating the experimentally measured velocity field. Coarse graining is further applied to generate a concentration field. The simulation is thus designed to mimic the experiment. We will incorporate the results of this work in the discussion of the physical experimental results.

Figure 2: A particular realization of the vorticity field, in the statistically steady state; the grey scale is linear in the vorticity.

3   The enstrophy cascade

The instantaneous vorticity field in the statistically stationary state is shown in Fig. 2. We see elongated structures, in the form of filaments or ribbons, some of them extending across a large fraction of the cell. At variance with the decaying regimes, and consistently with the above discussion, we have not seen any long lived vorticity concentration, i.e. persisting more than a few seconds. This is further confirmed by a measurement of the flatness of the vorticity distribution, a diagnostic previously introduced in [10] and which is found slightly above the Gaussian value in our case. The presence of coherent structures would have been associated to much larger values of this quantity. The isotropy of the vorticity field is not obvious from the inspection of a single realization, such as the one of Fig. 2; nonetheless, as will be shown later, the overall anisotropy level, obtained after statistical averaging, turns out to be reasonably small.

The spectrum of the velocity field, averaged over 200 realizations, in the statistically steady state, is shown in Fig. 3. The forcing wave-number $k_f \sim 0.6\text{cm}^{-1}$ corresponds to the location of the maximum of the energy spectrum; it is associated to an injection scale $l_f = 2\pi/k_f$ estimated at 10cm, a value consistent with the size of our permanent magnet clusters. The wave-number associated to the stratified fluid layer may be defined as $k_l = 2\pi/b \sim 12\text{cm}^{-1}$ (where b is the fluid thickness). This wave-number, together with the sampling wave-number, which is 25cm^{-1}, is well outside the region of interest. Fig. 3 shows that in the high wave-number region, i.e. above 9cm^{-1}, the spectrum is flat. This region is dominated by white noise; it reflects a limitation in the PIV technique to resolve low velocity levels at small scales.

The interesting feature is that there exists a spectral band, lying between k_f and $k_{\max} \sim 7\text{cm}^{-1}$, uncontaminated by a possible interaction with the layer wave-number, in which a power law behaviour is observed. The corresponding exponent is close to -3. A direct measurement of the exponent, performed by

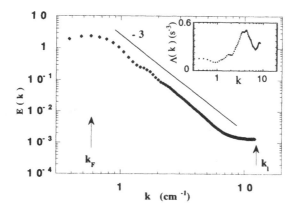

Figure 3: Energy spectrum of the velocity field, averaged over 200 realizations of the velocity field in the statistically stationary state; the insert shows the enstrophy transfer rate $\Lambda(k)$, calculated in similar experimental conditions.

using least square fit in the scaling region, leads to proposing the following formula for the spectrum:

$$E(k) \sim k^{-3.0\pm0.2} \tag{3.1}$$

The exponent we find is thus close to classical expectation. There is no steepening effect of the spectrum, which could be attributed, as in decaying systems, to the presence of coherent structures. Further analysis of the vorticity field shows homogeneity and stationarity of the process. Isotropy is also obtained, albeit only roughly, as shown in Fig. 4: to estimate the anisotropy level, we follow circles, embedded in the inertial range, in the spectral plane of Fig. 4, and determine by how much the spectral energy departs from a constant value along such circles. This leads to an anisotropy level on the order of 15% in the central region of the inertial range. Determining the Kraichnan–Batchelor constant is a delicate task, which entirely relies on the measurement of the enstrophy pumping rate η. The constant we discuss here, called C', is defined by expressing the energy spectrum in the form:

$$E(k) = C'\eta^{2/3}k^{-3} \tag{3.2}$$

To measure C', we measured the spectral enstropy transfer rate from below k to above k, namely $\Lambda(k)$; the result is shown in the insert of Fig. 3. $\Lambda(k)$ is found positive above 1cm^{-1}; this covers most of the range where the k^{-3} spectral law holds, and thus confirms the cascade is forward. To determine η, we further average out $\Lambda(k)$, between k_f and $k_{\max}$. This procedure provides

Figure 4: Iso levels of the energy spectrum in the wave-number space (k_x, k_y) defining, respectively, the horizontal and vertical axes of the plot. The boundaries of the rectangle along the x-axis correspond to $k_x = \pm 12\text{cm}^{-1}$ The grey scale is periodic. The two peaks at $k_x = \pm 0.6\text{cm}^{-1}$ around the centre signal the forcing.

the following estimate for the Kraichnan–Batchelor constant C':

$$C' \approx 1.4 \pm 0.3 \tag{3.3}$$

This estimate agrees with that found in the high resolution study of Ref [21], for which values ranging between 1.5 and 1.7 have been proposed. We provide here the first experimental measurement ever achieved for this constant.

We now turn to the intermittency problem. Here we consider the statistics of the vorticity increments, a central quantity considered in the recent analytical approaches to the enstrophy cascade [14, 15]. Fig. 5 shows a set of five distributions of the vorticity increments, obtained for different inertial scales, ranging between 2 and 9cm. As usual, in order to analyse shapes, the pdfs have been renormalized to require their variance be equal to unity. The shapes of the pdfs are not exactly the same, but it is difficult to extract a systematic trend with the scale. Within experimental error, they seem to collapse onto a single curve; the tails of such an average distribution are broader than a Gaussian curve, and the deviations have a moderate amplitude. It is difficult here, from the inspection of the distributions, to reveal the presence of intermittency in the enstrophy cascade.

The analysis of the structure functions of the vorticity, shown on Fig. 6, confirms this statement. These structure functions are defined by:

$$S_p(r) = \langle (\omega(\mathbf{x} + \mathbf{r}) - \omega(\mathbf{x}))^p \rangle \tag{3.4}$$

in which $\mathbf{x}$ and $\mathbf{r}$ are vectors, and r is the modulus of $\mathbf{r}$. The angle brackets mean double averaging, both in space, throughout the plane domain, and in time, between 20 and 280s. We use here 10^5 data points to determine the structure functions; this allows us to determine up to twelfth order, because of the near Gaussianicity of the pdfs. Fig. 6 thus represents a series of vorticity structure functions $S_p(r)$, obtained in such conditions, emphasizing the inertial domain, i.e. with r varying between 1 and 10cm. The structure functions weakly vary with the scale, indicating the exponents are close to zero.

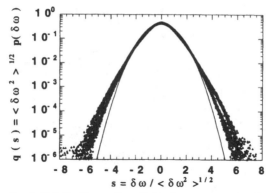

Figure 5: Normalized distributions of vorticity increments, for five separations r: 2, 3, 5, 7 and 9cm.

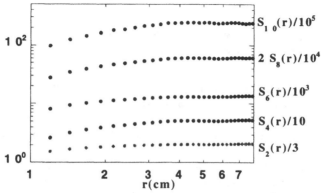

Figure 6: Structure functions of the vorticity increments, for various orders comprised between 2 and 10.

The corresponding values fall in the range –0.05, 0.15, for p varying between 2 and 10; owing to experimental uncertainty, this is indistinguishable from zero. We thus obtain here a result fully compatible with the classical theory, for which the exponents are predicted to be exactly zero at all orders. Concerning logarithmic deviations, such as those proposed by the theory [3, 15], it is difficult to draw out a firm conclusion at the moment.

4 The dispersion of tracers in the enstrophy cascade

Fig. 7 shows a typical evolution of a spot of fluorescein 5cm in diameter, released in the system at time $t = 0$. At early time, the blob is localized in the centre of the cell, and is slightly distorted. It is further vigorously advected and strained, under the action of the velocity field. At this stage, the striations display a broad range of sizes; it would be hard to extract a simple

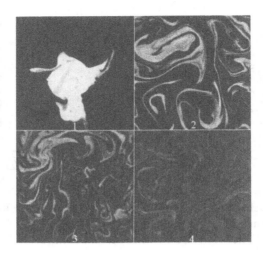

Figure 7: Time evolution of a blob of fluorescein of density $\rho = 1002\mathrm{gl}^{-1}$ in a 16cm×16cm region, at times $t = 1$, 12, 20, and 40s.

characteristic length from the inspection of the turbulent fields displayed in Fig. 7. After a period of 30s the concentration field is homogeneous; we have checked, by sampling the fluid, that the colorant remains confined in the upper layer, throughout the mixing process. In this system, mixing seems homogeneous and there is no evidence for the presence of any long lived structure trapping the tracer throughout the duration of the experiment. We believe this is due to the particular forcing we use, which disrupts the coherent structures, a requirement for the development of an enstrophy cascade (see Section 3 and [28]), and probably also for that of a Batchelor regime.

Fig. 8 shows the evolution of the dissipation spectrum $k^2 E_\theta(k)$ (where k is the wave-number and $E_\theta(k)$ is the variance spectrum of the concentration field), corresponding to the sequence of Fig. 7. Each dissipation spectrum is averaged over one second, then slightly smoothed. At early times, the dissipation is dominated by a bump, localized at a wave-number corresponding to the initial drop size. Between 8 and 18s, the spectra show the same characteristics: they gradually increase up to $k = 12\mathrm{cm}^{-1}$, then decrease. At larger time, the dissipation spectrum progressively collapses; it keeps evolving with time up to a final stage where equipartition is obtained, corresponding to complete mixing. The existence of a range of time, comprised between 8 and 18s, where the concentration field seems to reach a quasi-stationary state is confirmed by the inspection of the total dissipation $\int k^2 E(k)$ (see the insert of Fig. 8), which displays a broad maximum in this range of time. We further concentrate our attention on the analysis of this region, which we call the 'quasi-stationary domain'.

Fig. 9 displays the scalar variance spectra, now averaged over the entire

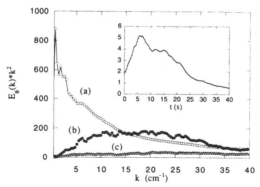

Figure 8: Evolution of the dissipation spectrum of a $\rho = 1002\mathrm{gl}^{-1}$ density pollutant, at times (a) $t = 1\mathrm{s}$, (b) $t = 12\mathrm{s}$, and (c) $t = 40\mathrm{s}$. Insert shows the total dissipation versus time.

quasi-stationary domain, along with the enstrophy cascade (indicated by arrows on Fig. 9). Beyond $\dot{k} = 7\mathrm{cm}^{-1}$, the spectrum drops. This defines an effective dissipative wave-number for our experiment. This wave-number is well beyond the Batchelor dissipation scale $k_B = (\gamma/\kappa)^{1/2}$, which in our case should be on the order of $2800\mathrm{cm}^{-1}$. The dissipation scale of the experiment originates in the fact that at each inversion of the electrical current, a transient shear develops in the upper layer, enhancing tracer diffusion in the horizontal plane. This cut-off can be substantially shifted (by a factor of 2, roughly), by working with tracer sheets of different thicknesses. The existence of an effective dissipative cut-off leads to us re-estimate the Péclet numbers of the experiment. By using classical estimates, one infers our effective Péclet number on the order of 10^5, which is lower than the molecular Péclet number, but still compatible with the existence of a Batchelor regime. The numerical simulation we performed, in which there is no such enhancement process, led to the same spectrum as in the experiment, indicating our results are probably not contaminated by this effect, in the inertial convective range.

We now turn to the measurement of the probability density functions (pdf) of scalar fluctuations $P(\theta(\mathbf{x}) - \langle\theta(\mathbf{x})\rangle)$, a crucial quantity to consider for comparing with theory. Fig. 10 displays the pdfs at $t = 8\mathrm{s}$ and $t = 16\mathrm{s}$, *i.e.* well within the quasi-stationary interval; the pdf is representative of the distributions observed throughout the interval. They are strikingly different from those observed at early times (which are bimodal) and at late times (which are Gaussian). The pdfs of the quasi-stationary domain are non-symmetric and non-Gaussian: as shown in Fig. 10, they develop exponential tails on the positive side, and Gaussian behaviour on the negative one. The tails seem controlled by the fronts, located at the edges of the tracer filaments. Here again, the numerical simulation shows the same results. These characteristics for the pdfs of the concentration field are in good qualitative agreement with

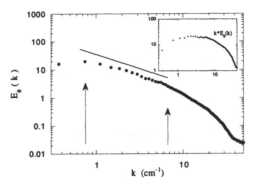

Figure 9: Variance scalar spectra with a pollutant of density $\rho = 1002\mathrm{gl}^{-1}$, the straight line has a k^{-1} slope. In insert are shown the compensated spectra $E_\theta(k) \times k$.

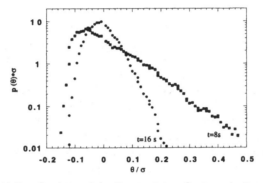

Figure 10: Rescaled distributions of the fluctuations of concentration at time $t = 8\mathrm{s}$ and $t = 16\mathrm{s}$. The passive scalar used has the density $\rho = 1002\mathrm{gl}^{-1}$.

the theoretical analysis of Pumir *et al.* [34] and Chertkov *et al.* [33].

Another quantity of interest is the standardized (i.e. rescaled so that the variance be unity) pdfs of the concentration increments defined by:

$$\Delta\theta = \theta(\mathbf{x} + \mathbf{r}) - \theta(\mathbf{x}) \tag{4.1}$$

where r is the scale. Fig. 11 displays such pdfs, at various times in the quasi-stationary domain. These pdfs appear self-similar in space, and roughly symmetric; the small increment region looks Gaussian while the tails, defined for increments above one standard type deviation, may be fitted by exponential functions. We thus have here a Gaussian bump with exponential tails, which agrees well with the theoretical preditions of [33]. In the insert, we show the corresponding distributions obtained by using a simulated concentration field; they are similar to the experiment.

As is traditionally done for turbulent signals, we measure the structure

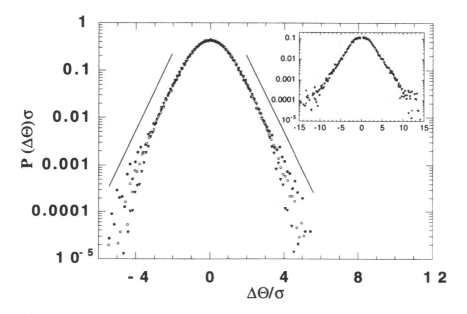

Figure 11: Rescaled distributions of the concentration increments at time $t = 16$s for three different increments in the Batchelor range for a $\rho = 1002$gl^{-1} density pollutant. Full squares correspond to $r = 1.165$cm, empty squares to $r = 4.078$cm and triangles to $r = 6.991$cm. In insert are shown the numerical result for the same increments at the same time.

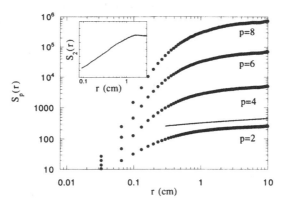

Figure 12: Structure functions of orders up to 8. Straight line is a logarithm fit for $S_2(r)$, $S_2(r) \approx 300 + 55\ln(r/L)$. The passive scalar used has the density $\rho = 1002$gl^{-1}. The numerical result for $S_2(r)$ is shown in insert with lin-log axis.

function of the concentration increments defined by

$$S_p = \langle (\theta(\mathbf{x} + \mathbf{r}) - \theta(\mathbf{x}))^p \rangle \qquad (4.2)$$

Fig. 12 displays such structure functions as functions of the scale r, up to 8th order. The scaling range corresponding to the inertial convective domain is difficult to determine accurately; it roughly lies between $r = 0.3$ and 5cm. In this range, there is no clear power law, and one may propose a logarithmic law for representing $S_2(r)$ (see Fig. 12). The logarithmic type behaviour is confirmed by the simulated particles experiment (see the insert of Fig. 12). The fit formula for $S_2(r)$ turns out to be consistent with the early proposals of Batchelor [27] and Kraichnan [31]. In the framework of Chertkov *et al.* [33] theory our fit formula corresponds to the case where, in the enstrophy cascade, the logarithmic corrections are small (which appears to be the case in our experiment (see Section 3 and [28])). We thus have consistency between the theory and the experiment. It is certainly difficult to say much more at the moment.

5 Conclusions

To summarize, we have performed, in a physical system, an extensive observation of the enstrophy cascade. Previous experiments inferred its existence from the interpretation of k^{-3} spectra. We provide here a complete observation, along with a measurement of the Kraichnan–Batchelor constant, and a determination of the high order vorticity statistics, a crucial quantity to consider in the intermittency problem. We obtain that classical theory is strikingly successful. There is no substantial small scale intermittency and the vorticity statistics depart only moderately from Gaussian. Because of these particular features, one may perhaps hope this problem will be brought to theoretical understanding. The role of coherent structures, long emphasized, is indeed important and interesting, but should probably be considered as a separate issue. Note finally these conclusions agree with a recent numerical study [26].

Concerning the dispersion problem, we have investigated the Batchelor regime, for a free decaying blob of a passive scalar, in a physical experiment. We have observed the k^{-1} spectrum, exponential tails for the distributions of the concentration and concentration increments, and logarithmic like behaviour for the second order structure functions. These findings have been confirmed by using simulated particles. Our observations agree well with the theoretical analysis of [33]. Although we stand at a qualitative level (unfortunately, there is no simple universal constant to measure in this problem), and we deal with freely decaying concentration fields, it is fair to say the experiment provides strong support to the theory. We thus report here the first case, in the domain of turbulence, where a physical experiment confirms a theoretical analysis, which has successfully calculated the two-point statistics of the problem, without calling for a closure scheme.

This work was supported by Ecole Normale Supérieure, Universités Paris 6 et Paris 7, Centre National de la Recherche Scientifique, and by EC Network Contract FMRX-CT98-0175. The authors wish to thank R. Benzi, G. Falkovitch, K. Gawedzki, V. Lebedev, C. Pasquero and A. Provenzalle, for enlightening discussions concerning this study.

References

[1] R.H. Kraichnan, *Phys. Fluids* **9**, 1937, (1966).

[2] G. Batchelor, *Phys Fluids Supp. II*, **12**, 233, (1969).

[3] R. Kraichnan, *J. Fluid Mech.*, **47**, 3, 525, (1971).

[4] P. Santangelo, B. Legras, R. Benzi, *Phys Fluids A*, **I**, 1027, (1989).

[5] M.E. Brachet, M. Meneguzzi, P.L. Sulem, *Phys. Rev. Lett.*, **57**, 683, (1986).

[6] B. Legras, P. Santangelo, R. Benzi, *Europhys. Lett.*, **5**, 37, (1988).

[7] K. Ohkitani, *Phys. Fluids*, **A3**, 1598, (1991).

[8] J.R. Herring, J. McWilliams, *J. Fluid Mech.*, **153**, 229, (1985).

[9] A. Babiano, B. Dubrulle, P. Frick, *Phys. Rev. E*, **52**, 4, 3719, (1995).

[10] M.E. Maltrud, G.K. Vallis, *J. Fluid Mech.*, **228**, 321, (1991).

[11] H.K. Moffatt, in *Advances in Turbulence*, edited by G. Comte-Bellot and J. Mathieu, Springer-Verlag, Berlin, (1986), p. 284.

[12] P.G. Saffman, *Stud. Appl. Math.*, **50**, 277, (1971).

[13] A. Polyakov, *Nucl. Phys.*, **B396**, 367, (1993).

[14] G. Eyink, *Physica D*, **91**, 97, (1996).

[15] G. Falkovitch, V. Lebedev, *Phys. Rev. E*, **52**, 49, 3, (1994).

[16] M. Gharib, P. Derango, *Physica D*, **37**, 406, (1989).

[17] B.K. Martin, X.-L. Wu, W.I. Goldburg, M.A. Rutgers, *Phys. Rev. Lett*, **80**, 18, 3964, (1995).

[18] H. Kellay, X.-L. Wu, W.I. Goldburg, *Phys. Rev. Lett.*, **74**, 20, 3975, (1995).

[19] K.G. Oetzel, G.K. Vallis, *Phys. Fluids*, **9**, 10, 2991, (1997).

[20] M. Rutgers, *Phys. Rev. Lett.*, **81**, 11, 2244, (1998).

[21] W. Borue, *Phys. Rev. Lett.*, **71**, 24, 3967, (1993).

[22] J. Paret, D. Marteau, O. Paireau, P. Tabeling, *Phys. Fluids*, **9**, (10), 3102, (1997).

[23] J. Paret, P. Tabeling, *Phys. Rev. Lett.*, **79**, 4162, (1997).

[24] J. Paret, P. Tabeling, *Phys. Fluids*, **10**, 12, 3126, (1998).

[25] A.E. Hansen, D. Marteau, P. Tabeling, *Phys. Rev. E*, **58**, 7261, (1998).

[26] C. Pasquero and A. Provenzalle have conducted a numerical study, whose conclusions agree well with the experiment. This work will be reported soon.

[27] G.K. Batchelor, *J. Fluid Mech.*, **5**, 113, (1959).

[28] J. Paret, M.-C. Jullien, P. Tabeling, *Phys. Rev. Lett.*, **83**, 3418, (1999).

[29] H.L. Grant, B.A. Hugues, R.B. Williams and A. Moillet, *J. Fluid Mech.*, **34**, 423, (1968).

[30] N.S. Oakey, *J. Phys. Oceanogr.*, **12**, 256, (1982).

[31] R. Kraichnan, *Phys. Fluids*, **10**, 1417, (1967).

[32] R. Kraichnan, *Phys. Fluids*, **11**, 945, (1968).

[33] M. Chertkov, G. Falkovich, I. Kolokolov, V. Lebedev, *Phys. Rev. E*, **51**, 5609, (1995). The generalization can be made in a Kraichnan framework (private communication).

[34] A. Pumir, B. Shraiman, E.D. Siggia, *Phys. Rev. Lett.*, **66**, 2984, (1991).

[35] G. Falkovich, I. Kolokolov, V. Lebedev, A. Migdal, *Phys. Rev. E*, **54**, 4896, (1996).

[36] C.H. Gibson, W.H. Scharz, *J. Fluid Mech.*, **16**, 365, (1963).

[37] A. Gargett, *J. Fluid Mech.*, **159**, 379, (1985).

[38] B.S. Williams, D. Marteau, J.P. Gollub, *Phys. Fluids*, **9** (7), 2061, (1997).

[39] P.L. Miller, P.E. Dimotakis, *J. Fluid Mech.*, **308**, 129, (1996).

[40] D. Bogucki, J.A. Domaradzki, P.K. Yeung, *J. Fluid Mech.*, **343**, 111, (1997).

[41] M. Holzer, E.D. Siggia, *Phys. Fluids*, **6** (5), 1820, (1994).

[42] M.-C. Jullien, J. Paret, P. Tabeling, *Phys. Rev. Lett.*, **82**, 2872 (1999).

On the Origins of Intermittency in Real Turbulent Flows

Michael Kholmyansky and Arkady Tsinober

Abstract

A set of data is obtained for all the nine components of the velocity gradient tensor $A_{ij} = \partial u_i / \partial x_j$ along with all the three velocity components in a field experiment at Reynolds numbers (based on the Taylor microscale) $\mathrm{Re}_\lambda \approx 10^4$. The data are used in order to demonstrate that the small scales – both from the inertial range and from the dissipation range – are *not decoupled* from the energy containing range even at Re_λ as large as 10^4. It is argued that this *direct* interaction/coupling of large and small scales seems to be a generic property of all turbulent flows and the main reason for small scale intermittency, non-universality, and quite modest manifestations of scaling.

1   Introductory notes, motivation

There is no concensus on the meaning of the term *intermittency* even in the community working in the field of fluid turbulence. We shall use it in connection with the small scale (SS) – another not well defined term – structure of turbulent flows and 'misbehaviors' of SS. We will understand the term 'scale' in its simplest geometrical meaning in physical space and/or use the field of velocity derivatives $A_{ij} = \partial u_i / \partial x_j$ as the one objectively (i.e. decomposition/representation independent) representing small scales, e.g. dissipation happens to occur on the scale$s^2 \equiv s_{ij} s_{ij}$, where $s_{ij} = \frac{1}{2}(A_{ij} + A_{ji})$ and vorticity is known as a basic SS quantity (Tsinober (1999b)).

There are several reasons for SS intermittency in turbulent flows (see, e.g. Sreenivasan and Antonia (1997), Tsinober (1993, 1998), Yeung *et al.* (1995), Warhaft (2000) and references therein).

Our concern in this article is with the *direct coupling* between the large and small scales. The evidence for such a coupling is massive (see references in Sreenivasan and Antonia (1997), Tsinober (1993, 1998), Warhaft (2000)). In particular, it goes back to the fact that the skewness of the derivative of temperature fluctuations is not small but is of order 1 (Stewart (1969); Gibson *et al.* (1970)). For example, in the experiments of Gibson *et al.* (1970) it was about -0.5, in conditions as in our experiment decsribed below, whereas for a locally isotropic flow it should be close to zero[1] Similar observations

---

[1]For additional references see Sreenivasan and Antonia (1997); also Figure 7 therein.

were made in laboratory flows at $Re_\lambda = 500$ (see Warhaft (2000) and references therein). Quite recently similar observations were made for the velocity derivatives in the direction of the shear (Garg and Warhaft (1998) and references therein).

An important observation was made by Praskovsky *et al.* (1993) (see also Sreenivasan and Dhruva (1998)). The specific feature of these works is the rather large (Taylor microscale) Reynolds number $Re_\lambda \sim 10^4$. In spite of such a large Reynolds number there is clear evidence of strong coupling between large and small scales, which according to 'common wisdom' should not exist. This fact only calls for experimental verification. However, our results – except those performed at $Re_\lambda \sim 10^4$ – have an additional important feature, namely, in our experiment, all the nine velocity derivatives $\partial u_i / \partial x_j$ were evaluated along with all the three velocity components u_i. This was done via implementation in a field experiment of the multi-hotwire technique used by Tsinober *et al.* (1992, 1997) in the laboratory. This allowed the use of invariant quantities (i.e. independent of the system of reference) such as energy, full dissipation (not its surrogate), enstrophy, enstrophy generation, etc., as being most appropriate to describe physical processes.

We briefly describe below the experiment followed by some results from it relevant to this paper. A more detailed description of the experiment along with various results will be published elsewhere.

2   The experiment

The choice of the site was one of the most complicated problems. A site was found after a year of searching that met most of the requirements. It is located in a field just outside, and belonging to, the Kfar Glikson kibbutz. It is flat grass-covered ground with 4.5km fetch downwind starting with a grove of trees of about 15m height. Concrete foundations were cast at the station for a mast 10m height, diesel generator, and a light prefabricated building for a laboratory with the equipment necessary for the experiment. This included a PC, sample and hold, anemometers and other electronics, various controls and auxiliary equipment. A special high precision calibration unit was designed and manufactured for the computer controlled three-dimensional calibration of the twenty hotwire probes consisting of five four-wire arrays forming a cross. Several essential technological innovations were introduced in the manufacturing process of the probe in view of specific requirements of a field experiment as compared with the probe used by Tsinober *et al.* (1992, 1997) in laboratory experiments. These innovations improved the reliability of the probes and hastened the production process. A considerable amount of work was required to prepare the software for data acquisition and calibration. The preparation of the experiment took 3.5 years. This is mainly due to the specific aspects of a field experiment, in general, and special require-

U_1	u_1'	u_2'	u_3'	λ	η	r_{uw}	C	Re_λ
m/s	m/s	m/s	m/s	m	m			
7.0	1.0	1.0	0.6	0.14	$8 \cdot 10^{-4}$	-0.33	0.53	10^4

Table 1. Basic information on the experimental run.

ments on the precision, in particular, which apart from huge investments in the 'hardware', required a much greater variety and amount of work.

The reported experiment involves recording and processing large amounts of data. The results presented here are based on one chosen measurement run, taken at the height of 10m in approximately neutral, slightly unstable conditions. The duration of the run was 15 min (i.e it was a 6.3km long sample) and it contained about 8.5×10^6 *simultaneous* samples *of 20 channels* taken at sampling rate nearly 10,000kHz per channel.

The basic data on this particular run are given in Table 1.

The notations are as follows: x_1 – horizontal streamwise, x_2 – horizontal spanwise, and x_3 – vertical coordinates respectively; u_i – corresponding components of velocity fluctuations, u_i' – their *rms* values; $r_{u_1 u_3} = \langle u_1 u_3 \rangle / \sigma_{u_1} \sigma_{u_3}$ – the correlation coefficient between the streamwise and vertical components of velocity fluctuations; C — Kolmogorov constant.

A number of tests were performed regarding the properties of the field of velocity fluctuations itself and the velocity derivatives. These include those as in Tsinober *et al.* (1992) as well as several additional ones. The spectra for all the three velocity components have a wide inertial range with a power law $(k\eta)^{-5/3}$ extending over about 3.5 decades, ending at longitudinal scale about 7cm, where the spectra deviate from the power law due to the influence of viscosity. It is noteworthy that the *compensated* spectra do not look that 'nice', so that the inertial range is considerably shorter. Similar behavior is observed when looking at the r dependence of structure functions. All this seems to be related to a much broader issue on the very existence of scaling in turbulent flows.

Velocity derivatives in the mean flow direction, x_1, were calculated according to the Taylor hypothesis. In calculating the x_1-derivative we used both the mean and the instantaneous velocity. The difference was insignificant.

The derivatives in the spanwise and vertical directions were calculated taking velocity differences between suitable arrays and dividing by the distance between these arrays. Data from five arrays allow us to calculate the spatial derivatives in several ways, so that we used nine possible combinations in order to check the reliability of evaluating velocity derivatives.

Skewness and flatness of velocity derivatives are shown in Table 2, where ÷ indicates the range of variation of the quantity in question.

Skewness

$\frac{\partial u_1}{\partial x_1}$	$\frac{\partial u_2}{\partial x_2}$	$\frac{\partial u_3}{\partial x_3}$	$\frac{\partial u_i}{\partial x_k}, i \neq k$	$\frac{\langle \omega_i \omega_k s_{ik} \rangle}{\langle \omega^2 \rangle^{3/2}}$	$\frac{\langle s_{ij} s_{jk} s_{ki} \rangle}{\langle s^2 \rangle^{3/2}}$
-0.73	-0.65	-0.65	$-0.05 \div -0.1$	0.18	-0.38
				$(0.21)_{S_{\frac{\partial u_1}{\partial x_1}} = -0.7}$	$(0.42)_{S_{\frac{\partial u_1}{\partial x_1}} = -0.7}$

Flatness

	$\frac{\partial u_i}{\partial x_k}$	$\frac{15}{7}\frac{\langle s^4 \rangle}{\langle s^2 \rangle^2}$	$\frac{9}{5}\frac{\langle \omega^4 \rangle}{\langle \omega^2 \rangle^2}$	$\frac{\langle \omega^2 s^2 \rangle}{\langle \omega^2 \rangle \langle s^2 \rangle}$	$3\frac{\langle (\omega_k s_{ik})^2 \rangle}{\langle \omega^2 \rangle \langle s^2 \rangle}$
Real	$20 \div 25$	17.5	27.6	6.7	3.6
Gaussian	3	3	3	1	1

Table 2. Skewness and flatness (kurtosis) of velocity derivatives. The third line in the table for *skewness* contains values of $\frac{\langle \omega_i \omega_k s_{ik} \rangle}{\langle \omega^2 \rangle^{3/2}}$ and $\frac{\langle s_{ij} s_{jk} s_{ki} \rangle}{\langle s^2 \rangle^{3/2}}$ obtained assuming isotropy and $S_{\frac{\partial u_1}{\partial x_1}} = -0.7$.

It is noteworthy that our result for the skewness $S_{\frac{\partial u_1}{\partial x_3}} \approx -0.05$ is significantly larger in magnitude than $S_{\frac{\partial u_1}{\partial x_2}} \approx -0.03$. Similarly $S_{\frac{\partial u_3}{\partial x_2}} \approx -0.2$ is even larger. This is consistent with the results on the anisotropy of small scales resulting from the presence of mean shear (Garg and Warhaft (1998) and references therein).

It is seen that the skewness of the derivatives $\partial u_2/\partial x_2$ and $\partial u_3/\partial x_3$ is close to that of $\partial u_1/\partial x_3$. Also, noteworthy is the agreement of these values ($\sim 0.6 \div 0.7$) and of the flatness ($\sim 20 \div 25$) with the one known from literature (e.g. see the review by Sreenivasan and Antonia (1997)).

3   Direct coupling of large and small scales

Here to a large extent we follow the approach of Praskovsky *et al.* (1993). The main difference is that we have access to invariant quantities such as energy, dissipation (not the usually used surrogate), enstrophy, etc.

We start with the correlation coefficients between the large scale quantities like u_i, v (the centered magnitude of the vector of velocity fluctuations) and the moments of different velocity differences from the inertial and dissipative range. Obviously we have the latter only in the streamwise direction with the exception of those used to estimate the derivative in the vertical and spanwise direction. It is noteworthy that the former are quite reliable. Some results are shown in Table 3.

It is seen that the correlation between the large and small scales definitely is not negligible. The results shown are in agreement with those obtained by Praskovsky *et al.* (1993).

r/η	1	35	110	350	1000
$\langle u_1 \cdot \delta u_1 \rangle / (\sigma_{u_1}\sigma_{\delta u_1})$	0.0028	0.0415	0.0646	0.0991	0.1424
$\langle u_2 \cdot \delta u_2 \rangle / (\sigma_{u_2}\sigma_{\delta u_2})$	0.0041	0.0528	0.0806	0.1231	0.1755
$\langle u_3 \cdot \delta u_3 \rangle / (\sigma_{u_3}\sigma_{\delta u_3})$	0.0065	0.0876	0.1345	0.2062	0.2974
$\langle v \cdot \delta u_1 \rangle / (\sigma_v\sigma_{\delta u_1})$	0.0001	−0.0023	−0.0033	−0.0033	−0.0012
$\langle v \cdot \delta u_2 \rangle / (\sigma_v\sigma_{\delta u_2})$	−0.0007	−0.0038	−0.0061	−0.0089	−0.0112
$\langle v \cdot \delta u_3 \rangle / (\sigma_v\sigma_{\delta u_3})$	−0.0001	0.0046	0.0064	0.0094	0.0045
$\langle u_1 \cdot \delta u_2 \rangle / (\sigma_{u_1}\sigma_{\delta u_2})$	−0.0003	−0.0001	0.0008	0.0026	0.0070
$\langle u_1 \cdot \delta u_3 \rangle / (\sigma_{u_1}\sigma_{\delta u_3})$	0.0009	0.0014	0.0024	0.0035	0.0066
$\langle u_1 \cdot (\delta u_1)^2 \rangle / (\sigma_{u_1}\sigma^2_{\delta u_1})$	0.0070	−0.0333	−0.0427	−0.0520	−0.0545
$\langle u_2 \cdot (\delta u_2)^2 \rangle / (\sigma_{u_2}\sigma^2_{\delta u_2})$	−0.0150	−0.0199	−0.0212	−0.0217	−0.0312
$\langle u_3 \cdot (\delta u_3)^2 \rangle / (\sigma_{u_3}\sigma^2_{\delta u_3})$	0.0506	0.0769	0.0817	0.0906	0.0706
$\langle v \cdot (\delta u_1)^2 \rangle / (\sigma_v\sigma^2_{\delta u_1})$	0.0282	0.0386	0.0403	0.0466	0.0531
$\langle v \cdot (\delta u_2)^2 \rangle / (\sigma_v\sigma^2_{\delta u_2})$	0.0202	0.0276	0.0308	0.0364	0.0485
$\langle v \cdot (\delta u_3)^2 \rangle / (\sigma_v\sigma^2_{\delta u_3})$	0.0222	0.0336	0.0388	0.0466	0.0635
$\langle u_1 \cdot (\delta u_2)^2 \rangle / (\sigma_{u_1}\sigma^2_{\delta u_2})$	−0.0099	−0.0564	−0.0575	−0.0574	−0.0630
$\langle u_1 \cdot (\delta u_3)^2 \rangle / (\sigma_{u_1}\sigma^2_{\delta u_3})$	−0.0061	−0.0429	−0.0479	−0.0527	−0.0534
$\langle u_1 \cdot (\delta u_1)^3 \rangle / (\sigma_{u_1}\sigma^3_{\delta u_1})$	0.0004	0.0223	0.0374	0.0610	0.0930
$\langle u_2 \cdot (\delta u_2)^3 \rangle / (\sigma_{u_2}\sigma^3_{\delta u_2})$	0.0009	0.0236	0.0419	0.0669	0.1119
$\langle u_3 \cdot (\delta u_3)^3 \rangle / (\sigma_{u_3}\sigma^3_{\delta u_3})$	0.0025	0.0436	0.0734	0.1092	0.1722
$\langle u_1 \cdot (\delta u_1)^4 \rangle / (\sigma_{u_1}\sigma^4_{\delta u_1})$	0.0026	−0.0054	−0.0124	−0.0199	−0.0231
$\langle u_2 \cdot (\delta u_2)^4 \rangle / (\sigma_{u_2}\sigma^4_{\delta u_2})$	−0.0022	−0.0015	−0.0074	−0.0071	−0.0225
$\langle u_3 \cdot (\delta u_3)^4 \rangle / (\sigma_{u_3}\sigma^4_{\delta u_3})$	0.0044	0.0153	0.0186	0.0118	0.0100

Table 3. Correlation coefficient between u_i, $\delta u_i = u_i(x+r)-u_i(x)$, and the centered magnitude of the vector of velocity fluctuations $v = u - \langle u \rangle$, $u^2 = u_1^2 + u_2^2 + u_3^2$.

The direct coupling of large and small scales is seen most clearly from the conditional averages of velocity differences conditioned on large scale quantities (see Figures 1 and 2). The main result is that all conditional statistics are *not independent* of large scale quantities as is expected without coupling between large and small scales.

In Figure 1 we show some results similar to ones obtained by Praskovsky *et al.* (1993) (in the left column)) in parallel with those conditioned on the *centered magnitude* of the vector of velocity fluctuations $v = u - \langle u \rangle$, where $u^2 = u_1^2 + u_2^2 + u_3^2$. Similar behavior is observed for the conditonal statistics of $\langle \delta u_i^n \rangle$ for all $i = 1, 2, 3$ and $n = 2, 3, 4$.

Two aspects deserve special comment.

First, there is a clear tendency of increase of the conditional averages of the structure functions with the *energy* of fluctuations as is seen from the right column of Figure 1. Second, such a tendency, that is the direct coupling, is observed also for the smallest distance $\sim \eta$, which was used for estimates of the derivatives in the streamwise direction. This result is quite reliable due to the absence of problems in estimating the derivatives in the streamwise direction like those in the other two directions.

In Figure 2 we show also similar conditional statistics for the enstrophy ω^2 and the total strain $s_{ij}s_{ij}$. The result is quite similar to the one shown in Figure 1 for the smallest distance $\sim \eta$. This result is definitely correct *qualitatively*.

Special mention is due the PDF of the angle between velocity and vorticity, since it reflects the correlation beween the two vectors (Figure 3). Though the alignment is weak, it is significant and is another manifestation of the direct coupling between large and small scales. It is not surprising that the correlation between a large scale and a small scale quantity is small, in our case $\langle \mathbf{u} \cdot \omega \rangle / u\omega = 5 \cdot 10^{-4}$. However, this does not mean that the coupling between them is small as well.

It is noteworthy that two correlations involving large and small scales are of particular importance for shear flows. These are $\langle \omega_2 u_3 \rangle$ and $\langle \omega_3 u_2 \rangle$, since they are directly related to the derivative[2] of the Reynolds stress:[3]

$$\frac{d \langle u_1 u_3 \rangle}{dz} \sim \langle \omega_2 u_3 \rangle - \langle \omega_3 u_2 \rangle$$

The important point is that the corresponding correlation coefficients, $C_{\omega_2 u_3} = 7.7 \cdot 10^{-3}$ and $C_{\omega_3 u_2} = 1.3 \cdot 10^{-2}$, are small, but significant: if they actually vanished the mean flow would not 'know' anything about the fluctuating part of the turbulent flow (Tsinober (1998, 1999a)).

4   Concluding remarks

The results obtained in this research are the first ones in which explicit information is obtained about the field of velocity derivatives (all nine components of the tensor $\partial u_i / \partial x_j$) and velocity differences along with all three components of the velocity fluctuations at $\mathrm{Re}_\lambda \sim 10^4$. Up to now, the field of velocity derivatives was accessible at $\mathrm{Re}_\lambda \sim 10^2$.

The general conclusion drawn from the whole experiment is that the basic physics of turbulent flow at high Reynolds number $\mathrm{Re}_\lambda \sim 10^4$, at least

---

[2] This is the quantity entering in the equation for the mean flow.
[3] This relation is precise if $(\partial/\partial x)\langle \cdots \rangle = 0$, e.g. in a channel flow.

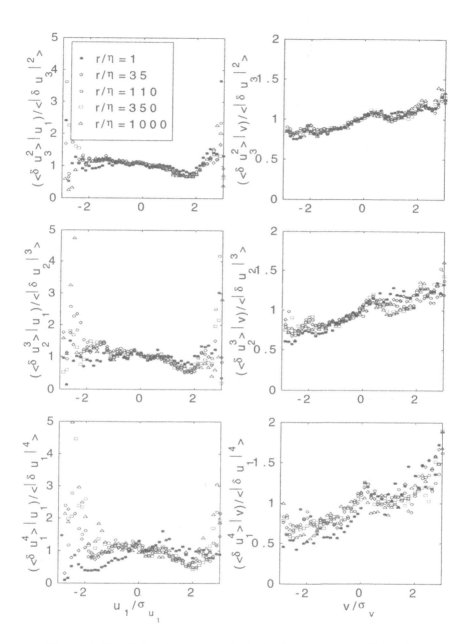

Figure 1. Conditional averages of velocity increments conditioned:
left – on the u_1 fluctuation, right – on the centred magnitude of
the vector of velocity fluctuations v.

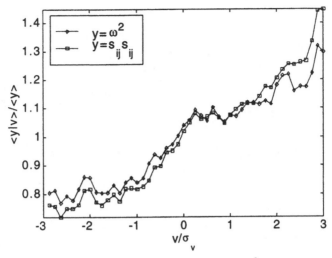

Figure 2. Conditional averages of enstrophy ω^2 and total strain $s_{ij}s_{ij}$ conditioned on the centred magnitude of the vector of velocity fluctuations v.

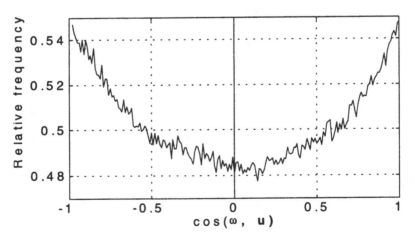

Figure 3. PDF of the cosine of the angle between the vectors of velocity and vorticity.

qualitatively, is the same as at moderate Reynolds numbers, $Re_\lambda \sim 10^2$. This is true of such basic processes as enstrophy and strain production, geometrical statistics, the role of concentrated vorticity and strain, and depression of nonlinearity.

One of the specific conclusions is that the small scales – both from the inertial range and from the dissipation range – are *not decoupled* from the energy containing range even at Reynolds numbers (based on the Taylor microscale) as large as 10^4. In other words the Galilean invariance is broken in the restricted sense that the properties of small scale turbulence are *not* independent of parameters characterizing the large scales, such as, e.g. the energy of velocity fluctuations. This *direct* interaction/coupling of large and small scales seems to be a generic property of all turbulent flows and one of the main reasons for small scale intermittency, non-universality, and quite modest manifestations of scaling.

This 'contamination' of small scales by the large ones seems to be unavoidable even in homogeneous and isotropic turbulence, since there are many ways to produce such a flow (i.e. many ways to produce the large scales).[4] It is the difference in the mechanisms of large scale production which 'contaminates' the small scales. Hence, non-universality.

The direct interaction/coupling of large and small scales is in full conformity and is the consequence of the generic property of Navier–Stokes dynamics (as well as some kinematics), that is, strong non-locality. This includes also another aspect of the coupling between the small and large scales, namely the bidirectional nature of this coupling, i.e. the 'reaction back' of the small scales (Tsinober (1998, 1999b)). But this is another issue.

Acknowledgements

This research was supported by the US–Israeli Binational Scientific Foundation, Belfer Center for Energy Research and The Israel Science Foundation. Participation of S. Yorish was indispensable: he manufactured the new improved version of the twenty-hotwire probe and took active part in the experiments. Our special thanks are to the people of Kfar Glikson kibbutz. Without their warm help the experiment would never been realized.

References

Garg, S. and Warhaft, Z. (1998) 'On the small scale structure of simple shear flow', *Phys. Fluids* **10**, 662–673.

Gibson, C.H., Stegen, G.S. and Williams, R.B. (1970) 'Statistics of the fine structure of turbulent velocity and temperature fields measured at high Reynolds numbers', *J. Fluid Mech.* **41**, 153–167.

---

[4]For this reason we did not attempt to introduce any corrections (as in Sreenivasan and Dhruva (1998)) due to the presence of shear (in our case $\sim 0.1\text{s}^{-1}$), since we don't see any rational way of doing so due to the nonlinear nature of the problem.

Praskovsky, A., Gledzer, E.B., Karyakin, M. Yu. and Zhou, Y. (1993) 'Fine-scale turbulence structure of intermittent shear flows', *J. Fluid Mech.* **248**, 493–511.

Sreenivasan, K.R. and Dhruva, B. (1998) 'Is there scaling in high-Reynolds number turbulence', *Prog. Theor. Phys., Suppl.* **130**, 103–120.

Stewart, R.W. (1969) 'Turbulence and waves in stratified atmosphere', *Radio Sci.* **4**, 1269–1278.

Sreenivasan, K.R. and Antonia, R. (1997), 'The phenomenology of small-scale turbulence', *Ann. Rev. Fluid Mech.* **29**, 435–472.

Tsinober, A. (1993) 'How important are direct interactions between large and small scales in turbulent flows?'. In *Proceedings of the Monte Veritá Colloquium on Turbulence, September 1991: New Approaches and Concepts in Turbulence*, T. Dracos and A. Tsinober (eds.), Birkäuser, Basel, 141–151.

Tsinober, A. (1998) 'Turbulence – beyond phenomenology'. In *Chaos, Kinetics and Nonlinear Dynamics in Fluids and Plasmas*, S. Benkadda and G.M. Zaslavsky (eds.), Lect. Notes in Phys. **511**, Springer, 85–143.

Tsinober, A. (1999a) 'On statisics and structure(s) in turbulence' Isaac Newton Institute for Mathematical Sciences, Preprint NI99010-TRB, 21 June 1999. A synopsis of lectures given at the IUTAM/IUGG Symposium, *Developments in Geophysical Turbulence, NCAR, Boulder, Colorado, USA, June 16–19, 1998* and at the Workshop on *Perspectives in Understanding of Turbulent Systems*, held at Isaac Newton Institute, Cambridge, January 13–22, 1999.

Tsinober, A. (1999b) 'Vortex stretching versus production of strain/dissipation', Isaac Newton Institute for Mathematical Sciences, preprint NI99010-TRB, 11 June 1999. To be published in *Turbulence Structure and Vortex Dynamics*, J.C.R. Hunt and J.C. Vassilicos (eds.), Cambridge University Press (2000)

Tsinober, A., Kit, E. and Dracos, T. (1992) 'Experimental investigation of the field of velocity gradients in turbulent flows', *J. Fluid Mech.* **242**, 169–192.

Tsinober, A., Shtilman, L. and Vaisburd, H. (1997) 'A study of vortex stretching and enstrophy generation in numerical and laboratory turbulence', *Fluid Dyn. Res.* **21**, 477–494.

Yeung, P.K., Brasseur, J. and Wang, Q. (1995) 'Dynamics of direct large-small-scale coupling in coherently forced turbulence: concurrent physical- and Fourier-space views' *J. Fluid Mech.* **283**, 743–95.

Warhaft, Z. (2000) 'Passive scalars in turbulent flows', *Ann. Rev. Fluid Mech.* **32**, in press.

Scaling and Structure in Isotropic Turbulence

Javier Jiménez, F. Moisy, P. Tabeling and H. Willaime

1 Introduction

An especially intriguing aspect of turbulence is the interplay between structure and statistics. It is tempting to conclude that turbulent flows, because of many degrees of freedom, should be studied statistically, and that the limit theorems for the probability distributions should apply. The earliest theoretical results on the subject were indeed based on postulating 'structureless' models and applying to them scaling laws (Kolmogorov 1941) but, while it works in some approximation, it is known that this description is incomplete. Deterministic structures exist both at the largest scales (Brown and Roshko 1974) and at the smallest ones (Batchelor and Townsend 1949). Less is known about the intermediate length scales, and the study of their structure has traditionally been done statistically, in terms of the 'anomalous' scaling of the difference of the velocities at two points. The basic model in this range is the self-similar multiplicative cascade, which we will briefly discuss below (see Frisch 1995 for a summary), but the main subject of this paper is the characterization of the inertial-range structures, if they can be shown to exist.

Anomalous scaling is, after all, probably the reflection of coherent structures within the inertial range of scales, but their study is complicated by experimental difficulties. The large-scale coherent structures are easily visualized, since they span the flow, while those at the smallest scales manifest themselves by strong gradients which can also be easily extracted from the velocity signals. None of those observational advantages are available in the inertial range, where structures are too small, and probably too numerous, to be readily apparent in visualizations of the velocity field, and also too weak to be seen in those of the gradients (see however the visualizations by Hosokawa, Oide and Yamamoto 1997, and by Porter, Woodward and Pouquet 1998).

In the next section we discuss our motivation for reconsidering the theory of the turbulent cascade. This is followed by the analysis of experimental velocity data, first of their coarse-grained dissipation, and later of the conditional distributions of the velocity increments and of band-pass filtered velocities. Discussions and conclusions are finally offered in §6.

2 Multiplicative cascades and blocking

The concept of cascade was probably first explicitly introduced by Richardson (1922) to describe high-Reynolds-number turbulence. Kolmogorov (1941) made it quantitative, on the implicit assumption that the velocity fluctuations were small enough for all the points in the flow to be described in terms of uniform characteristic scales. His later introduction of intermittency corrections (Kolmogorov 1962) improved on this approximation, but still assumed a uniform cascade in the sense that the only effect of local fluctuations was to introduce a locally variable velocity scale. It was pointed out by Jiménez and Wray (1998) and Jiménez (2000) that more complicated effects are possible.

Out of the many possible cascade models, self-similar multiplicative processes were first applied to turbulent flows by Gurvich and Yaglom (1967), and in more detail by Novikov (1971, 1990), although they were already implicit in the original Kolmogorov (1962) paper. Recent reviews are due to Meneveau and Sreenivasan (1991), Nelkin (1994) and Sreenivasan and Stolovitzky (1995). We will briefly summarize the basic ideas.

Consider a positive variable, v_n, that is assumed to cascade in discrete steps. Assume that the cascade is locally deterministic, which was defined by Jiménez (2000) as one in which the probability distribution of the cascading variable at one point, $p_n(v_n)$, depends only on its value at the previous step,

$$p_{n+1}(v_{n+1}) = \int p_t(v_{n+1}|v_n; n)p_n(v_n)\,\mathrm{d}v_n. \qquad (2.1)$$

This is in contrast to more complicated functional dependences, such as on the values of v_n in some extended spatial neighborhood, or on several previous cascade stages. This assumption intuitively implies that v_{n+1} evolves faster, or on a smaller scale, than v_n, and is in some kind of equilibrium with its precursor. If the cascade is deterministic in this sense, v_n can be represented as a product

$$v_n = x_n x_{n-1} \cdots x_1 u_0 \qquad (2.2)$$

of factors, $x_n = v_n/v_{n-1}$, which are statistically independent of one another.

If moreover the underlying process is invariant under scaling transformations of v it should also be true that the transition probability density function has the form

$$p_t(v_{n+1}|v_n) = v_n^{-1} w(v_{n+1}/v_n). \qquad (2.3)$$

It can easily be shown that local deterministic self-similar cascades lead naturally to intermittent distributions, in the sense that the high-order flatness factors for v_n become arbitrarily large as n increases, implying that arbitrarily strong, although rare, events inevitably appear at the later stages of the cascade.

If (2.3) does not depend explicitly on the cascade stage, these assumptions lead to power laws for the statistical moments of the pdfs, and to the theory of multifractal probability distributions. In particular the pdfs of v_n vary with the cascade step n in a well-defined manner that depends only on the distribution, w, of the 'breakdown' coefficients x. The application of this theory to turbulence is usually justified by the invariance of the Euler equations under scaling transformations (Frisch 1995) but, although this addresses (2.3), it says nothing about the locality hypothesis (2.1).

It was pointed out by Jiménez (2000) that locality is unlikely to be satisfied in multiplicative cascades involving fields. The reason is that each cascading step, presumably an instability of some structure, is bound to be controlled at least partly by the background fluctuations, which introduce a spatially global scale v'_n. The break-up of a given eddy is then controlled by two velocity scales, its own intensity and the global fluctuation level, and self-similarity is broken.

Consider for example the decay of a large-scale vortex in a turbulent flow. As long as its vorticity is of the same order as that of the background its decay is controlled by the outside perturbations, which for example fix the time scale of the break-up. One of the consequences of the random cascade process is however that some of the resulting vortices will be more intense than others and, as noted before, this will eventually lead to some structures which are much stronger than the average. Those strong structures are no longer subject to the influence of the background, and they in essence decouple from it. In some sense the cascade is 'blocked' for them, although all that we can conclude without going into the specific physics of a particular cascade process is that the breakdown of the weak and of the strong structures will be different, and that the flow will eventually differentiate into two distinct components. Simple models for cascades displaying such nonlinear blocking were given by Jiménez (1999), who also argued that, for the reasons sketched above, it is an almost inevitable consequence of applying random multiplicative cascades to fields. Such blocked cascades do not lead to power-law behaviors for the statistics.

In the particular case of vorticity in Navier-Stokes turbulence, it had been previously argued by Jiménez and Wray (1998) that the compact dissipation-range vortex filaments observed in many turbulent flows (Siggia 1981, Jiménez et al. 1993, Jiménez 1998) are in fact examples of such a blocked component of the standard Kolmogorov (or multifractal) cascade, which decouple from the background turbulent flow by virtue of their large internal vorticity.

3   Experimental set-up

To test these theoretical ideas we have analyzed experimental data of approximately isotropic turbulence at three different Reynolds numbers.

The set-up is the same as that described by Zocchi et al. (1994), Tabeling et al. (1996), Belin et al. (1997) and Moisy, Tabeling and Willaime (1999). The flow is confined to a cylinder limited axially by disks equipped with blades rotating in opposite directions at approximately equal angular velocities. The disks are driven by DC motors modified to work at low temperatures, whose rotation speeds are accurately measured by analyzing the frequency content of the electrical current supply. The cylindrical working volume in which

	Re_λ	L_ε/η	$\Delta x/\eta$	$u'/\overline{U}$	N	L/L_ε
----	155	2.5×10^2	1.6	0.20	1.4×10^7	8.8×10^4
—·—	760	2.8×10^3	1.8	0.22	3.7×10^7	2.5×10^4
——	1600	8.4×10^3	1.5	0.21	3.7×10^7	6.5×10^3

Table 1. Characteristics of the three data sets used in this paper; u' is the one-component rms velocity fluctuation intensity, and $\overline{U}$ is the mean longitudinal velocity. The total number of samples in the set is N, which corresponds to a total sample length L. The sampling distance, using Taylor's hypothesis, is Δx. The globally averaged energy dissipation ϵ is used to compute the integral length, $L_\varepsilon = \epsilon/u'^3$, and the Kolmogorov length scale η. Line types are consistently used in the figures.

turbulence takes place is 20cm in diameter, and 13.1cm in height. The whole system is enclosed in a larger cylindrical vessel in thermal contact with a liquid helium bath, and is filled with helium gas at controlled pressure. Its temperature is kept constant between 4.2 and 6.5K, with a long term stability better than 1mK. Under those conditions the kinematic viscosity of the gas is typically between $10^{-3} - 10^{-2}\text{cm}^2/\text{s}$, which is considerably below that of most conventional fluids. Far from the blades and away from the walls the flow behaves as a confined circular mixing layer (Zocchi *et al.* 1994), although the wall region is more complicated. For the data sets used in this paper the probe is located 4.7cm from the mid-plane of the system, and 6.5cm from the cylinder axis. This corresponds to the outer part of the mixing layer, which is more likely to have simple properties and where the presence of a substantial mean flow justifies the use of Taylor's hypothesis.

Velocity measurements are performed with a 'hot'-wire anemometer, whose sensor is a 7μm thick carbon fiber stretched across a rigid frame. A metallic layer, 1000A thick, covers the fiber everywhere except on a spot at the center, 7μm long, which defines the active length of the probe. The time response of the probe was analyzed by Tabeling *et al.* (1996). It depends on various factors, such as the overheat ratio, but it is typically on the order of 2μs. The sensor is unconventionally short but, owing to the thermal characteristics of carbon, no substantial enhancement of thermal inertia is to be expected. It was shown in the references given above that King's law accurately applies to the sensors in the range of operating conditions used in the experiments. Their directionality is bound to be poor but, because of the presence of a non-zero mean velocity, the measured fluctuations are predominantly longitudinal, and will be consider so in the rest of the paper.

The voltage from the anemometer is low-pass filtered, digitized and recorded. The reduction to velocity is done during the analysis, and time intervals are converted to lengths using Taylor's approximation with the overall mean

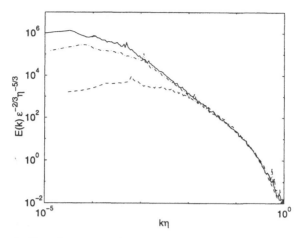

Figure 1. One-dimensional power spectra of the three data sets analyzed in this paper. Line types as in table 1.

velocity, $\Delta x = -\overline{U}\Delta t$. We consider three data sets, summarized in table 1. They are sampled at the Kolmogorov passing frequency, $\overline{U}/\eta$, and restricted to Reynolds numbers which are low enough for the dissipative range of scales to be well resolved. The three sets span a full decade of microscale Reynolds numbers, and the spectrum at the highest Reynolds number, shown in figure 1, presents almost three orders of magnitude of $k^{-5/3}$ power law. We will consider this to be the inertial range, although alternative characterizations are discussed by Moisy, Tabeling and Willaime (1999).

4 The breakdown coefficients for the dissipation

The usual scaling analysis of the coarse-grained surrogate dissipation,

$$\varepsilon_{\Delta x} = \frac{1}{\Delta x} \int_{x-\Delta x/2}^{x+\Delta x/2} (\partial_x u)^2 \, dx, \qquad (4.1)$$

was done first (Meneveau and Sreenivasan 1991). The pdfs of the centered breakdown coefficients,

$$q_{2\Delta x} = \frac{1}{2}\varepsilon_{\Delta x}/\varepsilon_{2\Delta x}, \qquad (4.2)$$

are bell-shaped in the inertial range, as previously reported by Van Atta and Yeh (1975) and by Chhabra and Sreenivasan (1992), but they become less so, and even concave, at the dissipative scales (see figure 2a). Somewhat surprisingly, the shapes of the distributions vary continuously with the averaging

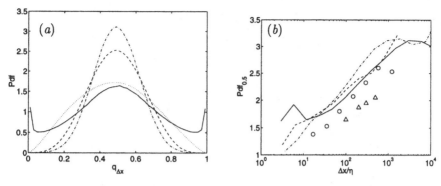

Figure 2. (*a*) Pdfs of the breakdown coefficients of the surrogate average dissipation, for several averaging lengths. $Re_\lambda = 1600$. ——— , $\Delta x/\eta = 3$; ········· , 24; ‐‐‐‐ , 380; —·— , 3000. (*b*) Midpoint value of the pdf as a function of averaging length. Lines are as in table 1; ○ , Van Atta and Yeh (1975). $Re_\lambda = 2600$; △ , Sreenivasan (private communication). $Re_\lambda = 15,000$.

scale, becoming more concentrated as the scale is made larger. This contradicts the conclusions of the two papers mentioned above, which were that the distributions are universal in the inertial range, and which have often been cited as experimental corroboration for the existence of a self-similar dissipative cascade. In fact, in spite of their own conclusions, the data of Van Atta and Yeh (1975) clearly show that their distributions vary with scale. They propose as a measure of the shape of the pdfs their value $p_{0.5}$ at $q = 0.5$. These values are plotted in figure 2(*b*), and it is clear that the same logarithmic trend is observed in their experiment as in ours. Chhabra and Sreenivasan's (1992) published plot is small and difficult to read, and suggests $p_{0.5} \approx 2$ across the inertial range. Newer data from Sreenivasan in the atmospheric boundary layer (private communication) are included in figure 2(*b*), and show variation with scale.

To get a clearer view of the cause of this variation, the pdfs of the breakdown coefficients $q_{\Delta x}$ of a 'parent' segment of length Δx were conditioned not only on Δx, but also on its integrated dissipation, which is expressed for convenience as a surrogate velocity difference

$$\delta u = (\varepsilon_{\Delta x} \Delta x)^{\frac{1}{3}}. \tag{4.3}$$

The conditioning cells are spaced logarithmically in both variables, by a factor of 2 in Δx, and by $\sqrt{2}$ in δu. The result is a two-dimensional array of pdfs, each of which can be characterized by its maximum. Contour plots of the distribution of the conditional $p_{0.5}$ are given in figure 3 for two Reynolds numbers. Both plots have been scaled so that the Kolmogorov length maps

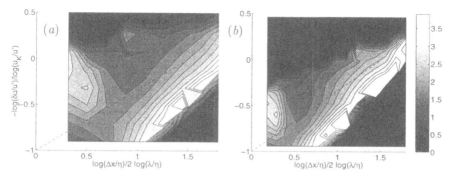

Figure 3. Midpoint value of the conditional pdfs of the breakdown coefficients for the surrogate average dissipation, as a function of the averaging length and of the surrogate velocity increment. (a) $\mathrm{Re}_\lambda = 155$. ($b$) $\mathrm{Re}_\lambda = 1600$.

to zero in the abscissae, and the integral length maps to 1.5. The abscissa 0.5 corresponds to the Taylor microscale λ. In the ordinates, zero corresponds to the large-scale fluctuation intensity u', and -1 to the Kolmogorov velocity scale $u_K = \nu/\eta$. The dashed diagonal line in each plot therefore represents the usual Kolmogorov cascade, $\delta u = (\varepsilon \Delta x)^{\frac{1}{3}}$. The black areas in the plots do not label cases in which the maximum value of the pdf vanishes, but cells for which there are not enough data to compile statistics. Not surprisingly, the cells containing more data cluster around the classical Kolmogorov line.

The behavior of $p_{0.5}$ is the same in both cases. The distributions of the breakdown coefficients of weaker fluctuations are, for a given length scale, more concentrated about the mean than those of stronger ones, and therefore have higher maxima. Concentrated distributions are the result of uncorrelated random processes. The integrated dissipation (4.1) is a sum of variables and, in the simplest hypothesis that the dissipation is only coherent over distances of the order of the Kolmogorov scale, it is easy to show that the distributions of $\varepsilon_{\Delta x}$ and of q should quickly become approximately Gaussian, with maxima increasing as $(\Delta x/\eta)^{\frac{1}{2}}$. The same conclusion would follow from a model in which a limited number of strong isolated small-scale structures account for the bulk of the dissipation. That this is not true is clear from figure 2(b), where the maxima increase only logarithmically with Δx, suggesting the presence of structure in the velocity field at all scales.

Low maxima and wide distributions are the results of coherence, or at least of local statistical inhomogeneity, since the distribution of the dissipation over the two halves of a segment Δx can take any value depending on where the midpoint lies with respect to the inhomogeneous structure. Over most of the plots in figure 3 it is clear that weaker fluctuations break more randomly,

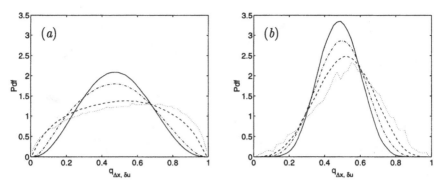

Figure 4. Conditional pdfs of the dissipation breakdown coefficients, for several dissipations and length scales. Re$_\lambda$ = 1600. (a)
$\Delta x/\eta$ = 23, corresponding to an abscissa of 0.5 in figure 3. Dissipations correspond to the following ordinates in figure 3: ——— ,
-0.94; —·— , -0.82; ---- , -0.59; ········ , -0.36. (b) $\Delta x/\eta$ = 1500
(abscissa, 1.2). Ordinates: ——— , -0.36; —·— , -0.24; ---- ,
-0.12; ········ , 0.

while stronger ones break more coherently. The pattern is apparently reversed
for distances in the dissipative range, but this can be shown to be due to a
different effect. The dissipation is effectively constant over those distances,
and segments are again most likely to contain one half of the dissipation of
their parents. In fact the dissipative distributions tend to be trimodal, with
a larger peak at $q = 0.5$ and weaker ones at $q = 0$ and $q = 1$.

Some examples of conditioned inertial pdfs are shown in figure 4. All the
distributions are roughly bell-shaped, but those which correspond to very
strong structures over short distances are almost flat, suggesting a breakdown
into random subintervals of single structures with sizes comparable to that of
the averaging segment. Note that the distributions are not exactly symmetric.
While the most probable value for the dissipation of the central half of a weak
fluctuation is weaker than that of its parent, that of a strong fluctuation tends
to be stronger.

This lack of universality of the breakdown coefficients weakens the experimental support for the self-similar cascade models, but it does not completely
disprove them. It has been noted on several occasions that the integrated dissipation is not a good quantity on which to base a local cascade theory since,
in the first place, it only represents the energy transfer rate in an averaged
sense and, in any case, the transfer rate is itself not a local quantity. It is easy
to see from simple dimensional considerations that the spatial energy fluxes
are of the same order as the energy transfer to smaller scales, and that the
kinetic energy of a fluid volume is as likely to diffuse to a neighboring struc-

ture of roughly the same scale, as to decay into smaller structures (see e.g. the discussion in Jiménez 2000). An experimentally confirmed consequence is that the local energy transfer to subgrid can be positive or negative with comparable probabilities (Piomelli *et al.* 1991).

4.1 The scaling exponents

That the pdfs of the breakdown coefficients are not found to be universal raises the question of why reasonably good power laws are experimentally found for the moments of the dissipation and for the structure functions of the velocity increments.

In the multiplicative model, and subject only to the statistical independence of consecutive cascade steps, the moments of the dissipation after n steps can be expressed as products of the moments of individual breakdown coefficients and, if we assume that Kolmogorov's detailed similarity hypothesis is satisfied, the scaling exponents of the structure functions of the absolute values of the velocity increments are given by (Frisch 1995)

$$S_p = \langle |\Delta u|_{\Delta x}^p \rangle \sim \Delta x^{\zeta_p}, \qquad \zeta_p = -\frac{1}{3}\log_2 \langle q_{\Delta x}^p \rangle, \tag{4.4}$$

where

$$\Delta u_{\Delta x} = u(x + \Delta x/2) - u(x - \Delta x/2). \tag{4.5}$$

The logarithm in (4.4) is one of the reasons why the slopes ζ_p appear to be relatively constant in experiments, even if the underlying distributions are not. Another reason, already noted by Sreenivasan and Stolovitzky (1995) and by Nelkin and Stolovitzky (1996), is that moments are poor indicators of the shape of probability distributions of bounded variables. The bell-shaped functions in figure 2(*a*) can be approximated by Beta distributions,

$$p_\alpha(q) = \frac{q^{\alpha-1}(1-q)^{\alpha-1}}{B(\alpha, \alpha)}, \tag{4.6}$$

where $B(\alpha, \alpha)$ is the Beta function, and $\alpha \approx$ 4-8 (see figure 5). The pth moment of (4.6) is

$$\langle q_{\Delta x}^p \rangle = \frac{B(\alpha + p, \alpha)}{B(\alpha, \alpha)}, \tag{4.7}$$

and is given, in the form of ζ_p, by the solid lines in figure 6(*a*). They are relatively insensitive to α in the range of interest, but they would still give a noticeable curvature in a logarithmic plot of S_p against Δx. A procedure to obtain better power laws was introduced by Benzi *et al.* (1993), by plotting S_p not against Δx (or α in our case), but against ζ_3, which should theoretically scale linearly with Δx in an ideal inertial range. This is equivalent to replacing the scaling exponents by

$$\zeta_p' = \zeta_p/\zeta_3, \tag{4.8}$$

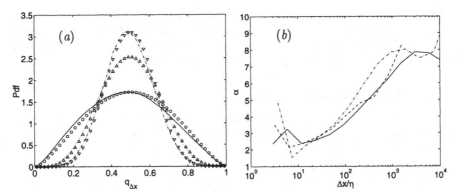

Figure 5. (a) Approximation of the experimental pdfs of the dissipation breakdown coefficients by symmetric Beta distributions having the same maxima. Re$_\lambda$ = 1600. ——— , $\Delta x/\eta$ = 23; ········· , 370; —·— , 6000. (b) Best value of α for the Beta approximation of the breakdown pdfs. Lines as in table 1.

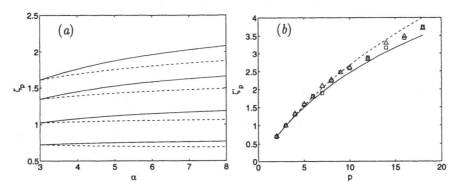

Figure 6. (a) Scaling exponents for the Beta distributions, as a function of the parameter α; p = 2, 4, 6, 8, in increasing order. ——— , true exponent; ---- , extended self-similarity. (b) Scaling exponents of the longitudinal velocity structure functions. 2 , Herweijer and van de Water (1994); △ , Anselmet et al. (1984). Re$_\lambda$ ≈ 500-800 in both cases; ——— , Beta distribution with α = 5; ---- , α = 8.

and has often been used since then to analyze experiments. These extended exponents are given as dashed lines in figure 6(a), and are even less sensitive than ζ_p to variations in α. Finally the results of (4.8) are compared in figure 6(b) with experimental data, with reasonable results.

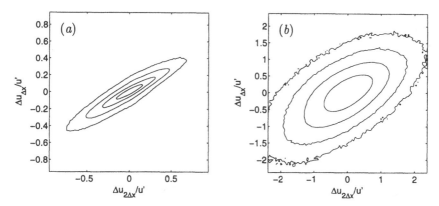

Figure 7. Joint pdfs of the velocity increments across centered segments whose lengths differ by a factor of 2. $Re_\lambda = 1600$. (*a*) $\Delta x/\eta = 6$. (*b*) $\Delta x/\eta = 90$. Contour lines are logarithmically spaced by factors of 10.

It should be emphasized how little this subsection has to do with physics. Its only input is that the pdfs of the breakdown coefficients are approximately bell-shaped, even if their maxima change by more than a factor of 2 in the scales being considered, but it ends with the analytic formula (4.4)–(4.7) for the scaling exponents. It was noted by Sreenivasan and Stolovitzky (1995) that similar results can be achieved with even less information.

Our purpose in this case cannot be to explain the physics of the cascade, even if we have already observed that there is some physical information in the relatively slow variation of the pdfs with the length scale, but to emphasize once more the insensitivity of the scaling exponents as a tool for this purpose. It can be noted in passing that the extended self-similarity procedure appears in this light as essentially a mathematical artifact, which hides, rather than highlighting, the underlying physical processes.

5 The velocity increments

To get more information on the inertial range than was possible from the coarse-grained dissipation, we now consider directly the velocity increments (4.5). It is unfortunately difficult to study them from the point of view of multiplicative cascades since they are not intrinsically positive and their most probable value is close to zero. It is easier to compute directly the joint probability function of the velocity increments at different scales, which contains the full two-point statistical information. The multiplicative cascade is just a particular model for this object. As in the previous section we will consider centred segments whose lengths differ by a factor of 2. The joint pdfs of the velocity increments were recently studied by Friedrich and Peinke (1997), who

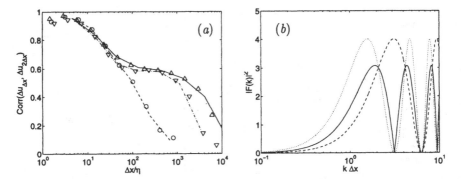

Figure 8. (*a*) Correlation coefficients for the joint pdfs in figure 7, as a function of the separation distance. Lines are as in table 1. (*b*) Transfer functions for the difference operator. ---- , $|F(k\Delta x)|^2$; ········ , $|F(2k\Delta x)|^2$; —— , $|F(k\Delta x)F^*(2k\Delta x)|$.

showed that they are consistent with a Markovian cascade characterized by a drift, depending linearly on the parent velocity difference, and by an additive noise with a parabolic variance. They analyzed data from the centerline of a jet at $\mathrm{Re}_\lambda \approx 600$, and their pdfs look similar to those in figure 7, which belong to our highest-Reynolds-number case. They are not Gaussian, and the correlation coefficient of the two increments, which is roughly proportional to the elongation of the ellipses, depends on the scale. It is given for our three data sets by the symbols in figure 8(*a*). It is close to unity for very short scales, and decays to almost zero near the integral length. For the higher Reynolds numbers the correlation has a plateau in a range that roughly coincides with the inertial range.

While that plateau looks interesting, it says more about our data analysis than about the structure of the flow. The reason for using velocity differences is to isolate particular length scales, which means that the difference operator is being used as a band-pass filter. As such it is far from ideal. If we consider the Fourier transform $\hat{u}(k)$ of the velocity, any homogeneous linear operator multiplies $\hat{u}$ by a transfer function which, in the particular case of (4.5), is

$$\widehat{\Delta u}/\hat{u} = F(k\Delta x) = 2\mathrm{i}\sin(k\Delta x/2). \tag{5.1}$$

It follows from Parseval's theorem that the covariance of two functions can be written in terms of their cospectrum,

$$\langle uv \rangle = \Re \int_0^\infty \hat{u}\hat{v}\,\mathrm{d}k, \tag{5.2}$$

which, for two velocity differences, takes the form

$$\langle \Delta_1 u\, \Delta_2 u \rangle = \int_0^\infty \Re\{F(k\Delta_1 x)F^*(k\Delta_2 x)\}E(k)\,\mathrm{d}k, \tag{5.3}$$

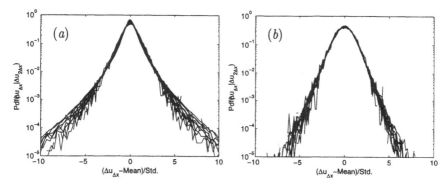

Figure 9. Normalized pdfs of $\Delta u_{\Delta x}$ conditioned on the velocity increment $\Delta u_{2\Delta x}$ of its parent interval. $\mathrm{Re}_\lambda = 1600$. (*a*) $\Delta x/\eta = 12$, $|\Delta u_{2\Delta x}/u_K| = 1\text{-}14$. (*b*) $\Delta x/\eta = 750$, $|\Delta u_{2\Delta x}/u_K| = 4\text{-}60$.

where the asterisk stands for complex conjugation and $E(k)$ is the energy spectrum of u. Since the variance of each velocity difference can also be computed in terms of spectrum and of the transfer function,

$$\langle \Delta u^2 \rangle = \int_0^\infty |F(k\Delta x)|^2 E(k)\,\mathrm{d}k, \tag{5.4}$$

the correlation coefficient

$$\mathrm{Corr}(\Delta_1 u\, \Delta_2 u) = \langle \Delta_1 u\, \Delta_2 u \rangle / \left(\langle \Delta_1 u^2 \rangle \langle \Delta_2 u^2 \rangle \right)^{\frac{1}{2}} \tag{5.5}$$

can be expressed solely in terms of the spectrum and of the filter. The key quantity is the overlap of the two transfer functions, which appears in (5.3). If the band-pass filters are narrower than the separation of the two filter widths, the correlation of the filtered velocities is small while, in the opposite case, it is large. The magnitude of (5.1) at separations spaced by a factor of 2 is given in figure 8(*b*), together with their product. The overlap is substantial, and the product has a wide support, implying that the correlation seen in figure 8(*a*) is mostly due to inadequacy of our filtering scheme. The lines in figure 8(*a*) are computed using (5.1)–(5.4) and the measured velocity spectra.

While the previous discussion is elementary, it should caution us against reading too much physics into the properties of the velocity differences since, at least in part, they may be due to the spectral spillage between different separation scales.

Consider for example the conditional pdfs of $\Delta u_{\Delta x}$, conditioned on the value of $\Delta u_{2\Delta x}$, some of which are given in figure 9. It is interesting that, once normalized with their own mean and standard deviation, they approximately collapse for a given Δx, independently of the conditioning velocity. They are however not Gaussian, and their shape depends on Δx, being more

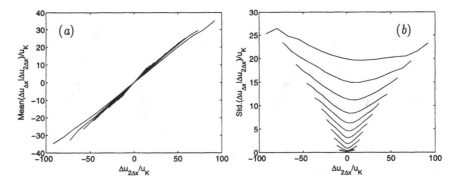

Figure 10. (*a*) Mean value of $\Delta u_{\Delta x}$, conditioned on the velocity increment $\Delta u_{2\Delta x}$ of its parent interval. (*b*) Standard deviations. In both figures $\Delta x/\eta$ ranges from 1.5 to 3000, increasing by factors of 2 in the direction of longer lines. $\mathrm{Re}_\lambda = 1600$.

intermittent for the smaller scales, and looking generally similar to the global pdfs of the velocity increments at the same length scale. Their conditional means and standard deviations are shown in figure 10, where each line corresponds to a given Δx, and where the quantities are plotted against the conditioning velocity difference. Note that the local self-similarity hypothesis for the cascade would imply that both the mean and the standard deviations should be proportional to the conditioning $\Delta u_{2\Delta x}$, and that, while this is approximately true for the former, it is not for the latter. This was also noted by Friedrich and Peinke (1997), and implies that self-similarity is incomplete, as discussed in §2, and that the weak perturbations are dominated by background fluctuations.

The minimum value of the conditional standard deviations for each Δx is plotted in figure 11(*b*), and is proportional to the Kolmogorov velocity at that scale $(\varepsilon\Delta x)^{1/3}$. The slope of the conditional mean with respect to the conditioning velocity is given in figure 11(*a*), and is close to 0.5, as it would be for the velocity differences of a smooth variable. This suggests that the behavior of the mean is a property of the filtering procedure rather than intrinsic to the velocity, which is not smooth at inertial scales. It is indeed easy to construct filters whose band-pass characteristics are sharp enough for the correlation of the filtered velocities to be much smaller than for the increments. We will discuss one such filter below, and we will see that the slope of the mean values can be made very small, or even negative, while the behavior of the standard deviations is robust. Note that the deviations of both the mean and the standard deviation from their 'inertial' values collapse well in Kolmogorov variables across different Reynolds numbers, and that there is no indication of other scales, such as the Taylor microscale, being important

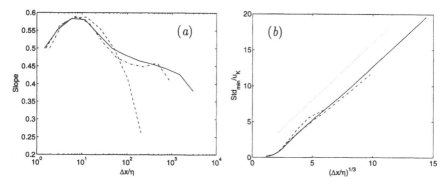

Figure 11. (a) Slope of the conditional mean values in figure 10(a), $\mathrm{d}\langle\Delta u_{\Delta x}|\Delta u_{2\Delta x}\rangle/\mathrm{d}\Delta u_{2\Delta x}$. ($b$) Minimum conditional standard deviations in figure 10(b). Lines as in table 1. The slope of the dotted line is 1.6.

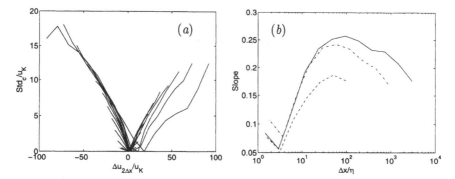

Figure 12. (a) 'Coherent' conditional standard deviation of the velocity increments, as a function of the conditioning velocity increments. Data as in figure 10. Each line represents a different Δx. $\mathrm{Re}_\lambda = 1600$. ($b$) Slope of the coherent standard deviation in (a) with respect to the conditioning velocity increments. Lines as in table 1. The slopes of the positive and negative branches of (a) have been averaged for this figure.

for the behavior of the pdfs.

The parabolic shape of the standard deviations in figure 10(b) suggests that the cascade process can be understood in terms of two processes: the random background noise discussed in the previous paragraph, and a more deterministic process which generates 'coherent' fluctuations which are proportional to those in the parent interval. Assuming both processes to be independent,

208   *Jiménez et al.*

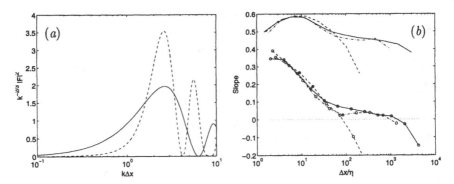

Figure 13. (*a*) Filtered inertial spectra $kE(k)|F(k)|^2$, with $E = k^{-5/3}$; ——— , for the difference operator (4.5); - - - - , for the 'sharp' filter (5.7). (*b*) Slope of the conditional mean values. Simple lines are the same as in figure 11(*a*). Those with symbols are computed with (5.7).

their variances add, and we can define a coherent standard deviation for the second process

$$\text{Std}_c = \left(\text{Std}^2 - \text{Std}_{min}^2\right)^{\frac{1}{2}}. \tag{5.6}$$

This is plotted in figure 12(*a*), and indeed varies approximately linearly with the parent velocity increment. Its slope, for our three data sets, is plotted in figure 12(*b*) as a function of the interval length. At least for the two higher Reynolds numbers it has a plateau at roughly 0.25 at the inertial scales, and falls to very low values in the dissipative range, as the flow becomes smooth and its stochastic component disappears. The location of the decay, 10-20η roughly corresponds to the end of the power-law behavior in the energy spectrum.

5.1 Sharper filters

We have noted that the properties of the velocity difference as a band-pass filter are far from ideal, which led to some spurious properties of the joint statistics of increments at different length scales. Sharper filters, with less spectral leakage, can be constructed, but generally need to use velocities from more points than the two used in (4.5), and they are most effective when tuned to a particular energy spectrum. In this section we will consider the next simplest filter, using four instead of two points. It can be written in the general form

$$\tilde{\Delta}u = a\left[u(x + \tilde{\Delta}x/2) - u(x - \tilde{\Delta}x/2)\right] - ab\left[u(x + \tilde{\Delta}x) - u(x - \tilde{\Delta}x)\right], \tag{5.7}$$

where the parameters a, b and p can be adjusted to get the desired transfer function. We will use

$$a = 1.075, \qquad b = 0.4, \qquad \tilde{\Delta}x = \frac{3}{2}\Delta x, \qquad (5.8)$$

which are chosen so that the peak of the transfer function is at the same location as in (5.1), and the correlation (5.3) vanishes when applied to an inertial power spectrum $E(k) = k^{-5/3}$ (see figure 13(a)). Note that, if (4.5) is a discrete approximation to the velocity gradient, (5.7), with the coefficients in (5.8), is very close to a discrete third derivative, and that it is also possible to think of it as an orthogonal wavelet basis in which orthogonality is defined by the integral in (5.3) with a weighting function which is the energy spectrum (Farge 1992)

When the analysis in the previous section is applied to velocities filtered in this way, instead of to the velocity differences, the general shape of the joint pdfs is the same, but the correlation of $\tilde{\Delta}u_{\Delta x}$ and $\tilde{\Delta}u_{2\Delta x}$ vanishes, and the slope of the conditional means is now much smaller than before, and even reverses in sign for the largest scales (figure 13(b)).

Other properties are more robust. The standard deviations of the conditional probabilities have the same general behavior as in the previous case. Its incoherent part is still proportional to the background fluctuation intensity, although with a slightly higher numerical factor, $\mathrm{Std}_{\min}(\varepsilon\Delta x)^{-1/3} \approx 1.8$, rather than the 1.6 in figure 11(b). A coherent part can also be defined, and it also depends approximately linearly on the filtered velocity of the parent interval, with a proportionality coefficient that reaches a plateau beyond $\tilde{\Delta}x/\eta \approx 10$, as in figure 12($b$), but which is now 0.3 instead of 0.25.

6 Discussion and conclusions

We have analyzed longitudinal velocity signals from three different data sets of approximately isotropic turbulence at low to moderate Reynolds numbers, the highest of which contains a well-developed inertial range over almost three decades. The classical analysis of the scaling of the coarse-grained dissipation does not support the often quoted assumption that the pdfs of the breakdown coefficients are universal in the inertial range, although their variation with scale is much slower than would correspond to a completely uncorrelated dissipation field.

A classification of the pdfs in terms of the length scale and of the dissipation of the parent interval shows the structure of the cascade, which is given in figure 2. Weaker fluctuations at each length scale break according to concentrated distributions, which correspond to uncorrelated processes, while stronger ones break more coherently, with broader pdfs. That separation becomes stronger at the smaller scales.

This interpretation, which is consistent with the theoretical ideas, summarized in §2, of how the strong fluctuations should decouple from the background, is confirmed by the conditional pdfs in §5, which are conditioned on the velocity increment of the parent interval. They do not support full self-similarity of the cascade, because the standard deviations of the 'children' increments are not proportional to the magnitude of their 'parents'. This is however approximately true of the stronger fluctuations, and the distributions can be separated into two components: a weaker one, which is dominated by the background, and whose standard deviation is independent of the intensity of the parent interval, and a stronger one, whose standard deviation is proportional to that intensity. The proportionality constant (~ 0.25) is approximately constant through the inertial range.

The velocity scale separating the two ranges is the Kolmogorov velocity at the particular length scale, $u'_{\Delta x} = (\varepsilon \Delta x)^{1/3}$. This contradicts the earlier assumption by Jiménez and Wray (1998), who had guessed that the relevant scale was the rms vorticity magnitude at the Kolmogorov scale $(\varepsilon/\nu)^{1/2}$, and implies that there is no preferred length scale in the cascade except for the Kolmogorov and the integral scales that bound it. The Reynolds number scaling of the observed distributions is consistent with this conclusion. It cannot however be determined from the present statistics whether the intermittent structures form a self-similar continuum across the inertial range, or are simply the reflection of the known coherence at the integral and Kolmogorov scales.

Another self-similar property, that of the mean values of the conditional distributions, was traced to the poor performance of the velocity increments as band-pass filters. It disappears when a sharper filter is used, but the behavior of the standard deviation is robust.

We have also noted that the lack of self-similarity found for the breakdown distributions is consistent with the experimentally observed approximate power-law behavior of the structure functions, and we have stressed, once more, that the latter are poor indicators of the self-similarity of the underlying processes, and should be used with care. We have shown this by computing the scaling exponents from a simple parametric approximation to the observed pdfs, and observed that, in this light, the better power laws obtained from extended self-similarity should be seen as hiding, rather than illuminating, the underlying physics of the cascade.

It should in any case be stressed that the analysis presented here, as most similar ones, is essentially kinematic, describing the state of the flow at a given time, and should not be confused with a dynamical description of a possible cascade process, which would require the study of the time evolution of individual structures.

This work was supported in part by the Spanish CICYT under contract PB95-0159, and by the Training and Mobility Programme of the EC under grant CT98-0175. Part of the work was carried out during a stay of J.J. in the Isaac Newton Institute of the University of Cambridge, whose hospitality is gratefully acknowledged. K.R. Sreenivasan and A.A. Wray read early versions of this manuscript and provided useful comments.

References

Anselmet, F., Gagne, Y., Hopfinger, E.J., Antonia, R.A. (1984) 'High-order structure functions in turbulent shear flow', *J. Fluid Mech.* **140**, 63–89.

Batchelor, G.K., Townsend, A.A. (1949) 'The nature of turbulent motion at large wave numbers', *Proc. Roy. Soc. Lond. A* **199**, 238–255.

Belin, F., Maurer, J., Tabeling, P., Willaime, H. (1997) 'Velocity gradient distributions in fully developed turbulence: experimental study', *Phys. Fluids* **9**, 3843–3850.

Benzi, R., Ciliberto, S., Tripiccione, R., Baudet, C., Massaioli, F., Succi, S. (1993) 'Extended self-similarity in turbulent flows', *Phys. Rev.* **48**, R29–R32.

Brown, G.L., Roshko, A. (1974) 'On the density effects and large structure in turbulent mixing layers', *J. Fluid Mech.* **64**, 775–816.

Chhabra, A.B., Sreenivasan, K.R. (1992) 'Scale-invariant multiplier distributions in turbulence', *Phys. Rev. Lett.* **68**, 2762–2765.

Farge, M. (1992) 'Wavelet transforms and their application to turbulence', *Ann. Rev. Fluid Mech.* **24**, 395–457.

Friedrich, R., Peinke, J. (1997) 'Description of a turbulent cascade by a Fokker-Planck equation', *Phys. Rev. Lett.* **78**, 863–866.

Frisch, U. (1995) *Turbulence. The legacy of A.N. Kolmogorov.* Cambridge University Press.

Gurvich, A.S., Yaglom, A.M. (1967) 'Breakdown of eddies and probability distributions for small-scale turbulence, boundary layers and turbulence', *Phys. Fluids Suppl.* **10**, S 59–65.

Herweijer, J.A., van de Water, W. (1995) 'Universal shape of scaling functions in turbulence', *Phys. Rev. Lett.* **74**, 4651–4654.

Hosokawa, I., Oide, S., Yamamoto, K. (1997) 'Existence and significance of "soft worms" in isotropic turbulence', *J. Phys. Soc. Japan* **66**, 2961–2964.

Jiménez, J. (1998) 'Small scale intermittency in turbulence', *Eur. J. Mech. B/Fluids* **17**, 405–419.

Jiménez, J. (1999) 'Self-similarity and coherence in the turbulent cascade', in *Proc. IUTAM Symp. Geometry and Statistics of Turbulence*, Hayama, Japan, Nov. 1–5, 1999.

Jiménez, J. (2000) 'Intermittency and cascades', *J. Fluid Mech.* **409**, 99–120.

Jiménez, J., Wray, A.A. (1998) 'On the characteristics of vortex filaments in isotropic turbulence', *J. Fluid Mech.* **373**, 255–285.

Jiménez, J., Wray, A.A., Saffman, P.G., Rogallo, R.S. (1993) The structure of intense vorticity in isotropic turbulence, *J. Fluid Mech.* **255**, 65–90.

Kolmogorov, A.N. (1941) 'The local structure of turbulence in incompressible viscous fluids at very large Reynolds numbers', *Dokl. Akad. Nauk SSSR* **30**, 301–305.

Kolmogorov, A.N. (1962) 'A refinement of previous hypotheses concerning the local structure of turbulence in a viscous incompressible fluid at high Reynolds number', *J. Fluid Mech.* **13**, 82–85.

Meneveau, C., Sreenivasan, K.R. (1991) 'The multifractal nature of the energy dissipation', *J. Fluid Mech.* **224**, 429–484.

Moisy, F., Tabeling, P., Willaime, H. (1999) 'Kolmogorov equation for a fully developed turbulence experiment', *Phys. Rev. Lett.* **82**, 3994–3997.

Nelkin, M. (1994) 'Universality and scaling in fully developed turbulence', *Adv. Phys.* **43**, 143–181.

Nelkin, M., Stolovitzky, G. (1996) 'Limitations of random multipliers in describing turbulent energy dissipation', *Phys. Rev. E* **54**, 5100–5106.

Novikov, E.A. (1971) 'Intermittency and scale similarity in the structure of a turbulent flow', *Prikl. Mat. Mekh.* **35**, 266–277. Translated in *Appl. Math. Mech.* **35**, 231–241.

Novikov, E.A. (1990) 'The effect of intermittency on statistical characterization of turbulence and scale similarity of breakdown coefficients', *Phys. Fluids A* **2**, 814–820.

Piomelli, U., Cabot, W.H., Moin, P., Lee, S. 1991 'Subgrid-scale backscatter in turbulent and transitional flows'. *Phys. Fluids A* **3**, 1766–1771.

Porter, D.H., Woodward, P.R., Pouquet, A. (1998) 'Inertial range structures in decaying compressible turbulent flows', *Phys. Fluids* **10**, 237–245.

Richardson, L.F. (1922) *Weather Prediction by Numerical Process*, p. 66, Cambridge University Press, reprinted by Dover.

Siggia, E.D. (1981) 'Numerical study of small scale intermittency in three dimensional turbulence', *J. Fluid Mech.* **107**, 375–406.

Sreenivasan, K.R., Stolovitzky, G. (1995) 'Turbulent cascades'. *J. Stat. Phys.* **78**, 311–333.

Tabeling, P., Zocchi, G., Belin, F., Maurer, J., Willaime, H. (1996) 'Probability density functions, skewness and flatness in large Reynolds number turbulence', *Phys. Rev. E* **53**, 1613–1621.

Van Atta, C.W., Yeh, T.T. (1975) 'Evidence for scale similarity of internal intermittency in turbulent flows at large Reynolds numbers', *J. Fluid Mech.* **71**, 417–440.

Zocchi, G., Tabeling, P., Maurer, J., Willaime, H. (1994) 'Measurement of the scaling of the dissipation at high Reynolds numbers', *Phys. Rev. E* **50**, 3693–3700.

Capturing Turbulent Intermittency

Willem van de Water, Adrian Staicu and Marie-Christine Guegan

Abstract

We analyze turbulent signals from an array of velocity sensors in terms of moments of velocity increments $u(x + r) - u(x)$ which depend on the angle θ between the measured velocity component and the separation vector r. For the transverse arrangement, $\theta = \pi/2$, we find a stronger intermittency than for the longitudinal $\theta = 0$ one which is reflected in a more strongly anomalous value of the scaling exponents. The greater efficiency of transverse measurements for capturing violent events in turbulence is further demonstrated by average velocity profiles of these events.

1   Introduction

Strong turbulence is characterized by scaling exponents $\zeta(p)$ which express the way statistical moments of a velocity increment

$$\langle \Delta u(r) \rangle^p = \langle [u(x + r) - u(x)]^p \rangle$$

increase with the distance over which it is measured. These scaling exponents are universal; they do not depend on the way turbulence is made. As strong turbulence is punctuated by exceptionally large velocity excursions, the scaling exponents ζ_p are anomalous, that is they differ from their self-similar value $\zeta_p = p/3$ that was predicted by Kolmogorov [1]. Intense slender vortex tubes have been proposed as the prime actors of turbulent intermittency. Evidence for these events was found in numerical simulations [2] and in experiments [3].

There is now growing evidence that the value of the scaling exponent may depend on the relative orientation of Δu and r. It was first found in laboratory turbulent flows [4], and subsequently verified in other experiments [5] and in numerical simulations [6, 7, 8]. In the customary longitudinal arrangement, Δu and r point in the same direction. It appeared that the scaling exponents for the transverse case, where Δu and r are perpendicular, deviated more strongly from the self-similar value $\zeta(p) = p/3$ than the longitudinal ones, pointing to a stronger intermittency of the transverse velocity increments.

Several explanations for this remarkable result have been proposed. First, it has been suggested that transverse probing is better suited for capturing slender vortex tubes [9]. In [8] it was concluded that different longitudinal and transverse exponents would be associated with different scaling behavior of energy dissipation and enstrophy (that is the squared vorticity). However in [10] this finding was contested with the argument that in the limit of infinite Reynolds number, no thinkable vortex structure would support different scaling of these quantities. Lastly, as almost all studied turbulent flows were not completely isotropic, it may be that different scaling anomaly for longitudinal and transverse velocity increments is inherited from the residual anisotropy.

In this paper we study the statistical moments of the velocity increments $G_p(r, \theta) = \langle [u(\boldsymbol{x}+\boldsymbol{r})-u(\boldsymbol{x})]^p \rangle$ as a function, both of the separation r and of the angle θ between the vector $\boldsymbol{r}$ and the x-axis. Throughout, we use a coordinate system with the x-axis pointing in the mean flow direction and with (u, v, w) the components of the fluctuating velocity $\boldsymbol{u}$. For $\theta = 0$ and $\theta = \pi/2$ the generalized structure functions $G_p(r, \theta)$ correspond to the longitudinal and transverse cases, respectively. There is a scaling exponent $\zeta(p, \theta)$ for each angle θ. We will demonstrate for the first time that $\zeta(p, \theta)$ is indeed more strongly anomalous for all $\theta > 0$ and changes rather suddenly at $\theta = 0$ to the accepted value of the longitudinal scaling exponents.

2   Experimental

Longitudinal velocity increments $\Delta u(x)$ can be measured with a single probe which registers a time-dependent velocity signal. By invoking Taylor's frozen turbulence hypothesis, temporal delays τ can be interpreted as spatial separations $x = U\tau$, with U the mean flow velocity. More complicated instrumentation is required for the transverse arrangement where Δu and r are perpendicular. With r pointing in the mean flow direction U, transverse velocity increments can be measured with the help of x-probes which are sensitive to the v-component of the velocity. Alternately, an array of probes oriented perpendicularly to the mean flow direction can be used to measure transverse increments $\Delta u(y_i)$ of the fluctuating u-component at discrete separations [11]. Using a non-intrusive optical technique Noullez et al. [9] obtained similar information, but at a much finer spatial resolution.

Our experiments were done in turbulence generated by a grid placed inside a closed windtunnel. The dimensions of the experiment section were 0.7 × 0.9m², and velocities were measured 2m downstream from the grid (which has mesh size 0.1m and solidity 0.34). As the turbulence intensity on the grid centerline was too small to observe a significant scaling range, measurements were done off centerline, 0.1m from the boundary. In this way a fairly large Reynolds number ($\mathrm{Re}_\lambda = 8.6 \times 10^2$) could be achieved, but at the expense of anisotropy of the velocity fluctuations. This anisotropy will be quantified precisely below. The mean velocity was $U = 12\mathrm{ms}^{-1}$, with rms turbulent velocity fluctuations of $u_{\mathrm{rms}} = 1.1\mathrm{ms}^{-1}$, which is a small fraction of U.

We employed an array of eight hot-wire probes at positions y_i oriented perpendicularly to the mean flow (x-direction) and parallel to the wall. With an array length of 0.07m, the flow is homogeneous over the sensor array. Each of the locally manufactured hot wires had a sensitive length of 200μm, which is comparable to the smallest length scale of the flow (the Kolmogorov scale is $\eta = 199\mu$m). They were operated at constant temperature using computerized anemometers that were also developed locally. The signals of the sensors were sampled exactly simultaneously at 20kHz, after being low-pass filtered at 10kHz (which equaled the Kolmogorov frequency). An accurate calibration of each of the wires is crucial, it was achieved in an automated procedure prior to each run.

If the turbulence intensity is small, the probes are mainly sensitive to the u-component of the velocity. Thus, by time-delaying the signals of the probe at y_i by $\tau_i = y_i \tan(\pi/2 - \theta)/U$, the direction of the separation vector $\boldsymbol{r}$ in $\Delta u(\boldsymbol{r}) = u(\boldsymbol{x} + \boldsymbol{r}) - u(\boldsymbol{x})$ can be set to θ. Here, we have used Taylor's frozen turbulence hypothesis to translate the time delay τ_i into a spatial separation $x_i = \tau_i U$. In this manner, the generalized structure function $G_p(r, \theta)$ can be measured at discrete separations $r = (y_i - y_j)/\cos(\theta)$. The purely longitudinal structure function $G_p^L(r)$ at $\theta = 0$ was measured from single probes using time delays only. An analogous measurement of $G_p(r, \theta)$ has been reported in [12], but using only two probes. In this case the separation r varies jointly with θ and the algebraic dependence of $G_p(r, \theta)$ on r at a given θ is not accessible.

3 Anisotropy

In the case of isotropic and incompressible turbulence, it is straightforward to derive an isotropy relation for $G_2(r, \theta)$,

$$G_2(r, \theta) = G_2^L(r) + \left(1 - \cos^2(\theta)\right) \frac{r}{2} \frac{\partial G_2^L(r)}{\partial r} \tag{3.1}$$

Satisfaction of the isotropy relation Eq. 3.1 was tested by computing $\tilde{G}_2(r, \theta)$ from the measured pure longitudinal structure function G_2^L and comparing the transverse $\tilde{G}_2(r, \theta)$ to the one that was actually measured. The result, shown in Fig. 1, illustrates that, whilst the scaling exponents of $G_2(r, \theta)$ and $\tilde{G}_2(r, \theta)$ are the same, the ratio $R(r) \equiv G_2(r, \theta)/\tilde{G}_2(r, \theta)$ deviates appreciably from unity. At $\theta = \pi/2$ the discrepancy is largest, but it is comparable to other work, both numerical [7] and experimental [12]. It should also be kept in mind that our measurement of the purely transverse $G_2(r, \theta = \pi/2)$ does *not* use Taylor's frozen turbulence hypothesis, in contrast with most other experiments. Clearly, the ratio $R(r)$ at $\theta = \pi/2$ is the most faithful measure of anisotropy.

A striking observation is that the generalized structure functions for $\theta > 0$ exhibit better scaling behavior at small separations than the pure longitudinal $G_p^L(r)$. Thus, although the dynamical range of distances r/η spanned

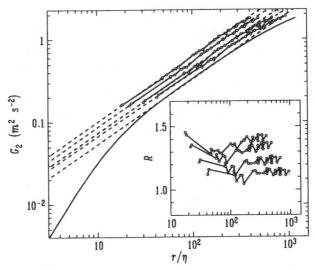

Figure 1: Second-order structure function. Full line, longitudinal $G_2^L(r)$; open circles, generalized structure functions $G_2(r,\theta)$, for $\theta = 90°, 54°, 36°$, and $25°$ from top to bottom, in order. Dashed lines, fit of r^ζ over the inertial range. Inset, ratio $R(r) = G_2(r,\theta)/\tilde{G}_2(r,\theta)$ of measured $G_2(r,\theta)$ structure functions to one computed from G_2^L and the isotropy relation Eq. (3.1). For isotropic turbulence, $R = 1$.

by the probe separations, namely $[20, 400]$, is not very large, it is possible to determine scaling exponents rather well.

An intriguing suggestion, raised in [13], is that neither the longitudinal nor the transverse structure function has pure scaling behavior, but rather the irreducible components of the tensor $S_{ij} = \langle [u_i(\boldsymbol{x} + r_j) - u_i(\boldsymbol{x})]^p \rangle$. Different irreducible components may then have different scaling exponents. In the case of isotropic and incompressible turbulence, those scaling exponents for $p = 2$ are bound to be the same, but this is no longer so in the presence of anisotropy. In [12] it was shown how to unravel $G_2(r,\theta)$ into its irreducible components and indeed two different exponents were found in slightly anisotropic atmospheric turbulence. It was concluded that especially the transverse structure function was contaminated by anisotropy. This contrasts with our observation of clear transverse scaling. In an isotropic velocity field, the second-order irreducible component of S vanishes at the magic angle $\theta_m = 54°$. If each irreducible component would contribute with its own scaling exponent to S, one would expect to see purer scaling at the magic angle. We have included G_2 at this angle and its complement $\theta_m - \pi/2$ in Fig. 1, but nothing special is seen.

Our earlier finding [4] that the transverse scaling exponents $\zeta_p(\theta = \pi/2)$ are

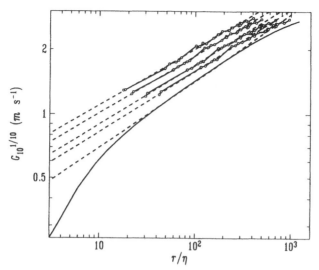

Figure 2: 10th-order structure function. Full line, longitudinal $[G_{10}^L(r)]^{1/10}$.
Open circles, generalized structure functions $[G_{10}(r,\theta)]^{1/10}$, for $\theta = 90°$, $55°$, $36°$, and $25°$ from top to bottom, in order. Dashed lines, fit of r^ζ over the inertial range.

more anomalous is illustrated in Fig. 2 for the tenth-order structure function $G_{10}(r,\theta)$. When $\theta \to 0$, $G_p(r,\theta)$ measured with the probe array approaches the longitudinal structure function $G_p^L(r)$ measured with a single probe, but the scaling exponent remains smaller than the slope of the longitudinal curve. The measured scaling exponents ζ_p^L/p and $\zeta_p(\theta)/p$ are shown in Fig. 3 in a way that most vividly demonstrates the deviation from Kolmogorov's 1941 [1] self-similar prediction (in which case $\zeta_p/p = 1/3$ for all p). For purely transverse structure functions $G_p(r,\theta = \pi/2)$, the velocity increments $\Delta u(y)$ and $\Delta u(-y)$ have identical statistics, and only even moments are non-zero. For all other angles $0 \le \theta < \pi/2$, odd-order moments also exist, but in a measurement with multiple probes these moments have relatively large errors as they are affected by slight systematic errors in the velocity readings that differ from probe to probe. Therefore, all results of Fig. 3 are obtained from moments of the absolute value of the velocity increments. For these structure functions, the scaling exponent $\tilde{\zeta}_3(\theta)$ is not necessarily 1. In fact, we find $\tilde{\zeta}_3(\theta) = 1.13$, for $\theta = 0$ and $\theta = \pi/2$, and we show the normalized exponents $\tilde{\zeta}_p(\theta)/\tilde{\zeta}_3(\theta)$ which are trivially 1 at $p = 3$. Whereas the longitudinal scaling anomaly $\zeta_p(\theta = 0)$ can be compared to the log-Poisson model of She and Lévêque [14], Fig. 3 demonstrates that all transverse exponents are more strongly anomalous.

Of interest is the manner in which the transverse ($\theta \ne 0$) exponents tend to

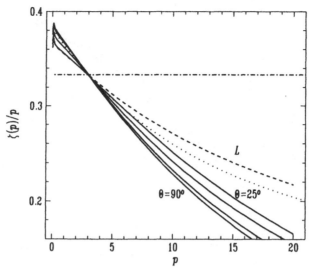

Figure 3: Full lines, scaling exponents of structure functions $G_p(r, \theta)$ for $\theta = 90°, 55°, 36°$, and $25°$, in order. Dashed line, scaling exponent of longitudinal structure function G_p^L. In all cases, moments are computed from absolute values of velocity increments. The measured scaling exponents have been normalized such that $\zeta_3(\theta) = 1$. Dash-dotted line, Kolmogorov's self-similar prediction [1]; dotted line, prediction of log–Poisson model [14].

the longitudinal one if $\theta \rightarrow 0$. Fig. 3 demonstrates that this transition is rather sudden. Even at a small angle $\theta = 25°$, the exponents measured with the probe array are significantly below the longitudinal ($\theta = 0$) exponents which are measured with a single probe. We therefore conclude that the probe array, which cuts a plane through the velocity field, captures stronger intermittency than a point measurement which cuts a line.

4   Strong events

In order to see whether a transverse measurement is indeed more sensitive to the large events which cause intermittency, we collected velocity profiles containing a large velocity difference $|\Delta u|$ across a separation of $y/\eta = 18$ (the smallest separation between the probes). The average over the 200 largest events, corresponding to a cumulative probability of 2×10^{-6}, is shown in Fig. 4a. The average was done by taking the inverse $-[u(x, y) - U]$ of the contributions with $\Delta u < 0$ (otherwise the average would be 0) and choosing the local maximum of Δu at $x = 0$.

A first striking observation is that over a separation of merely 3.7mm, the maximum average velocity increment is 4.3ms^{-1}, which is four times the size of the rms velocity fluctuations and can be compared to the mean velocity

$U = 12.1\text{ms}^{-1}$. A second observation is that Fig. 4a bears a striking resemblance to the velocity profile of a slender vortex tube with a core diameter smaller than the separation $y/\eta = 18$ of the closest probes. In Fig. 4a the signature $y/(x^2 + y^2)$ of an irrotational vortex with its axis pointing in the z-direction can be recognized. Vortex tubes are expected to come with random orientations and with a distribution of strengths and the question is how this distribution can be inferred from the conditional averages shown in Fig. 4a.

The result of a similar quest for large events, but now in the longitudinal x-direction, is shown in Fig. 4b. Large velocity differences $|\Delta u|$ were sought at a longitudinal separation $x/\eta = 24$, and the resulting velocity fields were averaged. Because of the negative skewness of longitudinal velocity increments, the x-coordinate in Fig. 4b is reversed. Compared to the transverse events of Fig. 4a, the maximum average longitudinal increment $\Delta u = 2.1\text{ms}^{-1}$ is a factor of 2 smaller than that of the transverse event. At first sight, Fig. 4 corroborates our hypothesis that the transverse setup is more sensitive to large events than the longitudinal one. At first sight it may also appear surprising that the average velocity profile of the longitudinal event is not of the form $y/(x^2 + y^2)$.

The problem is that conditional averages may be determined by the imposed condition, and not by an indigenous property of turbulence. A test is to perform the same conditional average on pseudo-turbulence, that is a random velocity signal which has the same (low-order) statistical properties as turbulence. Such pseudo-turbulence was constructed by Fourier transforming a measured turbulence signal, randomizing the phases (such that the Fourier transform remains real), and performing the inverse Fourier transform. The resulting velocity field has the same second-order structure function and the same turbulence characteristics as the original turbulence signal, but it is not turbulence [15].

For the transverse case this procedure gave a mean velocity profile similar to that of Fig. 4a, but with a maximum $\Delta u = 2.3\text{ms}^{-1}$, which is a factor 2 smaller than that of Fig. 4a. The pseudo-turbulence longitudinal profile has size $\Delta u = 1.5\text{ms}^{-1}$, which is comparable to that of the true turbulence signal. We therefore conclude that the average transverse velocity profile of Fig. 4a is genuine but that the longitudinal profile of Fig. 4b may just be an artifact of the conditional averaging procedure. Isotropy and incompressibility predict the ratio of velocity increments $[G_2(r, \theta = \pi/2)/G_2^L]^{1/2}$ to be $(4/3)^{1/2}$ in the inertial range and increasing to $2^{1/2}$ in the dissipative range. In our experiment we find it to be 1.5 at $r/\eta = 18$, due to small-scale anisotropy. Also in view of these numbers, the average of the rare transverse velocity increments of Fig. 4a is anomalously large.

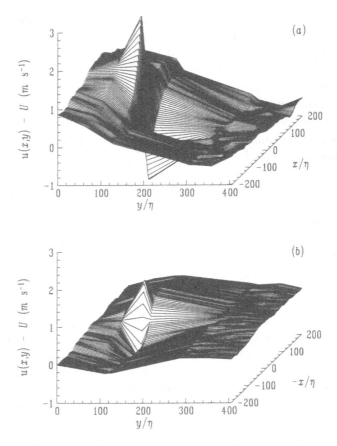

Figure 4: Average velocity profile of the 200 largest events in a time series of 10^8 velocity samples. *(a)* Conditioned on transverse $\Delta u \equiv u(x, y_4) - u(x, y_3)$ with $y_4 - y_3 = 3.7\text{mm} = 18\eta$ and with x shifted to $x = 0$. (b) Conditioned on longitudinal $\Delta u \equiv u(x + 24\eta, y_3) - u(x, y_3)$, with x shifted to $x = 0$. Note that the probes are not spaced equidistantly (which causes distances between them to be spread more evenly on a log axis as shown in Fig. 1).

5   Conclusion

In conclusion, for almost all non-zero angles between the measured velocity component u and the direction of the separation vector r, transverse velocity increments $u(x + r) - u(x)$ are more intermittent than longitudinal ones. This shows in a stronger anomaly of the transverse scaling exponents and is related to the structure of the most violent events in turbulence which come as slender vortex tubes. Further evidence for this is presented in the form of conditional averages of a two-dimensional cut of the turbulent velocity field. The question of anisotropy of scaling exponents, which was first raised in [12, 13], is intimately tied to the structure of the intermittent events. Clearly, higher Reynolds numbers, a more precise control of anisotropy and finer instrumentation are needed to further this interesting problem.

We gratefully acknowledge financial support by the "Nederlandse Organisatie voor Wetenschappelijk Onderzoek (NWO)" and "Stichting Fundamenteel Onderzoek der Materie (FOM)". We thank Gerard Trines and Gerald Oerlemans for technical assistance.

References

[1] A.N. Kolmogorov (1941), *Dokl. Akad. Nauk***26**, 115. This paper is reprinted in *Proc. Roy. Soc. London, Ser. A* **434**, 9–13 (1991).

[2] J. Jiménez, A.A. Wray, P.G. Saffman and R.S. Rogallo (1993), *J. Fluid Mech.* **255**, 65.

[3] S. Douady, Y. Couder and M.E. Brachet (1991), *Phys. Rev. Lett.* **67**, 983.

[4] First experimental evidence was given by J.A. Herweijer and W. van de Water, in *Advances in Turbulence* (ed. R. Benzi), 210–216 Kluwer (1995).

[5] B. Dhruva, B.Y. Tsuji and K.R. Sreenivasan (1997), *Phys. Rev. E* **56**, R4928.

[6] O.M. Boratav and R.B. Pelz (1997), *Phys. Fluids* **9**, 1400.

[7] S. Grossmann, D. Lohse and A. Reeh (1997), *Phys. Fluids* **9**, 3817.

[8] S. Chen, K.R. Sreenivasan, M. Nelkin and N. Cao (1997), *Phys. Rev. Lett.* **79**, 2253.

[9] A. Noullez, G. Wallace, W. Lempert, R.B. Miles and U. Frisch (1997), *J. Fluid Mech.* **339**, 287.

[10] G. He, S. Chen, R. Kraichnan, R. Zhang and Y. Zhou (1998), *Phys. Rev. Lett.* **81**, 4636.

[11] W. van de Water and J.A. Herweijer (1999), *J. Fluid Mech.* **387**, 3–37.

[12] I. Arad, B. Dhruva, S. Kurien, V.S. L'vov, I. Procaccia and K.R. Sreenivasan (1998), *Phys. Rev. Lett.* **81**, 5330.

[13] V. L'vov, E. Podivilow and I. Procaccia (1997), *Phys. Rev. Lett.* **79**, 2050.

[14] Z.-S. She, and E. Lévêque (1994), *Phys. Rev. Lett.* **72**, 336.

[15] A pseudo-turbulent velocity signal $u(x, y)$ was constructed by Fourier-transforming with respect to x (the time-like direction), and then taking the same random phases for each of the probes. This is not a complete randomization as it also left invariant the cross-correlation between any two probes.

The Exit-Time Approach for Lagrangian and Eulerian Turbulence

A. Vulpiani, L. Biferale, G. Boffetta, A. Celani, M. Cencini and D. Vergni

Abstract

Usually, intermittency in turbulence is studied by looking at the scaling of structure functions of different orders versus time and space separation (Eulerian statistics) or by looking at different moments of particle distance versus time (Lagrangian statistics). Here, we discuss an alternative approach in which one analyses the statistical properties of the time necessary to have a fixed velocity fluctuation in the Eulerian case or a fixed distance separation in the Lagrangian case. This method gives good results also at low Reynolds numbers, where the traditional approach for both Eulerian and Lagrangian descriptions is not very accurate. In addition, the approach here proposed is able to catch the statistical properties of laminar fluctuations in turbulent flows.

1 Introduction

The understanding of Eulerian intermittency, i.e. the corrections to the prediction obtained with dimensional arguments according to the Kolmogorov 1941 approach, is one of the main goals of theoretical investigation of fully developed turbulence (Frisch 1995). A related problem is the effect of Eulerian intermittency on the Lagrangian properties, i.e. the corrections to the Richardson law for the relative dispersion (Richardson 1926, Novikov 1989). For a detailed introduction to the statistical mechanics of Eulerian and Lagrangian turbulence see Monin and Yaglom (1975), Frisch (1995), Bohr *et al.* (1998). Eulerian turbulence has attracted most of the attention in the last 20 years, while there are relatively few studies for the corresponding Lagrangian statistics (Novikov 1989, Borgas 1993, Boffetta *et al.* 1999a, Paret *et al.* 1998).

The standard way to characterize Eulerian intermittency is the investigation of the scaling properties of the structure functions, i.e. the moments of the velocity difference as function of the space separation (or time delay according to the Taylor hypothesis, e.g. in the case of one point velocity measurement by an anemometer). Similarly for the Lagrangian aspects one looks at the moments of the relative particle pair distance as a function of time.

The above usual methods work very well in the cases of very extended inertial range. Unfortunately, in both experimental and numerical contexts one has to deal with a limited scale separation which entails several practical difficulties that may originate ambiguous results.

In the last few years, in order to characterize the non-asymptotic properties of transport in realistic systems, e.g. closed basins where the typical Eulerian length is not very small compared with the domain size, an alternative approach has been proposed (Artale *et al.* 1997, Boffetta *et al.* 1999a). The basic idea is to fix a certain separation for a particle pair and to analyse the statistical properties of the time necessary for doubling its separation. The fixed scale method is able to give the proper description also in non-ideal cases, while whenever large scale separations are involved it coincides with the usual analysis.

In Section 2 we introduce the concept of Finite Size Lyapunov Exponent and we show its application to relative diffusion in finite domains and for fully developed turbulence with the aid of synthetic velocity fields.

Section 3 is devoted to the investigation of the statistical properties of the time necessary to have a fixed velocity fluctuation of a turbulent signal, i.e. a sort of inverse structure function (Jensen 1999). This allows for an unambiguous detection of the intermediate dissipative range (Biferale *et al.* 1999).

For the sake of completeness in Appendix A we report the basic elements of the multifractal model. Appendix B contains some details about the Finite Size Lyapunov Exponent.

2 Exit times for Lagrangian dynamics

2.1 Finite Size Lyapunov Exponent

Understanding the statistics of particle pair dispersion is of fundamental interest in Lagrangian turbulence. At variance with absolute (one particle) dispersion, which is dominated by large scale flow, relative (two particle) dispersion is driven by the local velocity difference. Relative dispersion thus gives information on the velocity field structure at different scales. Nevertheless the reconstruction of the Eulerian properties from Lagrangian measurements is not a simple task (Monin and Yaglom 1975). This is due to the fact that, even in the presence of a simple Eulerian velocity field, Lagrangian trajectories can display a very complex behaviour due to the Lagrangian Chaos phenomenon (Ottino 1989).

By definition, chaotic motion means exponential separation of close trajectories. Therefore, in the presence of Lagrangian chaos we expect that the relative separation between advected tracers, $R(t)$, typically grows exponentially in time. The exponential regime is observed only for separation $R \ll \eta$, where η is the characteristic scale of the smallest Eulerian structures (i.e.

dissipative eddies in turbulence). For 3D homogeneous fully developed turbulence in the inertial range, $\eta \ll R(t) \ll L_0$ (where L_0 is the typical scale of energy injection), one has the Richardson law, $\overline{R^2(t)} \propto t^3$. Hereafter $\overline{(\cdots)}$ indicates the average over many particle pairs. For very large separations ($R \gg L_0$), the behaviour of $R(t)$ depends on the structure of the velocity field and one has the usual diffusive behaviour $\overline{R^2(t)} \sim Dt$, where D is the diffusion coefficient.

In real settings, e.g. relatively small inertial range, the standard analysis at fixed delay time, i.e. to look at $\overline{R^2(t)}$ as a function of t, can lead to ambiguous results. The possibility of having at the same time particle pairs in ranges of scales having different dynamical and statistical properties, e.g. in turbulence in the dissipative range and in the inertial one (or in the inertial range and in the diffusive one), can produce long crossover effects which can be wrongly interpreted (Artale *et al.* 1997).

In the last few years, in order to overcome such difficulties we have developed an alternative approach that is based on the study of the fixed-scale instead of the fixed-time statistics. In particular, we have introduced an indicator, the Finite Size Lyapunov Exponent (FSLE), which is an extension of the Lyapunov exponent by computing averaged quantities at fixed scale (Aurell *et al.* 1996, 1997). Let us now recall the basic idea.

We define a series of thresholds $R_n = \rho^n R_0$, $(n = 1, ..., N)$ and we measure the "doubling times", $T(R_n)$, it takes for the two particles' separation, initially of size $R(0) \leq R_0$, to grow from R_n to R_{n+1}, until it reaches the largest scale under consideration R_N. The threshold rate, ρ, has to be larger than 1. However, we choose ρ not too large in order to avoid contributions coming from different scales.

Performing $\mathcal{N} \gg 1$ experiments with different initial conditions, we define the Finite Size Lyapunov Exponent as

$$\lambda(R) = \frac{1}{\langle T(R) \rangle_e} \log \rho, \qquad (2.1)$$

where $\langle (\ldots) \rangle_e$ is the average on the doubling time experiments, which is different from the time average over a single realization (Appendix B). For very small separation (i.e. $R \ll \eta$) from (2.1) one recovers the standard Lagrangian Lyapunov exponent, i.e.

$$\lambda = \lim_{R \to 0} \lambda(R). \qquad (2.2)$$

See Aurell *et al.* (1996, 1997) for a detailed discussion about these points. At very large scales, if there is a standard diffusive regime ($\overline{R^2(t)} \sim Dt$), one has

$$\lambda(R) \approx \frac{D}{R^2}. \qquad (2.3)$$

Moreover, if the flow is turbulent, in the inertial range, between the two regimes (2.2) and (2.3) one has the Richardson law, i.e.

$$\lambda(R) \sim R^{-2/3}. \qquad (2.4)$$

The fixed scale analysis allows us to extract physical information at different spatial scales avoiding some unpleasant consequences resulting from working at a fixed delay time t, see Artale *et al.* (1997).

The FSLE analysis has been demonstrated to be a useful tool for the analysis of Lagrangian data in several situations. For example, in experiments on particle dispersion in closed basins of size L_B, the diffusive behaviour is observable only if $\eta \ll L_B$. Of course, this is not always the case (Boffetta *et al.* 1999b).

In the following we will describe the application of the FSLE analysis to the study of relative dispersion in turbulent flow and to the analysis of experimental data in a non-turbulent flow, in which complex Lagrangian trajectories appear due to chaotic advection.

2.2   Dispersion in synthetic turbulent fields

We consider now the relative dispersion of particle pairs advected by an incompressible, homogeneous, isotropic, fully developed turbulent field. The Eulerian statistics of velocity differences are characterised by the Kolmogorov scaling $\delta v(R) \sim R^{1/3}$, in the inertial range ($\eta \ll R \ll L_0$). Due to incompressibility, particles will typically diffuse away from each other (Cocke 1969, Orszag 1970). For separations less than η we have exponential separation of trajectories, whereas at separations larger than L_0 normal diffusion takes place. In the inertial range the average pair separation is not affected either by large scale components of the flow, which simply sweep the pair, or by small scale ones, which act incoherently and with low intensity. Accordingly, the separation $R(t)$ feels mainly the action of velocity differences $\delta v(R(t))$ at scale R. As a consequence of the Kolmogorov scaling the separation grows with the *Richardson law* (Richardson 1926, Monin and Yaglom 1975)

$$\overline{R^2(t)} \sim t^3 \,. \tag{2.5}$$

However, some unclear behaviours are observed when the Reynolds number is not large enough. As a matter of fact, even at very high Reynolds numbers, the inertial range is still too small to observe the scaling (2.5) without any ambiguity (Fung *et al.* 1992). On the other hand, we shall show that FSLE statistics are effective already at relatively small Reynolds numbers.

In order to investigate the problem of relative dispersion in turbulent flows a practical tool is the use of synthetic turbulent fields (Elliott and Majda 1996, Fung *et al.* 1992, Fung and Vassilicos 1998). In fact, by means of stochastic processes it is possible to build a velocity field which reproduces some of the statistical properties of velocity differences observed in real fully developed turbulence (Biferale *et al.* 1998). To overcome the difficulties related to the sweeping, we limit ourselves to a correct representation of two-point velocity differences. In this case, if one adopts the reference frame in which one of the two tracers is at rest at the origin (the so-called quasi-Lagrangian frame of reference), then the motion of the second particle is ruled by the velocity difference in this frame of reference, which has the same single time statistics as the Eulerian velocity differences (L'vov *et al.* 1997).

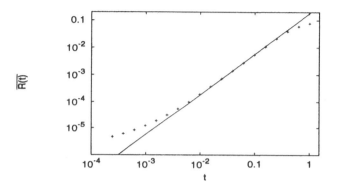

Figure 1: Relative dispersion $\overline{R(t)}$ versus time. Synthetic turbulent simulation averaged over 10^4 realizations. The line is the theoretical Richardson scaling $t^{3/2}$.

In Figure 1 we show the results of numerical simulations of pair dispersion in a synthetic turbulent field with the Kolmogorov scaling (i.e. with no intermittency) at Reynolds number Re $\simeq 10^6$ (Boffetta *et al.* 1999a). The expected power law (2.5) can be observed without ambiguity only for huge Reynolds numbers. To explain the depletion of scaling range for the relative dispersion, let us consider a series of experiments, in which a pair of particles is released at a separation R_0 at time $t = 0$. At a fixed time t, as customarily is done, we perform an average over all different experiments to compute $\overline{R^2(t)}$. But, unless t is large enough that all particle pairs have "forgotten" their initial conditions, the average will be biased. This is at the origin of the flattening of $\overline{R^2(t)}$ for small times, which we can call a crossover from initial condition to self similarity. In an analogous fashion there is a crossover for large times (of the order of the integral time-scale) since some pairs might have reached a separation larger than the integral scale, and thus diffuse normally, meanwhile other pairs still lie within the inertial range, biasing the average and, again, flattening the curve $\overline{R^2(t)}$. This correction to a pure power law is far from being negligible in experimental data, where the inertial range is generally limited due to the relatively small value of the Reynolds number and the experimental apparatus. For example, Fung and Vassilicos (1998) and Fung *et al.* (1992) show quite clearly the difficulties that may arise even in numerical simulations with the standard approach.

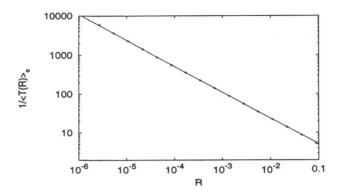

Figure 2: Average inverse doubling time $1/\langle T(R)\rangle_e$ for the same simulation as in Figure 1. Observe the enhanced scaling region. The line is the theoretical Richardson scaling $R^{-2/3}$.

To overcome these difficulties we exploit the approach based on the fixed scale statistics. The outstanding advantage of averaging at a fixed scale separation is that it removes all crossover effects, since all sampled pairs belong to the inertial range. The expected scaling properties of the doubling times are obtained by a simple dimensional argument. The time it takes for particle separation to grow from R to $2R$ can be estimated as $T(R) \sim R/\delta v(R)$; we thus expect the scaling

$$\frac{1}{\langle T(R)\rangle_e} \sim R^{-2/3}. \tag{2.6}$$

In Figure 2 the great enhancement of the scaling range achieved by using the doubling times is evident. In addition, by using the FSLE it is possible to study in detail the effect of Eulerian intermittency on the Lagrangian statistics of relative dispersion. See Boffetta *et al.* (1999a) for a detailed discussion in the framework of the multifractal model.

In conclusion we have that in this case doubling time statistics allows us a much better estimate of the scaling exponents with respect to the standard, fixed time, statistics.

2.3 Analysis of experimental data using the FSLE

Let us now discuss the use of the FSLE for the analysis of experimental Lagrangian data in a convective flow (Boffetta *et al.* 1999b). The experimental apparatus is a rectangular convective tank $L = 15.0$cm wide, 10.4cm deep and $H = 6.0$cm high filled with water. The upper and lower surfaces are kept at constant temperature and the side walls can be considered as adiabatic.

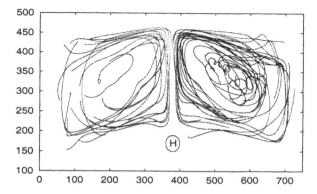

Figure 3: An example of trajectories reconstructed by the PVT technique (pixel units). The circle on the bottom represents the heater.

Convection is generated by an electrical circular heater, 0.8cm in diameter, located in the mid-line of the tank, just above the lower surface (see Figure 3). The heater furnish a constant heat flux controlled by a feedback on the power supply.

The control parameter of the experiment is the Rayleigh number, Ra, which varies over a wide range of values. The geometrical configuration constrains the convective pattern to two counter-rotating rolls divided by an oscillating thermal plume (Miozzi *et al.* 1998). The Eulerian velocity field is thus, basically, two-dimensional and time periodic.

Lagrangian data are obtained by the Particle Tracking Velocimetry (PTV) technique. Figure 3 shows an example of the output of the PTV analysis.

We report the result of the FSLE analysis applied to six different runs at different Rayleigh numbers. Each run consists of 22500 frames containing 900 simultaneous trajectories on average. In Figure 4 we show the FSLE for the run at $Ra = 2.39 \times 10^8$. In order to increase the statistics at large separations we performed the analysis with different initial thresholds R_0. For small R, we observe the collapse of $\lambda(R)$ to the value of the Lyapunov exponent, independent of R_0.

For larger separation $\lambda(R)$ decreases to smaller values, indicating a slowing down in the separation growth due to the presence of boundaries. The behaviour of $\lambda(R)$ is well described by assuming exponential relaxation of $R(t)$ to the saturation value $R_{\max}$ (uniform distribution in the tank). The prediction is (Artale *et al.* 1997)

$$\lambda(R) \simeq \frac{1}{\tau_R} \frac{R_{\max} - R}{R}, \tag{2.7}$$

where τ_R is the characteristic relaxation time (see Appendix B).

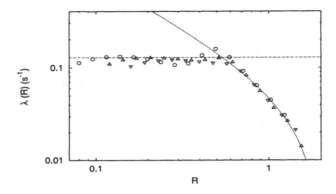

Figure 4: $\lambda(R)$ versus R for the run at Ra $= 2.39 \times 10^8$ and different initial thresholds $R_0 = 0.4$cm (o), $R_0 = 0.6$cm ($\triangle$) and $R_0 = 0.8$cm ($\triangledown$). The straight line is the Lyapunov exponent $\lambda = (0.12 \pm 0.01\text{s}^{-1}$ and the curve represents the saturation regime (2.7).

We observe that in this experiment we cannot expect to observe diffusive behaviour because the size of Eulerian structure is of the same order as the tank size. Let us conclude by observing that in this case the FSLE analysis has given clear evidence of Lagrangian chaos (i.e. $\lambda > 0$) and, moreover, it has given a good description of the separation evolution on all the observable scales.

3 Exit times for turbulent signals and the intermediate dissipative range

The most studied statistical indicators for intermittency in homogeneous isotropic turbulence are the longitudinal structure functions, i.e. moments of the velocity increments at distance R in the direction of $\hat{\mathbf{R}}$:

$$S_p(R) = \left\langle \left[(\mathbf{v}(\mathbf{x} + \mathbf{R}) - \mathbf{v}(\mathbf{x})) \cdot \hat{\mathbf{R}} \right]^p \right\rangle . \tag{3.1}$$

Basically in typical experiments one is forced to analyse a one-dimensional string of data, e.g. the output of a hot-wire anemometer. In such cases the Taylor Frozen-Turbulence Hypothesis is used to link measurements in space with measurements in time. Within the Taylor Hypothesis, one has the large-scale typical time, $T_0 = L_0/U_0$, and the dissipative time, $t_d = \eta/U_0$, where U_0 is the large scale velocity field at the scale of the energy injection. As a function of time increment, τ, structure functions (3.1) assume the form $S_p(\tau) = \langle [(v(t+\tau) - v(t)]^p \rangle$. It is well known that in the inertial range, $\tau_d \ll$

$\tau \ll T_0$, the structure functions develop an anomalous scaling behaviour, $S_p(\tau) \sim \tau^{\zeta(p)}$, where $\zeta(p)$ is a nonlinear function, while inside the dissipative range, $\tau \ll \tau_d$, they show the laminar scaling $S_p(\tau) \sim \tau^p$.

Beside the huge amount of theoretical, experimental and numerical studies devoted to the understanding of velocity fluctuations in the inertial range (Frisch 1995, Bohr *et al.* 1998), only a few –mainly theoretical– attempts have focused on the Intermediate Dissipation Range (IDR), we just mention Frisch and Vergassola (1991), Jensen *et al.* (1991), Gagne and Castaing (1991), L'vov *et al.* (1997), Benzi *et al.* (1999). By IDR we mean the range of scales, $\tau \sim \tau_d$, between the inertial and the dissipative range (see Appendix A).

The very existence of the IDR is relevant for the understanding of many theoretical and practical issues. Among them we cite: the modelizations of small scales for optimizing Large Eddy Simulations and the validity of the Refined Kolmogorov Hypothesis, i.e. the link between inertial and dissipative statistics.

Non-trivial IDR properties are connected to intermittent fluctuations in the inertial range. That is, anomalous scaling laws can be explained by assuming that velocity fluctuations in the inertial range are characterized by a spectrum of different local scaling exponents $\delta_\tau v = v(t + \tau) - v(t) \sim \tau^h$ with the probability of observing at scale τ a value h given by $P_\tau(h) \sim \tau^{3-D(h)}$. This is the celebrated multifractal picture of the energy cascade (see Appendix A) which has been confirmed by many independent experiments (Frisch 1995, Bohr *et al.* 1998). Non-trivial dissipative statistics are generated by fluctuations of the dissipative cut-off τ_d (see Appendix A):

$$\tau_d(h) \sim \nu^{1/(1+h)} . \tag{3.2}$$

Here we present the measurement in both experimental and synthetic data of a set of observables which are able to highlight the IDR properties. The main idea is to take a one-dimensional string of turbulent data, $v(t)$, and to analyze the statistical properties of the exit times from a set of defined velocity thresholds. Roughly speaking we are working with a kind of Inverse Structure Function (Jensen 1999). This approach is rather naturally related to the fixed scale method discussed in Section 2 for studying the particle separation statistics.

This analysis allows us to give the first clear evidence of non-trivial intermittent fluctuations of the dissipative cut-off in turbulent signals.

Fluctuations of viscous cut-off are particularly important for all those regions in the fluid where the velocity field is locally smooth, i.e. the local fluctuating Reynolds number is small. In this case, the matching between non-linear and viscous terms happens at scales much larger than the Kolmogorov scale, $\tau_d \sim \nu^{-3/4}$. It is natural, therefore, to look for observables which detect mainly laminar events. A possible choice is to measure the exit-time moments through a set of velocity thresholds. More precisely, given a reference initial time t_0 with velocity $v(t_0)$, we define $\tau(\delta v)$ as the first time necessary to have an absolute variation equal to δv in the velocity data, i.e. $|v(t_0) - v(t_0 + \tau(\delta v))| = \delta v$. By scanning the whole time series we recover the probability density functions of $\tau(\delta v)$ at varying δv from the typical large scale values down to the smallest dissipative values. Positive moments of $\tau(\delta v)$ are dominated by events with a smooth velocity field, i.e. laminar bursts in the turbulent cascade. Let us define the Inverse Structure Functions (Inverse-SF) as

$$\Sigma_p(\delta v) \equiv \langle \tau^p(\delta v) \rangle . \tag{3.3}$$

It is necessary to perform the average over the time-statistics in a weighted way. This is due to the fact that by looking at the exit-time statistics we are not sampling the time series uniformly, i.e. the higher the value of $\tau(\delta v)$ is, the longer it is detectable in the time series.

It is easy to realize (Aurell *et al.* 1996, 1997 and Appendix B) that the sequential time average of any observable based on exit-time statistics, $\langle \tau^p(\delta v) \rangle_e$, is connected to the uniformly-in-time multifractal average by the relation

$$\langle \tau^p(\delta v) \rangle = \frac{\langle \tau^{p+1} \rangle_e}{\langle \tau \rangle_e} . \tag{3.4}$$

According to the multifractal description we suppose that, for velocity thresholds corresponding to inertial range values of the velocity differences, $\delta_{\tau_d} v \equiv v_m \ll \delta v \ll v_M \equiv \delta_{T_0} v$, the following dimensional relation is valid:

$$\delta_\tau v \sim \tau^h \quad \rightarrow \quad \tau(\delta v) \sim \delta v^{1/h} ,$$

where the probability of observing a value τ for the exit time is given by inverting the multifractal probability, i.e. $P(\tau \sim \delta v^{1/h}) \sim \delta v^{[3-D(h)]/h}$. Given this ansatz, the prediction for the Inverse-SF, $\Sigma_p(\delta v)$, evaluated for velocity thresholds within the inertial range is

$$\Sigma_p(\delta v) \sim \int_{h_{min}}^{h_{max}} dh \; \delta v^{[p+3-D(h)]/h} \sim \delta v^{\chi(p)} \tag{3.5}$$

where the RHS has been obtained by a saddle-point estimate:

$$\chi(p) = \min_h \left\{ [p + 3 - D(h)]/h \right\} . \tag{3.6}$$

Let us now consider the IDR properties.

For each p, the saddle-point evaluation (3.6) selects a particular $h = h_s(p)$ where the minimum is reached. Let us also remark that from (3.2) we have an estimate for the minimum value assumed by the velocity in the inertial range given a certain singularity h: $v_{min}(h) = \delta_{\tau_d(h)} v \sim \nu^{h/(1+h)}$. Therefore, the smallest velocity value at which the scaling (3.5) still holds depends on both ν and h. That is, $\delta v_m(p) \sim \nu^{h_s(p)/1+h_s(p)}$. The most important consequence is that for $\delta v < \delta v_{\mathrm{m}}(p)$ the integral (3.5) is not any more dominated by the saddle-point value but by the maximum h-value still dynamically alive at that velocity difference, $1/h(\delta v) = -1 - \log(\nu)/\log(\delta v)$. This leads for $\delta v < \delta v_{\min}(p)$ to a pseudo-algebraic law

$$\Sigma_p(\delta v) \sim \delta v^{[p+3-D(h(\delta v))]/h(\delta v)} . \tag{3.7}$$

The presence of this p-dependent velocity range, intermediate between the inertial range, $\Sigma_p(\delta v) \sim \delta v^{\chi(p)}$, and the dissipative scaling, $\Sigma_p(\delta v) \sim \delta v^p$, is the IDR signature.

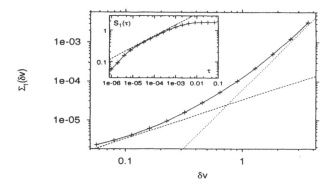

Figure 5: Inverse Structure Functions $\Sigma_1(\delta v)$. The straight lines show the dissipative range behaviour (dashed) $\Sigma_1(\delta v) \sim \delta v$, and the inertial range non intermittent behaviour (dotted) $\Sigma_1(\delta v) \sim (\delta v)^3$. The inset shows the direct structure function $S_1(\tau)$ with superimposed the intermittent slope $\zeta(1) = .39$.

Then, it is easy to show that Inverse-SF should display an enlarged IDR. Indeed, for the usual *direct* structure functions the saddle-point $h_s(p)$ value is reached for $h < 1/3$. This pushes the IDR to a range of scales very difficult to observe experimentally (Gagne and Castaing 1991). On the other hand, as regards the Inverse-SF, the saddle-point estimate of positive moments is always reached for $h_s(p) > 1/3$. This is an indication that we are probing the laminar part of the velocity statistics. Therefore, the presence of the IDR must be felt much earlier in the range of available velocity fluctuations. Indeed, if $h_s(p) > 1/3$, the typical velocity field at which the IDR shows up is given by $\delta v_{\mathrm{m}}(p) \sim \nu^{h_s(p)/(1+h_s(p))}$, that is much larger than the Kolmogorov value $\delta v_{\tau_d} \sim \nu^{1/4}$.

In Figure 5 we plot $\Sigma_1(\delta v)$ evaluated on a string of high Reynolds number experimental data as a function of the available range of velocity thresholds

δv. This data set has been measured in a wind tunnel at $Re_\lambda \sim 2000$. One can see that the scaling is very poor. Indeed, it is not possible to extract any quantitative prediction about the inertial range slope. For this reason, we have only drawn the dimensional non-intermittent slope and the dissipative slope as possible qualitative references. On the other hand (inset of Figure 5), the scaling behaviour of the direct structure functions $\langle|\delta v(\tau)|\rangle \sim \tau^{\zeta(1)}$ is quite clear in a wide range of scales. This is a clear evidence of IDR's contamination in the whole range of available velocity values for the Inverse-SF cases. Similar results (not shown) are found for higher order Σ_p structure functions.

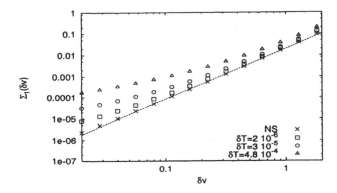

Figure 6: Inverse Structure Function $\Sigma_1(\delta v)$ versus δv for the synthetic signals not smoothed (NS) and smoothed with time windows: $\delta T = 4.8 \times 10^{-4}, 3 \times 10^{-5}, 2 \times 10^{-6}$, the straight line is obtained from the inverse multifractal prediction (3.6).

In order to better understand the scaling properties of $\Sigma_p(\delta v)$ we investigate a synthetic multiaffine field obtained by combining successive multiplications of Langevin dynamics (Biferale *et al.* 1998). The advantage of using a synthetic field is that one can control analytically the scaling properties of direct structure functions in order to have the same scaling laws observed in experimental data. An IDR can be introduced in the synthetic signals by smoothing the original dynamics on a moving time window of size δT. Imposing a smoothing time window is equivalent to fixing the Reynolds number, $Re \sim \delta T^{-4/3}$. The purpose of introducing this stochastic multi-affine field is twofold. First, we want to reach Reynolds numbers high enough to test the *inverse* multifractal formula (3.6). Second, we want to test that the very extended IDR observed in the experimental data, see Figure 5, is also observed in this stochastic field. This would support the claim that the experimental result is the evidence of an extended IDR.

In Figure 6 we show the Inverse-SF, $\Sigma_1(\delta v)$, measured in the multiaffine synthetic signal at high Reynolds numbers. The observed scaling exponent, $\chi(1)$, is in agreement with the prediction (3.6). The same agreement also holds for higher moments.

In Table 1, we compare the best fit to the $\Sigma_p(\delta v)$ measured on the synthetic field with the inversion formula (3.6). As for the comparison between the theoretical expectation (3.6) and the synthetic data let us note the following points. First, in Biferale *et al.* (1998), it was proved that the signal possesses the correct direct structure function exponents for positive moments, i.e. the $\zeta(p)$ exponents are in a one-to-one correspondence with the $D(h)$ curve for $h < 1/3$. Nothing was proved for observables that detect the $h > 1/3$ interval. Therefore the agreement between the inversion formula (3.6) and the numerical results cannot be found analytically. Second, because the synthetic signal is defined by using Langevin processes, the least singular h-exponent expected to contribute to the saddle point (3.6) is $h = 0.5$. Therefore, the theoretical prediction, $\chi_{\text{th}}(q)$, in Table 1 has been obtained by imposing $h_{\max} = 0.5$.

Let us now go back to the most interesting question about the statistical properties of the IDR. In order to study this question we have smoothed the stochastic field, $v(t)$, by performing a running-time average over a time window, δT. Then we compare Inverse-SF obtained for different Reynolds numbers, i.e. for different dissipative cut-offs: $\text{Re} \sim \delta T^{-4/3}$.

The expression (3.7) predicts the possibility of obtaining a data collapse of all curves with different Reynolds numbers by rescaling the Inverse-SF as follows (see Frisch and Vergassola (1991) and Jensen *et al.* (1991))

$$-\log(\Sigma_p(\delta v))/\log(\delta T/\delta T_0) \quad \text{vs.} \quad -\log(\delta v/U)/\log(\delta T/\delta T_0)\,, \qquad (3.8)$$

where U and δT_0 are adjustable dimensional parameters.

Within the same experimental (or synthetic) set-up they are Reynolds number independent (i.e. δT independent).

The rationale for the rescaling (3.8) stems from the observation that, in the IDR, $h_s(p)$ is a function of $\log(\delta v)/\log(\nu)$ only. Therefore, identifying $\text{Re} \propto \nu^{-1}$, the relation (3.8) directly follows from (3.7). This rescaling was originally proposed as a possible test of IDR for direct structure functions in Frisch and Vergassola (1991), but, as already discussed above, for the latter observable it is very difficult to detect any IDR due to the extremely small scales involved, as in Gagne and Castaing (1991).

Figure 7 shows the rescaling (3.8) of the Inverse-SF, $\Sigma_1(\delta v)$, both for the synthetic field at different Reynolds numbers and for the experimental signals. As one can see, the data collapse is very good. This is clear evidence that the poor scaling range observed in Figure 5 for the experimental signal can be explained as the signature of the IDR. The same behaviour holds for higher moments (not shown). It is interesting to remark that for a self-affine signal $(D(h) = \delta(h - 1/3))$, the IDR is greatly reduced and the Inverse-SF, scaling trivially as $\Sigma_p(\delta v) \sim (\delta v)^{3p}$, do not bring any new information.

p	1	2	3	4	5
$\chi_{\text{syn}}(p)$	2.32(4)	4.40(8)	6.38(8)	8.3(1)	10.1(2)
$\chi_{\text{th}}(p)$	2.32	4.34	6.34	8.35	10.35

Table 1: Comparison between the Inverse-SF scaling exponents $\chi_{\text{syn}}(p)$ measured in the synthetic signal and the inversion of the theoretical multifractal prediction (3.6), $\chi_{\text{th}}(p)$. The synthetic signal has been defined such that the $D(h)$ function leads to the same set of experimental $\zeta(p)$ exponents for the direct structure functions.

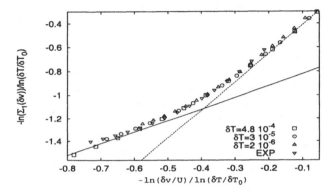

Figure 7: Data collapse of the Inverse-SF, $\Sigma_1(\delta v)$, obtained by the rescaling (3.8) for the smoothed synthetic signals (with time windows: $\delta T = 4.8 \times 10^{-4}$, 3×10^{-5}, 2×10^{-6}) and the experimental data (EXP). The two straight lines have the dissipative (solid line) and the inertial range (dashed) slope.

4   Concluding remarks

A new approach to looking at Lagrangian and Eulerian statistical properties of turbulent flows has been reviewed. The main idea consists in looking at *inverse* scaling properties, i.e. looking at the time necessary to reach a certain (fixed) separation between advected particles instead of fixing the time and looking at the separation distribution for Lagrangian aspects; or, for Eulerian aspects, fixing the velocity increments and looking at the distribution of spatial separations where fluctuations are detected.

In both cases some new and/or better statistical properties have been detected. First, for the Lagrangian properties, we have shown that the inverse statistics, based on the definition of 'doubling time' dramatically improve the scaling properties characterizing the statistics of multipoint particle correlations in both ideal turbulent flows and realistic domain-bounded flows. Second, for the Eulerian properties, the inverse statistics based on the concept of Inverse Structure Functions have revealed themselves to be able to highlight the properties of the Intermediate Dissipative Range, a range of scales almost inaccessible by the usual direct Structure Functions.

Many questions are still open. As far as the Lagrangian approach is concerned, one can think of extending the preliminary attempts here presented to investigate realistic flows in a variety of different cases, going from geophysical flows to flows in closed laboratories. A sensible improvement in the quality of data analysis must certainly be expected.

In the Eulerian case, the analysis of a wider set of experimental data could make it possible to quantify the agreement of the data collapse with the prediction based on (3.2) and (3.7). Indeed, it is easy to realize that, by using different parameterization for the onset of the viscous range, one would have predicted the existence of an extended IDR for $\Sigma_p(\delta v)$ but with a slightly different rescaling procedure (Benzi *et al.* 1996). The quality of experimental data available to us is not high enough to distinguish between the two different predictions. Analyzing different experimental data sets, at different Reynolds numbers, could also make it possible to better explore $D(h)$ for $h > 1/3$ and thus to enquire into possible non-universalities of these $D(h)$ values. For example, as discussed above, in the Langevin synthetic data a good agreement between the multifractal prediction and the numerical data is obtained by imposing $h_{max} = 0.5$, similarly in true turbulent data other h_{max} values could appear depending on the physical mechanism driving the energy transfer at large scales.

Acknowledgements

We thank U. Frisch, M.H. Jensen and M. Vergassola for useful suggestions and discussions. This work has been partly supported by INFM (*PRA-TURBO*) and by the European Network *Intermittency in Turbulent Systems* (contract number FMRX-CT98-0175).

A   Multifractal Model for Turbulence

In the Kolmogorov (1941) theory of fully developed turbulence a global invariance with a fixed exponent $h_{k41} = 1/3$, is assumed:

$$|(\mathbf{v}(\mathbf{x} + \mathbf{R}, t) - \mathbf{v}(\mathbf{x}, t)) \cdot \mathbf{R}| \equiv \delta v_R(\mathbf{x}, t) \sim R^{h_{k41}}. \tag{A.1}$$

Relaxing this restrictive hypothesis, in the framework of the multifractal model (Frisch 1995) one assumes that the velocity field possesses a local scale invariance with a continuous spectrum of exponents, each of which belongs to a given fractal set, Ω_h, with dimension $D(h)$

$$\delta v_R(\mathbf{x}, t) \sim R^h \quad \text{for } \mathbf{x} \in \Omega_h \tag{A.2}$$

where $h \in (h_{\min}, h_{\max})$.

The probability of having a given scaling exponent h at the scale R is $P_R(h) \sim R^{3-D(h)}$, so the scaling of the structure function assumes the form

$$S_p(R) = \langle (\delta v_R)^p \rangle \propto \int_{h_{\min}}^{h_{\max}} R^{hp} R^{3-D(h)} \mu(h) \mathrm{d}h \sim R^{\zeta(p)}, \qquad (A.3)$$

where $\mu(h)$ is a smooth function independent of R. If $R \ll 1$, using a saddle-point estimate one obtains

$$\zeta(p) \simeq \min_h \{hp + 3 - D(h)\} = h^* p + 3 - D(h^*), \qquad (A.4)$$

where h^* is a solution of the equation $D'(h^*(p)) = p$ and $D''(h^*(p)) < 0$. The Kolmogorov four-fifths law implies $\zeta(3) = \min_h \{hp + 3 - D(h)\} = 1$. For the other moments, $\zeta(p)$, depends on the shape of $D(h)$, which cannot be obtained with simple arguments.

Frisch and Vergassola (1991) have shown that, in the context of the multifractal model of fully developed turbulence, there exists a non-trivial behaviour of the structure functions in the region between the dissipative and the inertial range. In this range of scale, called the Intermediate Dissipative Range (IDR), the structure functions show a pseudo-algebraic scaling law, due to the fluctuations of the viscous cut-off. For a given scaling exponent h we can find the viscous cut-off, $\eta(h)$, for which the local Reynolds number is of the order of unity:

$$\mathrm{Re}(R) = \frac{R \delta v_R}{\nu} \sim O(1) \Rightarrow \eta(h) \sim \nu^{\frac{1}{(1+h)}}. \qquad (A.5)$$

Taking into account the fluctuations of the cut-off the formula (A.3) assumes the form

$$S_p(R) \sim \int_{h_{\min}}^{h(R)} R^{hp+3-D(h)} \mu(h) \mathrm{d}h \qquad (A.6)$$

and therefore

$$S_p(R) \propto \begin{cases} R^{\zeta(p)} & \text{if } R \gg \eta^* \\ R^{h(R)p+3-D(h(R))} & \text{if } \eta^* \gg R \gg \eta_{\min} \end{cases} \qquad (A.7)$$

where η^* is the length scale related to the $h^*(p)$ (given by the saddle-point evaluation) via $\eta^* = \nu^{\frac{1}{(1+h^*(p))}}$, $h(R)$ is the maximum exponent still present at the scale R for $\eta^* \gg R \gg \eta_{\min}$ given by $\nu^{\frac{1}{(1+h(R))}} = R$ and η_{min} is the viscous cut-off related to the strongest singularity $h_{\min}$. If the saddle-point estimation, $h^*(p)$, is between $h(R)$ and $h_{\min}$ then we have the usual scaling relation. Conversely, if $h(R) > h^*(p)$ we are in the intermediate dissipative range and we have the pseudo-algebraic scaling law.

A simple calculation shows that one has a universal function both in the inertial and in the IDR by introducing the multiscaling transformation. Let us discuss this for the energy spectrum $E(k) \sim k^{-1}S_2(R = k^{-1})$. Introducing the rescaled variables

$$F(\theta) = \frac{\log E(k)}{\log(1/\nu)} \text{ and } \theta = \frac{\log k}{\log(1/\nu)} \tag{A.8}$$

from (A.7) one obtains

$$F(\theta) \propto \begin{cases} -(1 + \zeta(2))\theta & \text{in the inertial range} \\ -2 - 2\theta + \theta D(h = -1 + 1/\theta) & \text{in the IDR.} \end{cases} \tag{A.9}$$

The presence of the multiscaling behaviour was first experimentally verified by Gagne and Castaing (1991). However, since for $S_p(R)$ the extension of the IDR is very small, it is extremely difficult to have a strong evidence.

B   Finite Size Lyapunov Exponent

In this appendix we discuss the practical method for computing the Finite Size Lyapunov Exponent. For a norm defined or the distance $R(t)$ between two trajectories (two particles for relative dispersion), one defines a series of thresholds $R_n = \rho^n R_0$ ($n = 1, \ldots, N$). Then one measures the time $T_\rho(R_n)$ that a perturbation of size R_n takes to grow up to R_{n+1}. The threshold rate ρ should not be taken too large to avoid growth on different scales. On the other hand, ρ cannot be too close to 1, otherwise $T_\rho(R_n)$ would be of the order of the time step in the integration. Typically, one uses $\rho = 2$ or $\rho = \sqrt{2}$. For simplicity T_ρ is called "doubling time" even if $\rho \neq 2$.

The doubling times $T_\rho(R_n)$ are measured by following the evolution of the trajectory's separation from its initial value $R_{min} \ll R_0$ up to the largest threshold R_N. One must choose $R_{min} \ll R_0$ in order to allow the direction of the initial perturbation to align with the most unstable direction in the phase-space. Moreover, one must pay attention to keep R_N smaller than the saturation distance, $R_N < R_{sat}$, so that all thresholds can be attained (R_{sat} is the typical distance of two uncorrelated trajectories).

The evolution of the error from the initial value, R_{min}, up to the largest threshold, R_N, carries out a single error-doubling experiment. At the end of each error-doubling experiment one rescales one trajectory at the initial distance R_{min} with respect to the other and starts another experiment. After $\mathcal{N}$ error-doubling experiments, one can estimate the expectation value of some quantity A as:

$$\langle A \rangle_e = \frac{1}{\mathcal{N}} \sum_{i=1}^{\mathcal{N}} A_i . \tag{B.1}$$

The average $\langle(\ldots)\rangle_e$ is not a time average: different error-doubling experiments may take different times. Indeed we have

$$\langle A \rangle_t = \frac{1}{T} \int_0^T A(t)\, \mathrm{d}t = \frac{\sum_i A_i \tau_i}{\sum_i \tau_i} = \frac{\langle A\tau \rangle_e}{\langle \tau \rangle_e}. \tag{B.2}$$

In the particular case in which A is the inverse of them doubling time itself (as when computing the Lyapunov exponent) we have from (B.2)

$$\lambda(R_n) = \frac{1}{\langle T_r(R_n)\rangle_e} \log \rho. \tag{B.3}$$

The above described method assumes that the distance between the two trajectories is continuous in time. For maps or for discrete sampling in time the method has to be slightly modified; see Aurell *et al.* (1997).

Before concluding this Appendix, we discuss the behaviour of $\lambda(R)$ near the saturation, i.e. for distance of the order of the domain size. This behaviour mainly stems from the assumption that for large times the tracers tend to uniformly distribute in the domain and that small deviations from the asymptotic uniform distribution relax exponentially to that. This assumption is usually satisfied in generic chaotic dynamical systems even if it is difficult to prove. In the language of dynamical systems exponential relaxation to asymptotic distribution means that the second eigenvalue, α, of the Perron-Frobenius operator is inside the unit circle, the relaxation time is $\tau_k = -\log|\alpha|$ (see Ott (1993) for an introduction).

If the distribution relaxes exponentially to the uniform, the same should hold for the moment of the distribution. Therefore, for the large time evolution of the distance between two trajectories, $R(t)$, one expects

$$R(t) \approx R_{\max} - A e^{-t/\tau_R} \tag{B.4}$$

where A and τ_R are system dependent. For $t \ll \tau_R$ or equivalently for $(R_{\max} - R(t))/R(t) \ll 1$, one has:

$$\frac{\mathrm{d}}{\mathrm{d}t} \log R = \lambda(R) = \frac{1}{\tau_R} \frac{R_{\max} - R}{R}. \tag{B.5}$$

For an exact computation of (B.4–B.5) in a particular system see the appendix in Artale *et al.* (1997).

References

Artale, V., Boffetta, G., Celani, A., Cencini, M., Vulpiani, A. (1997) 'Dispersion of passive tracers in closed basins: beyond the diffusion coefficient', *Phys. Fluids* **9**, 3162.

Aurell, E., Boffetta, G., Crisanti, A., Paladin, G., Vulpiani, A. (1996) 'Growth of non-infinitesimal perturbations in turbulence', *Phys. Rev. Lett.* **77**, 1262.

Aurell, E., Boffetta, G., Crisanti, A., Paladin, G., Vulpiani, A. (1997) 'Predictability in the large: an extension of the concept of Lyapunov exponent', *J. Phys. A* **30**, 1.

Benzi, R., Paladin, G., Parisi, G., Vulpiani, A. (1984) 'On the multifractal nature of fully developed turbulence and chaotic systems', *J. Phys. A* **17**, 3521.

Benzi, R., Biferale, L., Ciliberto, S., Struglia, M.V., Tripiccione, R. (1996) 'A generalized scaling of fully developed turbulence', *Physica D* **96**, 162.

Benzi, R., Biferale, L., Ruiz-Chavarria, G., Ciliberto, S., Toschi, F. (1999) 'Multiscale velocity correlations in turbulence: experiments, numerical simulations, synthetic signals', *Phys. Fluids* **11**, 2215.

Biferale, L., Boffetta, G., Celani, A., Crisanti, A., Vulpiani, A. (1998) 'Mimicking a turbulent signal: sequential multiaffine processes', *Phys. Rev. E* **57**, R6261.

Biferale, L., Cencini, M., Vergni, D., Vulpiani, A. (1999) 'Exit time of turbulent signals: a way to detect the intermediate dissipative range', chao-dyn/9904029.

Boffetta, G., Celani, A., Crisanti, A., Vulpiani, A. (1999a) 'Relative dispersion in fully developed turbulence: Lagrangian statistics in synthetic flows', *Europhys. Lett.* **46**, 177

Boffetta, G., Cencini, M., Espa, S., Querzoli, G. (1999b) 'Chaotic advection and relative dispersion in a convective flow', in preparation.

Bohr, T., Jensen, M.H., Paladin, G., Vulpiani, A. (1998) *Dynamical Systems Approach to Turbulence* (Cambridge University Press).

Borgas, M.S. (1993) 'The multifractal Lagrangian nature of turbulence', *Philos. Trans. Roy. Soc. London* **A342**, 379.

Cocke, W.J. (1969) 'Turbulent hydrodynamic line stretching: consequences of isotropy', *Phys. Fluids* **12**, 2488.

Elliott, F.W. Jr, Majda, A.J. (1996) 'Pair dispersion over an inertial range spanning many decades', *Phys. Fluids* **8**, 1052.

Frisch, U. (1995) *Turbulence. The Legacy of A.N. Kolmogorov* (Cambridge University Press).

Frisch, U., Vergassola, M. (1991) 'A prediction of the multifractal model: the intermediate dissipation range', *Europhys. Lett.* **14**, 439.

Fung, J.C.H., Vassilicos, J.C. (1998) 'Two-particle dispersion in turbulent-like flows', *Phys. Rev. E* **57**, 1677.

Fung, J.C.H., Hunt, J.C.R., Malik, N.A., Perkins, R.J. (1992) 'Kinematic simulation of homogeneous turbulence by unsteady random Fourier modes', *J. Fluid Mech.* **236**, 281.

Gagne, Y., Castaing, B. (1991) 'Une représentation universelle sans invariance globale d'échelle des spectres d'énergie en turbulence développée' *C. R. Acad. Sci. Paris* **312**, 441.

Jensen, M.H. (1999) 'Multiscaling and structure functions in turbulence: an alternative approach', *Phys. Rev. Lett.* **83**, 76.

Jensen, M.H., Paladin, G., Vulpiani, A. (1991) 'Multiscaling in multifractals', *Phys. Rev. Lett.* **67**, 208.

L'vov, V.S., Podivilov, E., Procaccia, I. (1997) 'Temporal multiscaling in hydrodynamic turbulence', *Phys. Rev. E* **55**, 7030.

Miozzi, M., Querzoli, G., Romano, G.P. (1998) 'The investigation of an unstable convective flow using optical methods', *Phys. Fluids* **10**, 2995.

Monin, A., Yaglom, A. (1975) *Statistical Fluid Mechanics*, Vol. 2, (MIT Press).

Novikov, E.A. (1989) 'Two-particle description of turbulence, Markov property and intermittency', *Phys. Fluids A* **1**, 326.

Orszag, S.A. (1970) 'Comment on: Turbulent hydrodynamic line stretching: consequences of isotropy', *Phys. Fluids* **13**, 2203.

Ott, E. (1993) *Chaos in Dynamical Systems* (Cambridge University Press).

Ottino, J.M. (1989) *The Kinematics of Mixing: Stretching, Chaos and Transport* (Cambridge University Press).

Paret J., Tabeling P. (1998) 'Intermittency in two-dimensional inverse cascade of energy : experimental observations', *Phys. Fluids* **10**, 3126.

Richardson, L.F. (1926) 'Atmospheric diffusion shown on a distance-neighbour graph', *Proc. Roy. Soc.* **A110**, 709.

Statistical Geometry and Lagrangian Dynamics in Turbulence

Michael Chertkov, Alain Pumir and Boris I. Shraiman

Abstract

It may be argued that the dynamically significant inertial scale object is not the conventional two-point velocity difference, but the velocity difference tensor constructed from the velocity measurement at at least four different points. The minimal 'tetrad' of measurement points also defines the moment of inertia type tensor characterizing the shape of the fluid volume under consideration. This tensor together with the velocity difference tensor are used to construct a phenomenological model describing the (short time) Lagrangian evolution of the tetrad. The model and its predictions are compared with the results of the direct numerical simulation of the Navier–Stokes equation with the focus on the geometric structure of turbulent fluctuations. We find that large energy transfer events are associated with strain dominated sheets with vorticity along the intermediate positive strain axis. The statistical measures we consider are also accessible experimentally and their investigation would be very instructive.

1 Introduction

The study of turbulence has relied for many years on measurements and analysis of a signal (such as a velocity component) at one or two points. Much effort has been devoted to the investigation of the nth order structure function, defined as the nth moment of velocity difference, and its scale dependence (Frisch, 1995). Although much valuable insight has been obtained by studying structure functions, it has become increasingly clear that deeper understanding of turbulence requires more information about the spatial organization of underlying flow than the two-point measurement can provide. In fact the 'structure function' tells us virtually nothing about the *structure* of turbulent fluctuations if the latter term is interpreted in the geometrical sense, i.e. refers to an instantaneous configuration of the velocity field over some region. It is clear that at least the largest fluctuations can be associated with well defined structures transiently appearing in the flow. For example, large values of local vorticity are due to vortex filaments readily identifiable in experimental flows (Douady *et al.*, 1991) and numerical simulations (Siggia, 1981; Jiménez *et al.*, 1993). An interesting compromise between the two-point measurements and the 'snapshots' of the velocity field is furnished by the correlation functions of the velocity taken at multiple points, which can be studied as functions of point configuration. To the extent that high order multipoint correlators pick out large fluctuations they can also serve as probes of the geometrical structure characteristic of the dominant events. Yet these new more complex objects not only require more elaborate measurement techniques but also need a novel theoretical framework to support their interpretation.

The importance of looking at the full correlation function became particularly evident in the study of passive scalar mixing by the white (in time) Gaussian random velocity field – the Kraichnan model (Kraichnan, 1968 and 1994). In this problem, the N-point correlation functions, $\langle \theta(x_1)\theta(x_2) \cdots \theta(x_N) \rangle$, satisfy a closed set of partial differential equations (Kraichnan, 1974; Shraiman and Siggia, 1994). These so-called Hopf equations arise from considering the temporal evolution of the multipoint correlators. Because the scalar is conserved along trajectories of the flow (up to a correction due to molecular diffusivity) the time evolution is determined entirely in terms of the dependence of the correlator on the configuration of measurement points. By contrast, an evolution equation for the structure function, where the configuration dependence has been lost, cannot be written down without additional assumptions (Kraichnan, 1974). Although the exact Hopf equation may only be derived for the white noise velocity ensemble, it is plausible to infer that the dynamical significance of the multipoint correlators is more general, and it is sensible to employ them, instead of the structure function, as the fundamental object in the quantitative modeling of the passive scalar in more realistic flows. In this approach one seeks to construct, on phenomenological grounds, an *effective* evolution operator governing multipoint correlators (Shraiman and Siggia, 1995). In what follows we shall transplant this general notion into the much more complex context of modeling the statistical structure of the velocity field itself.

The need for multipoint correlation functions in the study of the turbulent velocity seems intuitively obvious. Local dynamics of the velocity field requires at the very least the knowledge of the local velocity gradient tensor $m_{ab} \equiv \partial_a v_b$, which cannot be estimated by measuring velocity at only two points. For example, the evolution of vorticity, universally believed to be the essential aspect of 3D turbulence, involves the subtle properties of alignment between the vorticity vector and the eigenvectors of the rate of strain tensor $(s_{ab} \equiv (m_{ab} + m_{ba})/2)$ even in the homogeneous isotropic problem (Ashurst *et al.*, 1987), implying that the full structure of the m tensor is important, not just one component.

These observations have led us to focus on the Lagrangian dynamics of a cluster of points (Chertkov *et al.*, 1999). The smallest cluster necessary to estimate the velocity difference tensor is a *tetrad*: i.e. $n = 4$ points (in d dimensions, it would be $n = d + 1$). We think of a tetrad as a poor man's representation of a coarse grained volume of the fluid as it evolves in the flow, and seek to write down on phenomenological grounds a physically sensible model of its dynamics. The desired model would correctly implement the incompressibility and circulation constraints on local evolution and represent the coupling to the neighboring volumes in the sense of Mean Field theory. In this paper we discuss the first steps in the development and analysis of such a model. In Section 2 we introduce the tetrad variables which allow one to follow the evolution of the full velocity difference tensor, at the cost of additional degrees of freedom, characterizing the size and the shape of the tetrad. In Section 3 we formulate the stochastic evolution model and discuss the semiclassical approximation which may be invoked for its analysis. Section 4 contains a number of numerical results and comparisons between DNS and the model which demonstrate encouraging semiquantitative agreement for the statistics of certain interesting physical quantities, such as enstrophy, enstrophy production and energy dissipation. The tetrad dynamics which we consider is geometrical in nature and the statistical measures we introduce and analyze provide direct insight into the geometric structure of underlying turbulent fluctuations, e.g. we will see that vortex filaments evolve into sheets and that most of the energy dissipation occurs in strain ribbons where vorticity is aligned with the intermediate stretching direction (Ashurst *et al.*, 1987). This emphasis on the geometrical structure explains the 'statistical geometry' term which we used in the title. Section 5 presents the summary and the discussion of further prospects and explains how the statistical quantities which we introduce may be obtained from multipoint experimental measurements.

2 The tetrad approach

The idea of following a set of points comes from our goal of constructing the *coarse grained* approximation of the velocity field over a finite size region of fluid comoving with the flow.

Consider a set of N points $(r_1, r_2, r_3, \cdots, r_N)$, whose velocities are $(v_1, v_2, v_3, \cdots, v_N)$. The center of mass, $\vec{\rho}_0 \equiv (r_1 + r_2 + r_3 + \cdots + r_N)/N$ is not interesting for investigating the small scale properties of turbulence, since it merely reflects the large scale advection of a parcel of fluid. We focus here on differences, and define

$$
\begin{aligned}
\rho_1^a &= (r_2^a - r_1^a)/\sqrt{2}, \\
\rho_2^a &= (2r_3^a - r_1^a - r_2^a)/\sqrt{6}, \\
\rho_3^a &= (3r_4^a - r_1^a - r_2^a - r_3^a)/\sqrt{12},
\end{aligned}
\tag{2.1}
$$

$$
\begin{aligned}
u_1^a &= (v_2^a - v_1^a)/\sqrt{2}, \\
u_2^a &= (2v_3^a - v_1^a - v_2^a)/\sqrt{6}, \\
u_3^a &= (3v_4^a - v_1^a - v_2^a - v_3^a)/\sqrt{12}, \quad \text{etc.}
\end{aligned}
\tag{2.2}
$$

Let us now decompose the velocities of the particles in the N-point 'swarm' into the smoothly varying coherent component and the fluctuating residual. We shall approximate the coherent component by the best linear fit which defines the coarse grained velocity gradient tensor M_{ab} via least squares,

$$u_i^a = \rho_i^b M_{ba} + \zeta_i^a, \qquad (2.3)$$

with $\mathrm{tr} M = 0$ constraint. For $N > 4$ the M-tensor is defined via the pseudoinverse $k_i^a = g_{ab}^{-1} \rho_i^b$ (where $g_{ab} = \rho_i^a \rho_i^b$ is the moment of inertia tensor for the 'swarm'),

$$M_{ab} = k_i^a u_i^b - \delta_{ab} \frac{k_i^c u_i^c}{3}. \qquad (2.4)$$

The minimal number of points for which the procedure is defined is $N = 4$, in which case the 3×3 tensor ρ_i^a is in general invertible and $k = \rho^{-1}$. We shall often think of this minimal case and talk of *tetrads* instead of the larger 'swarms'. Note that M does not depend on the precise definition of ρ and u, taken as independent combinations of the coordinates and velocity differences, as in eqs. (2.1) and (2.2). In the case where $(r_2 - r_1, r_3 - r_1, r_4 - r_1)$ are three vectors of size ρ, orthogonal to one another, the definition (2.4) reduces to the usual finite difference approximation of the velocity gradient tensor. In general (2.4) provides the most sensible definition of a velocity *difference* tensor.

We are interested in the dynamics of the tetrad. With the definition already introduced, the equations of evolution for the difference vectors spanning the tetrad are

$$\frac{d}{dt} \rho_i^a = v_i^a = \rho_i^b M_{ba} + \zeta_i^a, \qquad (2.5)$$

where $\zeta_i^a = (\delta_{ij} - \rho_i^a k_j^a) v_j^a + \frac{1}{3} \rho_i^a k_j^b v_j^b$ is the fluctuating component of the velocity. To the extent that this fluctuating velocity component is associated with the length scales smaller than the blob size it should also fluctuate faster than the large scale coherent component described by M. Hence we shall model the small scale fluctuations as Gaussian random and white in time.

$$\langle \zeta_i^a(\rho, t) \zeta_j^b(0, 0) \rangle = C_u (\delta_{ij} \delta_{ab} \rho^2 - \rho_i^a \rho_j^b) \sqrt{\mathrm{tr}(MM^T)} \delta(t). \qquad (2.6)$$

The evolution of the velocity at a Lagrangian point r_i is given in the large Reynolds number limit by

$$\frac{d}{dt} V_i = -\nabla p_i + f_i \qquad (2.7)$$

where p_i is the pressure at point r_i and f_i the external force. Defining the $\delta \nabla p_i$ in the same way as ρ and u, see eqs. (2.1) and (2.2), the equation of evolution for the matrix M becomes

$$\frac{d}{dt} M_{ab} + M_{ab}^2 - \frac{\delta_{ab}}{3} \mathrm{tr} M^2 = k_i^a (\delta \nabla p)_i^b - \frac{\delta_{ab}}{3} k_i^c (\delta \nabla p)_i^c + \xi_{ab}' \qquad (2.8)$$

where the last term combines the external force and the reaction of small scales (i.e. the contribution of incoherent fluctuations, ζ_i^a). The pressure term on the right-hand side of eq. (2.8) is the finite difference generalization of the pressure Hessian, and the key question is how $\delta\nabla p$ is correlated with the velocity difference tensor.

If the right hand side of (2.8) were set to zero, the resulting dynamical equation for M would reduce to the one obtained by Vieillefosse (Vieillefosse, 1984) for the velocity gradient proper. The Vieillefosse, or so-called Restricted Euler, dynamics is derived directly from the Euler equation by making an isotropic approximation for the pressure $\partial_{ab}p \rightarrow -\frac{1}{3}\text{tr}(m^2)\delta_{ab}$ (where $m_{ab} = \partial_a v_b$). This results in a closed set of ODE for the m-tensor (Vieillefosse, 1984; Cantwell, 1992) which may be readily integrated and exhibit a finite time singularity.

In contrast to the Vieillefosse approach, the present model deals not with velocity gradient but with velocity difference tensor and contains an additional tensorial object, g_{ab}, describing the 'shape' of the tetrad which allows one to introduce a more complex, anisotropic model of pressure. Thus we propose (see Chertkov *et al.* (1999) for details) the following stochastic dynamics:

$$\frac{d}{dt}M_{ab} + \alpha(M_{ab}^2 - \kappa_{ab}\text{tr}M^2) = \xi_{ab} \qquad (2.9)$$

which involves a tetrad anisotropy tensor $\kappa_{ab} \equiv k_i^a k_i^b / k^2$ (note $\text{tr}\kappa = 1$) and invokes a 'renormalization' of the dynamics by factor $\alpha < 1$ due to the pressure. This renormalizing effect was first noted in the numerical study of the correlation between the pressure Hessian and the velocity gradient tensor (Borue and Orszag, 1997) and holds also for finite differences (Chertkov *et al.*, 1999) considered here. The remainder of the right hand side of (2.8) is lumped into ξ and treated as Gaussian random and white in time

$$\langle\xi_{ab}(0)\xi_{cd}(t)\rangle = C_\eta \varepsilon \rho^{-2}\delta(t) \qquad (2.10)$$

obeying Kolmogorov's scaling with ε, the energy dissipation rate.

The anisotropy tensor κ was introduced into the pressure Hessian model to insure that the local pressure term drops out of the energy balance equation (Chertkov *et al.*, 1999). It plays the crucial role in eliminating the finite time singularity in the dynamics described by (2.9). Anisotropy of the tetrad depends (through (2.5)) on the history of strain which has been stretching it and increases rapidly with increasing M. As a result the singularity is cut off. The unphysical blow-up of M is replaced by the plausible tendency of the initially isotropic volumes to distort into pancakes and ribbons!

Eqs. (2.9), (2.5), (2.6), (2.10) completely specify our model of Lagrangian dynamics of the tetrad. The philosophy behind its construction was an attempt to identify the correct non-linear dynamics of strain and vorticity coarse

grained over a Lagrangian volume, and treat all the non-local pressure medi-
ated forces coupling the volume in question to the rest of the fluid as random
and obeying K41 scaling. As we shall see next the model makes non-trivial
predictions concerning the statistics of the velocity difference tensor.

3 The Lagrangian and Eulerian Probability Distribution

The tetrad dynamics equations (2.5), (2.9) define a well-posed stochastic
problem. It is possible to write down the corresponding Fokker–Planck equation for the probability distribution $P(M, \rho, t)$:

$$\partial_t P = \mathbf{L} P \tag{3.1}$$

with the evolution operator

$$
\begin{aligned}
\mathbf{L} =\ & (1-\alpha)\frac{\partial}{\partial M_{ab}}\left(M_{ab}^2 - \kappa_{ab}\,\mathrm{tr}\mathbf{M}^2\right) - \frac{\partial}{\partial \rho_b^i}\,\rho_a^i M_{ab} \\
& + C_\eta \frac{\varepsilon}{\rho^2}\left(\frac{\partial^2}{\partial M_{ab}\partial M_{ab}} - \frac{1}{3}\frac{\partial^2}{\partial M_{aa}\partial M_{bb}}\right) \\
& + C_u \sqrt{\mathrm{tr}\mathbf{M}\mathbf{M}^\dagger}\,\frac{\partial}{\partial \rho_i^a}\left(\rho^2 \frac{\partial}{\partial \rho_i^a}\right).
\end{aligned}
\tag{3.2}
$$

It is instructive to consider the resulting energy balance from an *Eulerian*
point of view. To this end, we consider a fixed shape of the tetrad, and
average with the help of the Fokker–Planck equation over M, to get the
following relation in the case of isotropic tetrads:

$$\partial_t \langle V^2 \rangle = -\partial_{\rho_i^a}\langle V_i^a \mathrm{tr}(V^2)\rangle + \alpha\langle \mathrm{tr}(M M^T \rho^T \rho M)\rangle \tag{3.3}$$

where $V_i^a \equiv \rho_i^b M_{ba}$.

Eq. 3.3 has a rather clear physical interpretation. The first term in the
right-hand side is an energy flux term, analogous to the flux term appearing in
the von Kármán–Howarth analysis. The velocity field V considered in Eq. 3.3
however is only the coarse grained velocity field at scale R. As such, the flux
term balances with an eddy damping term, of the form $\langle V_a V_b \partial_a V_b \rangle$ (stress-
strain correlation), as expected from standard arguments. Interestingly, the
term appearing on the right-hand side of eq. (3.3) has the same structure
as the term proposed by Bardina *et al.*, 1980, in the context of Large Eddy
Simulation (see also Borue and Orszag, 1998). The stress–strain correlation
reduces, for an isotropic tetrad, to $\mathrm{tr}(M^2 M^T) = \mathrm{tr}(s^3) + (\omega \cdot s \cdot \omega)/4$; its
structure will be investigated in the next section.

The Eulerian probability distribution $P(M \mid \rho)$ must satisfy the $LP = 0$
stationarity condition and the normalization $\int dM\,P(M \mid \rho) = 1$. In addition

one should also specify boundary conditions at large scale, $\rho^2 = L^2$, where L is the integral length scale. We shall rely on the well known fact that the velocity is Gaussian at large scales and simply impose that $P(M \mid \rho = L)$ is a Gaussian distribution.

The solution can be looked for by using a path integral formulation. Specifically, one defines the Lagrangian transition probability (effectively the Green's function) by

$$G_t(M, \rho \mid M', \rho') = \int \mathcal{D}M \int \mathcal{D}\rho \exp -S(M, \rho) \qquad (3.4)$$

where S is the usual action:

$$S = \int_0^t dt \left(\frac{\operatorname{tr}(\dot{M} + M^2 - \kappa \operatorname{tr}(M^2))^2}{C_\eta \epsilon \rho^{-2}} + \frac{\operatorname{tr}(\dot{\rho} - \rho M)^2}{C_u \|M\| \rho^2} \right). \qquad (3.5)$$

The solution of the problem can be expressed in terms of Green's theorem

$$P(M \mid \rho) = \int dM' \int dt G_t(M, \rho \mid M', \rho') P(M' \mid \rho = L) \qquad (3.6)$$

which allows us to express the solution at a scale ρ by using the knowledge of the solution at the integral scale, and the transition probability G_t. Schematically, the probability distribution function of having M at scale ρ at time $t = 0$ is related to the probability of getting there starting from M' at a scale L at some earlier time $-T$. A full analysis of this formal solution (3.6) is feasible in the limit $C_\eta \to 0$, $C_u \to 0$ using the semiclassical approximation. In the semiclassical approximation $G_T(M, \rho \mid M', L) \approx \exp[-S_{sc}(M, \rho \mid M', L; T)]$ with the action corresponding to the least action trajectory connecting the specified initial and final configurations in time T. We restrict ourselves here to an even simpler 'classical', or zero-action, approximation, which consists of finding the most probable path passing through (M, ρ): the one with $S = 0$. This path is the classical Lagrangian trajectory, obtained by setting $C_u = 0$, $C_\eta = 0$ in the equation of motion. Within this approximation $P(M \mid \rho)$ is given by the probability of the Lagrangian pre-image $M'(M, \rho)$ evaluated at the integral scale. Crude as it is, this approximation allows us to compute a number of quantities, and leads to qualitatively interesting results we will present in the next section.

There are several ways to analyze how the velocity difference tensor evolves. We use in this work the notion of conditional flow in the (R, Q) plane. That is, we investigate the evolution of the invariants (R, Q), and look at the conditional averages of $\langle \frac{dR}{dt} \mid R, Q \rangle$ and $\langle \frac{dQ}{dt} \mid R, Q \rangle$.

A number of properties of the statistical distribution of (R, Q) has been investigated in the case of the velocity gradient tensor (i.e. in our formulation, $\rho \rho^T \to 0$). The isoprobability lines in the (R, Q) plane have a very skewed petal shape. It is clearly of interest to investigate the properties of the distribution of the (R, Q) invariants as a function of the size of the tetrad, since it gives an indication about the topology of the flow as a function of scale ρ.

4 Results

We present now the results obtained both from direct numerical simulations (DNS) and from the classical approximation outlined in the previous section. Before we proceed however we must remind the reader of a few kinematic facts about velocity gradient (or difference) tensors and about Restricted Euler dynamics.

In order to characterize the tensor M, it is useful to introduce the invariants (Vieillefosse, 1984; Cantwell, 1992)

$$Q \equiv -\frac{1}{2}\text{tr}(M^2), \quad R \equiv -\frac{1}{3}\text{tr}(M^3). \tag{4.1}$$

Because M is a matrix of rank 3, and $\text{tr}(M) = 0$, by definition, all the invariants $\text{tr}(M^n)$ can be expressed in terms of Q and R only. The two invariants Q and R provide a characterization of the local topology of the flow (Cantwell, 1992), since the eigenvalues of M are the roots of

$$\lambda^3 + Q\lambda + R = 0. \tag{4.2}$$

The zero discriminant line $27Q^3 + 4R^2 = 0$ separates the (R, Q) plane into two regions: for $27Q^3 + 4R^2 > 0$ (upper part of the (R, Q) plane), the flow is elliptic, with locally swirling streamlines, whereas for $27Q^3 + 4R^2 < 0$ (lower part of the (R, Q) plane), strain dominates and the flow is locally hyperbolic.

Using the invariants R and Q defined above, eq. (4.1), the Restricted Euler (RE) dynamics (Vieillefosse 1984) $\frac{d}{dt}M_{ab} + M_{ab}^2 - \frac{1}{3}\delta_{ab}\text{tr}M^2 = 0$ (i.e. the left-hand side of eq. (2.8)) reduces to

$$\frac{dQ}{dt} = -3R, \quad \frac{dR}{dt} = \frac{2}{3}Q^2. \tag{4.3}$$

These two equations can be integrated; the resulting streamlines are shown in Fig. 1. Interestingly, the zero discriminant line $4Q^3 + 27R^2 = 0$ is invariant with respect to the flow, and the solutions asymptote with t to the $R \rightarrow +\infty$ limit of the zero discriminant line. In fact, it is easy to see that R, Q diverge in a finite time implying a finite time singularity. This disturbing prediction points to a flaw in the isotropic pressure approximation (Vieillefosse, 1984), but it turns out, remarkably, that the general structure of the phase space flow governed by (4.3) is consistent with the shape of the isoprobability lines in the (R, Q) plane as observed in the numerical simulations (Cantwell 1992; Borue and Orszag 1998), see Fig. 2. It is also possible to test the validity of the model directly by comparing the predictions of the Lagrangian dynamics defined by (2.5), (2.9) with the short time Lagrangian statistics of point trajectories in the DNS of the Navier–Stokes equation.

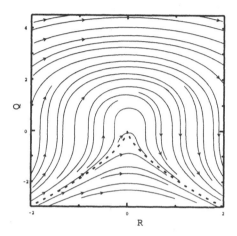

Figure 1: The streamline of the flow resulting from the isotropic pressure approximation (Vieillefosse flow)

Briefly, our numerical results were obtained by integrating the Navier–Stokes equations by using a pseudo-spectral methods. With a resolution of up to $(128)^3$, the highest Reynolds number we could reach, while maintaining adequate resolution, was $R_\lambda \approx 85$. The quantities presented were obtained by averaging over both space and time, with a sampling time larger than three eddy turnover times. More detail on the numerical code used may be found elsewhere (Pumir, 1994).

The first question we address here is how close is the flow resulting from the Navier–Stokes equations to the model dynamics, eq. (2.9). To this end, we consider isotropic (regular) tetrads, with a size ρ, and we compute numerically the conditional averages: $\langle dR/dt \mid R, Q \rangle$ and $\langle dQ/dt \mid R, Q \rangle$. These conditional values define the 'velocity field' in the (R, Q) phase plane, and allow us to construct an effective flow, which can then be compared with the Restricted Euler approximation, Fig. 1.

The results, obtained for values of ρ both in the dissipative range, $\rho = 2\eta$, and in the inertial range, $\rho \approx 30\eta$, are shown in Fig. 2. The reconstructed flow streamlines at small scales, Fig. 2a, show some significant difference

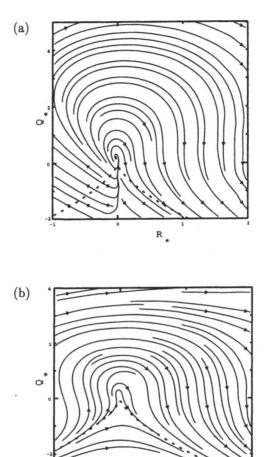

Figure 2: Streamlines of the flow of the M invariants (R, Q) constructed from the conditional averages $\langle \dot{Q} \mid Q, R \rangle$ and $\langle \dot{R} \mid Q, R \rangle$ measured from DNS at different length scales: (a) dissipative range $\rho = 2\eta = L/32$, (b) upper end of the inertial range $\rho = L/2 = 32\eta$.

with RE case. This is not so surprising, since viscous effects, which were never taken into account in the theoretical considerations leading to (2.9) or (4.3) are expected to play an important role in the dissipative range (see also Ooi *et al.*, 1999). In contrast, the picture obtained at larger scales shows a qualitatively much better agreement with the Vieillefosse flow. The flows are topologically similar. In addition, the flow asymptotes the separatrix, $4Q^3 + 27R^2 = 0$, $R \geq 0$, marked by the dashed lines in Fig. 2. It must be however remembered, that similarity between Fig. 1 and Fig. 2 does not imply the validity of RE dynamics, but merely the similarity between the short time evolution of (R, Q) invariants for an initially *isotropic* tetrad and RE. According to (2.9) we expect that tetrad dynamics is only tangent to the RE and at longer times (R, Q) invariants of a given Lagrangian volume must deviate from RE streamlines.

We next turn to the probability distribution function (pdf) of (R, Q), and its dependence as a function of the overall scale of the tetrad, ρ. Again, we restrict ourselves to regular tetrads. The numerical results are shown for ρ in the dissipative range ($\rho \sim 2\eta$), Fig. 3a, and in the inertial range ($\rho \sim 30\eta$), Fig. 3b. Fig. 3a is very similar to the pdf already obtained in the inertial range (Cantwell, 1993); it shows a very skewed shape, with a large probability along the separatrix $4Q^3 + 27R^2 = 0$; $R \geq 0$, and in the upper left quadrant. At higher values of ρ, the pdf is more symmetric with respect to the $R = 0$ line. In particular, the isoprobability lines do not extend as far along the separatrix. The systematic evolution of the shape of the pdf as a function of the overall scale can be easily inferred from these figures. Note that for a strictly Gaussian velocity field, the pdf of (R, Q) are strictly symmetric with respect to the $R = 0$ line.

An example of pdf computed in the classical approximation, see the previous section, is shown in Fig. 4. Although there are clear similarities with the shape of the pdf obtained from numerical simulations, there are also some significant differences. Some of the latter, like the low probability 'gulf' along the $R > 0$ semiaxis, are clearly artifacts of the approximation which, by neglecting the fluctuations (i.e. setting $C_u = 0$, $C_\eta = 0$), underestimates the probability of large fluctuations.

It is interesting to investigate more physical quantities. We begin with the density of enstrophy in the (R, Q) plane: $\langle \omega^2 \mid R, Q \rangle \times P(R, Q)$. Fig. 5a shows the DNS results, whereas Fig. 5b shows the model predictions (using again the semiclassical approximation). The qualitative agreement between the two sets of pictures is rather good. The enstrophy density is largest around $(R, Q) = (0, 0)$. Note that this point corresponds to a locally 2D configuration. The enstrophy is largest in the upper part of the plane ($Q \geq 0$). The model suggests that a small, although significant, part of the enstrophy comes from the separatrix $4Q^3 + 27R^2 = 0$. In this picture, the strong vortex

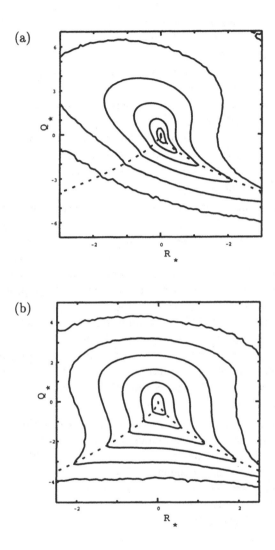

Figure 3: The pdf of Q_*, R_* invariants normalized to the variance of strain, $Q_* \equiv Q/\langle s^2 \rangle$ and $R_* \equiv R/\langle s^2 \rangle^{3/2}$ ('star' denotes normalization), obtained from DNS at $R_\lambda = 85$ measured at different length scales: (a) dissipation range $\rho = 2\eta$; (b) inertial range $\rho = L/2 = 32\eta$. The isoprobability contours are logarithmically spaced, and are separated by factors of 10. The dashed line corresponds to the separatrix: $4Q^3 + 27R^2 = 0$.

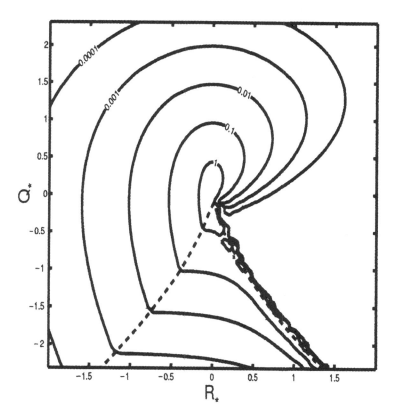

Figure 4: The pdf of Q_*, R_* invariants (normalized as in Fig. 3) calculated for the tetrad model in the deterministic approximation for $\rho/L = .5$.

filaments are located higher up in the upper half-plane.

Enstrophy production density, defined by $\langle \omega.s.\omega \mid R, Q \rangle \times P(R, Q)$, obtained from numerical simulations, see Fig. 6a, and from the model calculation, Fig. 6b, is another important physical quantity. Vortex stretching comes mostly from regions with $Q \geq 0$, $R \leq 0$, where vorticity is aligned with the (single) positive strain direction. (Note that strain has two positive eigenvalues for $R > 0$ and one positive eigenvalue for $R < 0$.) Vorticity is also stretched in the $R > 0$ separatrix region, where it is aligned with the intermediate stretching axis. Also note a region of negative enstrophy production in the $R, Q > 0$ quadrant corresponding to the 'collapse' of nearly 2D vortex filaments.

The energy flux density, which, for the isotropic tetrads considered here

256 *Chertkov, Pumir & Shraiman*

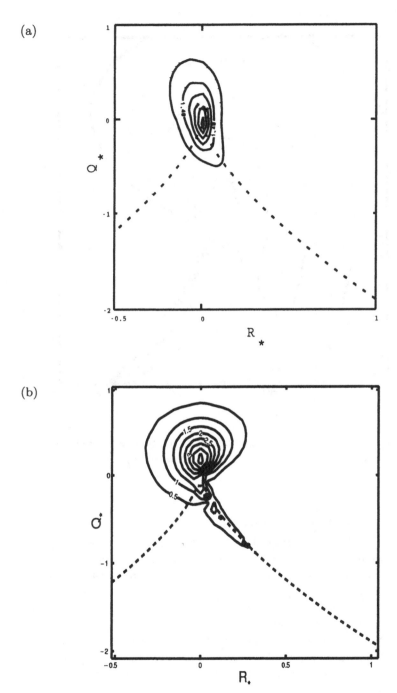

Figure 5: Enstrophy density in (R, Q) plane measured from DNS at $R_\lambda = 85$, at $\rho/L = 0.125$ (a) and computed from the tetrad model at $\rho/L = .5$ (b).

(a)

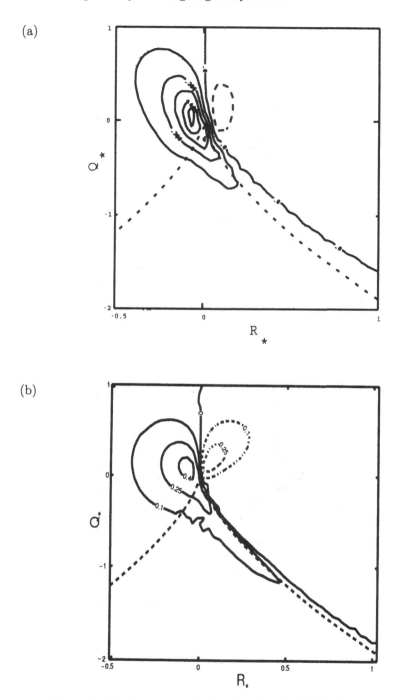

Figure 6: Enstrophy production density in (R, Q) plane measured from DNS at $R_\lambda = 85$, at $\rho/L = 0.125$ (a) and computed from the tetrad model at $\rho/L = .5$ (b).

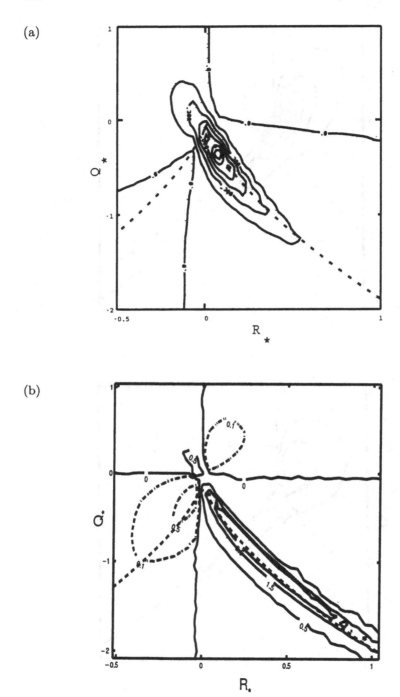

Figure 7: Energy flux density in the (R, Q) plane measured from DNS at $R_\lambda = 85$, at $\rho/L = 0.125$ (a) and computed from the tetrad model at $\rho/L = .5$ (b).

reads $\langle \mathrm{tr}(M^2 M^T) \mid R, Q \rangle \times P(R, Q)$, also shows an interesting structure. We find that both the DNS (Fig. 7a) and the calculations with the model (Fig. 7b) predict that the energy flux comes mostly from the $R > 0$ separatrix region, and certainly *not* from the high enstrophy or enstrophy production regions (Tsinober, 1998). Interestingly, some regions of negative energy flux ('backscattering of energy') are also found in the $(R \geq 0, Q \leq 0)$ and in the $(R \geq 0, Q \leq 0)$ quadrants. The intensity of backscattering remains weak.

We have also investigated the alignment of vorticity and strain in different regions of the (R, Q) plane and the distribution of strain eigenvalues. The great advantage of analysis statistics of the velocity difference tensor *conditioned* by the (R, Q) invariants is that it allows us to dissect the topologically different regions: e.g. identify the region of vortex filaments and strain sheets. The result is an intuitively plausible geometric picture where vortex filaments are formed in the $(R < 0, Q > 0)$ quadrant and achieve maximum enstrophy as they follow the RE streamlines to the $R = 0$ axis, then collapse and turn into weak vorticity sheets aligned with the intermediate stretching axis of the strain. The latter sheets are responsible for large fluctuations in energy transfer.

5 Conclusion

We have introduced the velocity difference tensor as a novel and interesting observable and discussed its statistical properties on the basis of the DNS and as predicted by the Lagrangian dynamics model. Considering the full tensorial quantity defined by four-point measurement, renders considerable insight into the geometric structure of dominant fluctuations in turbulence and the role of vorticity and strain. On the other hand it is evident that modeling Lagrangian dynamics of tetrads (or 'swarms') of points allows us to address the same issues in quantitative detail. The tetrad model which we defined above of course is only phenomenological and its validity is to be ascertained by exploring its consequences and through detailed comparison with experiment and numerical simulations. Our present results although based on a crude approximation are quite encouraging in their semiquantitative agreement with the DNS. Further work will improve the solution of the model and extend the comparison with the DNS for both the model equations and the statistical predictions.

The utility of working with the velocity difference tensor and (R, Q) conditional statistics is evident. We emphasize, that these are not just theoretical or numerical constructs, but are eminently experimentally *measurable*. Although multiple (at least three) three-wire probes are required, they need not be placed close together. The *difference* tensor measurement in the inertial range is *as* or *more* interesting than the local velocity gradient! There is much to be learned from its study.

We have benefitted from discussions with E. Siggia and A. Tsinober. AP and BIS acknowledge the hospitality of the Isaac Newton Institute for Mathematical Science during the turbulence program. We acknowledge a grant of computer time from IDRIS (France).

References

Ashurst, W., Kerstein, A., Kerr, R. and Gibson, C. (1987) 'Alignment of Vorticity and Scalar Gradient with Strain rate in Simulated Navier-Stokes Turbulence', *Phys. Fluids* **30**, 2343–2353.

Bardina, J., Ferziger, J. and Reynolds, W. (1980) 'Improved Subgrid Scale Models for Large Eddy Simulation', *AIAA Paper 80–1357*.

Borue, V. and Orszag, S.A. (1998) 'Local energy flux and subgrid scale statistics in three-dimensional turbulence', *J. Fluid Mech.* **336**, 1.

Cantwell, B.J. (1992) 'Exact Solution of a Restricted Euler Equation for the Velocity Gradient Tensor', *Phys. Fluids* **A4**, 782–793.

Cantwell, B.J. (1993) 'On the behavior of the velocity gradient tensor invariants in DNS of turbulence', *Phys. of Fluids* **A5**, 2008–2013.

Chertkov, M., Falkovich, G., Kolokolov, I. and Lebedev, V. (1995) 'Normal and anomalous scaling of the fourth-order correlation function of randomly advected passive scalar', *Phys. Rev.* **E52**, 4924.

Chertkov, M., Pumir, A. and Shraiman, B.I. (1999) 'Lagrangian tetrad dynamics and the phenomenology of turbulence', *Phys. Fluids* **11**, 2394.

Douady, S., Couder, Y. and Brachet, M. E. (1991) 'Direct observation of Intense Vortex Filaments in Turbulence', *Phys. Rev. Lett.* **67**, 983–986.

Frisch, U. (1995) *Turbulence: the legacy of A.N. Kolmogorov*, Cambridge University Press, Cambridge.

Gawędzki, K. and Kupiainen, A. (1995) 'Anomalous scaling of the passive scalar', *Phys. Rev. Lett.* **75**, 3834.

Jiménez, J., Wray, A., Saffman, P. and Rogallo, R. (1993) 'The structure of intense vorticity in homogeneous isotropic turbulence', *J. Fluid Mech.* **225**, 65.

Kraichnan, R.H. (1968) 'Small-Scale Structure of a Scalar Field Convected by Turbulence', *Phys. Fluids* **11**, 945.

Kraichnan, R.H. (1974) 'Convection of a passive scalar by a quasi-uniform random straining field', *J. Fluid Mech.* **64**, 737.

Kraichnan, R. H. (1994) 'Anomalous scaling of a randomly advected passive scalar', *Phys. Rev. Lett.* **72**, 1016–1019.

Mydlarski, L., Pumir, A., Shraiman, B.I., Siggia, E.D. and Warhaft, Z. (1998) 'Struture and Multipoint Correlators for Turbulent Advection', *Phys. Rev. Lett.* **81**, 4373.

Ooi, A., Martin, J., Soria, J. and Chong, M. S. (1999) 'A study of the evolution and characteristics of the invariants of the velocity gradient tensor in isotropic turbulence', *J. Fluid Mech.* **381**, 141.

Pumir, A. (1994) 'A numerical study of the mixing of a passive scalar in the presence of a mean gradient', *Phys. Fluids* **6**, 2118.

Shraiman, B.I. and Siggia, E.D. (1994) 'Lagrangian path integrals and fluctuations in random flow', *Phys. Rev.* **E49**, 2912.

Shraiman, B.I. and Siggia, E.D. (1995) 'Anomalous scaling of a passive scalar in a turbulent flow', *C.R. Acad. Sci.* **321**, 279.

Siggia, E.D. (1981) 'Numerical Study of Intermittency in 3D Turbulence', *J. Fluid Mech.* **107**, 375–406.

Tsinober, A. (1998) 'Is concentrated vorticity that important?', *Eur. J. Mech., B Fluids* **17**, 421–449.

Vieillefosse, P. (1984) 'Internal Motion of a Small Element of a Fluid In inviscid Flow', *Physica* **125A**, 150–162.

Flow Structure Visualization by a Low-Pressure Vortex

Shigeo Kida, Hideaki Miura and Takahiro Adachi

1 Introduction

Fluid motions of practical interest have in general complicated spatio-temporal structures which are three-dimensional and time-dependent. Understanding of the dynamics of fluid motion may be difficult without knowing the structure. It is a long time since the coherent tubular vortical motions have been recognized as common structures in turbulence such as ribs and braids in free shear flows, streamwise vortices in wall turbulence, worms in isotropic turbulence. They play important roles in turbulence dynamics in diffusion, mixing, transport, instability, and so on. Thus the coherent tubular vortical motions are essential ingredients in turbulence, and visualization of their three-dimensional structures is entirely desirable.

There have been lots of attempts to visualize them, either by using streamlines, pathlines, vorticity, pressure, the Laplacian of pressure, rate-of-strain, or whatever. We may recall, for example, the Schlieren picture for mixing layer turbulence (Bernal & Roshko 1986), aluminous powder for Kármán vortex streets (Taneda 1988), sectional streamlines for a cross-section of streamwise vortices in a wake (Wu *et al.* 1996), aluminous powder for low-speed streaks in wall turbulence (Cantwell *et al.* 1978), hydrogen bubbles for streamwise vortices in boundary layer turbulence (Kim *et al.* 1971), iso-surfaces of vorticity for streamwise vortices in a channel flow (Miyake *et al.* 1995), bubbles for low-pressure vortices in a shearing flow in a circular cylinder (Bonn *et al.* 1993), the Laplacian of pressure for tubular vortices in simple shear turbulence (Tanaka & Kida 1993), vorticity lines around a strong tubular vortex in isotropic turbulence (She *et al.* 1990).

Historically the existence of tubular vortices in turbulence has been recognized as high-vorticity concentrated regions by plotting iso-surfaces of magnitude or a single component of vorticity (see Lesieur 1997). However, high-vorticity regions do not always pick up swirling tubular vortices, but often vortex layers as well. A simple way to distinguish vortex tubes and layers was proposed by Tanaka & Kida (1993) by the use of relative magnitude of vorticity and strain rate. Vorticity is dominant in tubes, while strain is so in layers. Vorticity-dominant regions also correspond to a high Laplacian of pressure since $\nabla^2 p = \frac{1}{2}|\omega| - |\underline{s}|$. However, the critical ratio of $|\omega|/|\underline{s}|$ which discriminates tubes and layers is still unknown. Actually, structures of tubular vortices are different depending on the threshold. An identification method which is free from arbitrariness of the choice of threshold is waited for.

In this article we present a method of visualization of flow structure without ambiguity by a low-pressure vortex. In §2, the notion of a low-pressure vortex is introduced and a detection algorithm for their axes is formulated. The swirl condition which characterizes the low-pressure vortex is compared with the Δ-definition. In §3, an automatic tracing scheme for a low-pressure vortex axis is developed and applied to a field of the interaction of two orthogonal vortices. In §4, several interesting phenomena are discussed which are discovered by the skeleton representation by the axis of low-pressure vortices. Section 5 is devoted to concluding remarks.

2   Low-Pressure Vortex

Pressure is generally lower at the center of a swirling tubular vortex because it is maintained by pressure gradient against centrifugal force (figure 1). Pressure may take a minimum somewhere across a vortex. This is called the sectional minimum of pressure and is used for the definition of a swirling vortex as explained below (Miura & Kida 1997).

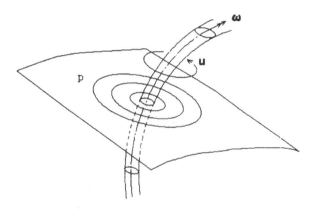

Figure 1. A low-pressure vortex. The axis and the core are defined by a line of sectionally local minimum of pressure and swirl condition.

In order to detect the position of a pressure minimum we first calculate the pressure hessian,

$$H_{ab} = \frac{\partial^2 p}{\partial x_a \partial x_b} \qquad (a, b = 1, 2, 3), \qquad (2.1)$$

at every grid point of numerical turbulence. The eigen-vectors $\{e_a^{(i)}\}$ $(i, a = 1, 2, 3)$ of this symmetric tensor are obtained by solving the eigen-equation,

$$H_{ab}e_b^{(i)} = \lambda^{(i)}e_a^{(i)}, \tag{2.2}$$

where $\lambda^{(i)}$ $(i = 1, 2, 3)$ are eigen-values which are real and their magnitudes are assumed as $\lambda^{(1)} \geq \lambda^{(2)} \geq \lambda^{(3)}$ without loss of generality. Summation for repeated subscripts is implicit. Eigen-vectors are normalized as

$$e_a^{(i)}e_a^{(i)} = 1. \tag{2.3}$$

The condition that pressure takes a sectional minimum is then given by

$$\lambda^{(1)} \geq \lambda^{(2)} > 0 \tag{2.4}$$

and

$$(e^{(i)} \cdot \nabla)p = 0 \qquad (i = 1, 2) \tag{2.5}$$

on some spatial point. Pressure takes a minimum at this point on a plane spanned by $e^{(1)}$ and $e^{(2)}$, both of which are orthogonal to $e^{(3)}$.

Since low-pressure regions are not always accompanied with swirl motions (Jeong & Hussain 1995), we impose a swirl condition as follows. For a two-dimensional compressible stagnation flow $u'(x')$ $(u'(o) = o)$, streamlines around the origin forms spirals if

$$D \equiv \tfrac{1}{4}(W_{11}' - W_{22}')^2 + W_{12}'W_{21}' < 0, \tag{2.6}$$

where

$$u_i' = W_{ij}'x_j', \qquad W_{ij}' = \frac{\partial u_i'}{\partial x_j'} \qquad (i = 1, 2) \tag{2.7}$$

(McWilliams 1984, Brachet *et al.* 1988, Weiss 1991). According as the sign of $W_{11}' + W_{22}'$ is positive or negative, streamlines spiral out of or into the origin (figure 2).

For a general three-dimensional incompressible flow a swirl condition is imposed on a cross-section of a low-pressure vortex. If we take the third eigen-vector of the pressure hessian as the x_3^*-axis and the (x_1^*, x_2^*)-plane perpendicular to it, the swirl condition in the (x_1^*, x_2^*)-plane is written as

$$
\begin{aligned}
D(\theta, \phi) \\
= \tfrac{1}{4}(W_{11}^* - W_{22}^*)^2 + W_{12}^*W_{21}^* \\
= \tfrac{1}{4}[(\cos^2\theta\cos^2\phi - \sin^2\phi)W_{11} + (\cos^2\theta\sin^2\phi - \cos^2\phi)W_{22} + \sin^2\theta W_{33} \\
\quad + (\cos^2\theta + 1)\sin\phi\cos\phi(W_{12} + W_{21}) - \sin\theta\cos\theta\sin\phi(W_{23} + W_{32}) \\
\quad - \sin\theta\cos\theta\cos\phi(W_{31} + W_{13})]^2 \\
\quad + [\cos\theta\cos\phi\sin\phi(W_{22} - W_{11}) + \tfrac{1}{2}\cos\theta(\cos^2\phi - \sin^2\phi)(W_{12} + W_{21}) \\
\quad - \tfrac{1}{2}\sin\theta\cos\phi(W_{23} + W_{32}) + \tfrac{1}{2}\sin\theta\cos\phi(W_{31} + W_{13})]^2 \\
\quad - \tfrac{1}{4}[\cos\theta(W_{12} - W_{21}) + \sin\theta\cos\phi(W_{23} - W_{32}) + \sin\theta\sin\phi(W_{31} - W_{13})]^2 \\
< 0, \tag{2.8}
\end{aligned}
$$

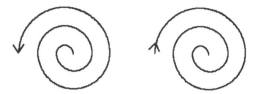

Figure 2. Swirling motion. If $D \equiv \frac{1}{4}(W'_{11} - W'_{22})^2 + W'_{12}W'_{21} < 0$, streamlines spiral out of or into the origin according as $W'_{11} + W'_{22}$ is positive or negative.

where θ and ϕ are the Eulerian angles of the vortex coordinate (x_1^*, x_2^*, x_3^*) system (figure 3). Note that $D(\theta, \phi)$ depends on the orientation angle (θ, ϕ) of a vortex. This means that a point can belong to cores of more than two vortices simultaneously. If a flow is irrotational, then $D > 0$. On the other hand, if a flow is rotational, there exists such an angle (θ, ϕ) that $D(\theta, \phi) < 0$. Furthermore, we calculate $D(\theta, \phi)$ around an axis of a low-pressure vortex and define the core as regions of negative $D(\theta, \phi)$.

The above swirl condition is reminiscent of the so-called Δ-definition of topology of streamlines around a stagnation point of a three-dimensional incompressible flow field (Dallmann 1983, Chong *et al.* 1990). Let the origin be a stagnation point, in the vicinity of which a velocity field is expressed as

$$u_i = W_{ij}x_j, \qquad W_{ij} = \frac{\partial u_i}{\partial x_j} \qquad (i = 1, 2, 3). \tag{2.9}$$

Eigen-values λ of the velocity gradient tensor $\{W_{ij}\}$ satisfy

$$\lambda^3 + Q\lambda + R = 0, \tag{2.10}$$

where

$$Q = -\tfrac{1}{2}W_{ij}W_{ji} = \tfrac{1}{2}\nabla^2 p \tag{2.11}$$

and

$$R = -\tfrac{1}{3}W_{ij}W_{jk}W_{ki}. \tag{2.12}$$

If the discriminant,

$$\Delta = \left(\tfrac{1}{3}Q\right)^3 - \left(\tfrac{1}{2}R\right)^2, \tag{2.13}$$

of the cubic algebraic equation (2.10) is positive (Δ-definition), two of the three eigen-values are complex and the other is real, and streamlines are generally helical. Note that if $Q > 0$, then $\Delta > 0$, and if a flow is irrotational, then $\Delta \leq 0$.

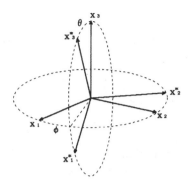

Figure 3. Eulerian angles θ and ϕ.

It is obvious that the Δ-definition and the D-definition are conceptually different. The former is a spiral condition on streamlines in the three-dimensional space, while the latter is that for those projected on a plane. To exemplify the difference, we draw in figure 4 streamlines of a velocity field given by

$$\left.\begin{array}{rl} u_1 = & 2x_1 - 2x_2, \\ u_2 = & -2x_1 - x_2 - x_3, \\ u_3 = & x_2 - x_3. \end{array}\right\} \tag{2.14}$$

Although a three-dimensional perspective of streamlines viewed along the x_1-axis does not show any swirl motions, those projected on the (x_2, x_3)-plane spiral into the origin. Actually, $\Delta = -24 < 0$ and $D = -1 < 0$ (in the direction of the x_1-axis) at the origin. Note that this is different from the Q-definition and the Δ-definition except for two-dimensional incompressible flows in which $D = -Q = -\frac{1}{2}\nabla^2 p$.

Now let us examine what the low-pressure vortex looks like for a stationary Burgers vortex tube whose vorticity is written in a cylindrical polar coordinate (r, θ, z) system as

$$\left.\begin{array}{l} \omega_r = 0, \\ \omega_\theta = 0, \\ \omega_z = (\Gamma/2\pi r_B)\exp\left[-(r^2/r_B^2)\right], \end{array}\right\} \tag{2.15}$$

where

$$r_B = \sqrt{2\nu/\alpha} \tag{2.16}$$

is the $(1/\sqrt{e})$-radius. The velocity field is expressed as

$$\left\{\begin{array}{l} u_r = -\frac{1}{2}\alpha r, \\ u_\theta = (\Gamma/2\pi r)\left\{1 - \exp\left[-(r^2/r_B^2)\right]\right\}, \\ u_z = \alpha z, \end{array}\right. \tag{2.17}$$

where $\alpha\ (>0)$ is the strain rate and Γ is the circulation of the vortex. The associated pressure is

$$p = -\tfrac{1}{2}\alpha^2 z^2 - \tfrac{1}{8}\alpha^2 r^2$$
$$- \left(\frac{\Gamma}{2\pi}\right)^2 \int_0^r \frac{1}{r_1^3} \left\{1 - \exp\left[-\frac{r_1^2}{r_B^2}\right]\right\}^2 \mathrm{d}r_1 + \text{const.}$$
$$\approx -\tfrac{1}{2}\alpha^2 z^2 + \tfrac{1}{8}\alpha^2 \left\{\left(\frac{\Gamma}{4\pi\nu}\right)^2 - 1\right\} r^2 + \cdots \qquad (r \ll \sqrt{\nu/\alpha}). \qquad (2.18)$$

If $\Gamma/\nu > 4\pi$, a line of the sectional minimum of pressure lies on the z-axis. The discriminant is given by

$$D = \left(\frac{\Gamma}{2\pi}\right)^2 \frac{1}{r^4} \left\{1 - \exp\left[-\frac{r^2}{r_B^2}\right]\right\} \left\{1 - \exp\left[-\frac{r^2}{r_B^2}\right] - \frac{r^2}{r_B^2} r^2 \exp\left[-\frac{r^2}{r_B^2}\right]\right\},$$
$$(2.19)$$

which is negative where

$$r < r_* \approx 1.58 r_B. \qquad (2.20)$$

Thus, the core radius r_* has been determined unambiguously.

(a)   (b)

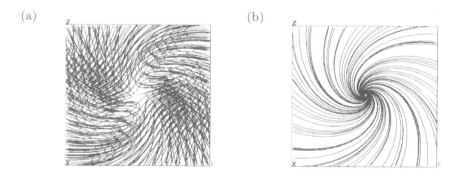

Figure 4. Comparison of Δ-definition and D-definition. (a) A perspective view and (b) projection on the (x_2, x_3)-plane of the velocity field $(u_1, u_2, u_3) = (2x_1 - 2x_2, -2x_1 - x_2 - x_3, x_2 - x_3)$. They do not form any spirals in (a) but they do in (b).

The axes and cores of low-pressure vortices have already been successfully visualized for isotropic turbulence and the existence of swirling motions around each axis was confirmed (Kida & Miura 1998a,b). The probability density function of core radius distributes around five times the Kolmogorov length, which is compatible with Jiménez *et al.* (1993) and Belin *et al.* (1996).

3   Automatic Chase of Low-Pressure Vortex Axes

Since there are numbers of vortices in a turbulent flow, it would be useful to trace an arbitrarily selected vortex automatically. Here, we introduce our new scheme for it. Suppose two velocity fields, one of which has evolved from the other. All the vortices in each field are educed by the method described in the preceding section. To make the identification scheme of corresponding vortices in the two fields effective we first calculate the traveling velocity of the axis of a low-pressure vortex and find an approximate position of it at a later time. The predicted vortices are then compared with the ones educed beforehand.

The traveling velocity of the axis of a low-pressure vortex is calculated as follows. The third eigen-vector of the pressure hessian is parallel in general to the pressure gradient on the vortex axis so that

$$e_a^{(3)} = \frac{\partial \hat{p}}{\partial x_a}, \tag{3.1}$$

where $\hat{p}$ is the normalized pressure ($|\nabla \hat{p}| = 1$). By taking a Lagrangian derivative of it,

$$\left(\frac{\partial}{\partial t} + \frac{\mathrm{d}x_b}{\mathrm{d}t} \frac{\partial}{\partial x_b} \right) e_a^{(3)} = \left(\frac{\partial}{\partial t} + \frac{\mathrm{d}x_b}{\mathrm{d}t} \frac{\partial}{\partial x_b} \right) \frac{\partial \hat{p}}{\partial x_a}, \tag{3.2}$$

we obtain

$$\frac{\mathrm{d}x_b}{\mathrm{d}t} \left(\frac{\partial e_a^{(3)}}{\partial x_b} - H_{ab} \right) = \frac{\partial}{\partial t} \left(\frac{\partial \hat{p}}{\partial x_a} - e_a^{(3)} \right), \tag{3.3}$$

where

$$\frac{\partial e_a^{(3)}}{\partial x_b} = \sum_{j=1}^{2} \left(\frac{1}{\lambda^{(39)} - \lambda^{(j)}} \frac{\partial H_{mn}}{\partial x_b} e_m^{(j)} e_n^{(3)} \right) e_a^{(j)}. \tag{3.4}$$

Equation (3.3) is composed only of information on the pressure field. The temporal derivative on the right-hand side is calculated by a finite difference scheme. The traveling velocity $\mathrm{d}x/\mathrm{d}t$ is obtained by solving this equation.

Now we apply this chasing scheme for the axis of a low-pressure vortex to a numerical flow of the interaction of two vortices placed orthogonally at the initial instant (see figure 5).

The initial vorticity field is given by

$$\left. \begin{array}{l} \omega_x = \omega_0 \exp[-a\{y^2 + (z + z_0)^2\}], \\ \omega_y = -\omega_0 \exp[-a\{x^2 + (z - z_0)^2\}], \\ \omega_z = 0, \end{array} \right\} \tag{3.5}$$

where $a = 50$ is the inverse of the square of the core radius, $\omega_0 = 100/\pi$ is the vorticity magnitude on the central axis, $z_0 = 3\Delta x$ is the half distance between the two vortices, and $\Delta x = 2\pi/64$ is the grid width. The vortex Reynolds number is $Re_\Gamma = \Gamma/\nu = 200$, where $\Gamma = \omega_0\pi/a = 2$ is the circulation, and $\nu = 0.01$ is the kinematic viscosity.

This configuration was studied extensively before by Melander & Zabusky (1988) and Boratav *et al.* (1992). Here, we give a simple skeleton representation of the interaction of vortices by the use of the axes of low-pressure vortices.

In figure 6, we show the time series of iso-surfaces of vorticity magnitude at the level of $0.5|\omega|_{max}$. The originally straight vortex tubes are bent by the induction velocity of each other. The bended tubes are then twisted by their own self-induction velocity. The two tubes approach each other anti-parallel. In the mean time there appear small bridges (or fingers) of high-vorticity regions which have been intensified by vortex stretching (figure 6(d), Kida & Takaoka 1988, Melander & Zabusky 1988). Eventually, the main vortices are reconnected through these bridges.

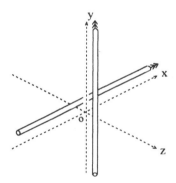

Figure 5. Two vortex tubes placed orthogonally to each other. A double arrow denotes the direction of vorticity.

A skeleton representation of these vortex tubes in terms of the axes of low-pressure vortices is shown in figure 7. Thick lines show the vortex tubes traced from the beginning, while thin lines are the ones generated at a later time. It is seen that the vortex axes are successfully traced and the reconnection is taking place through the bridges. This reconnection of vortex axes is a kind of T-type reconnection which will be discussed in the next section.

4 Vortex Interaction

As described in the preceding section, the axes of low-pressure vortices may serve as an efficient method of representation of the vortical structure in a

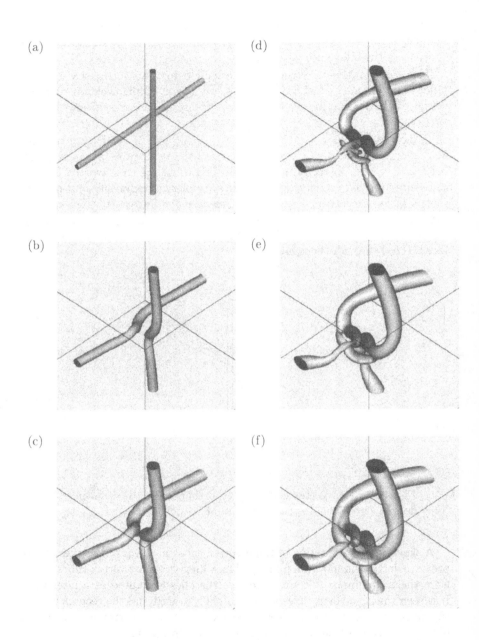

Figure 6. Time series of iso-surfaces of vorticity magnitude at the level of $0.5|\omega|_{\max}$. (a) $t = 0$, (b) $t = 1$, (c) $t = 2$, (d) $t = 3$, (e) $t = 5$, (f) $t = 10$.

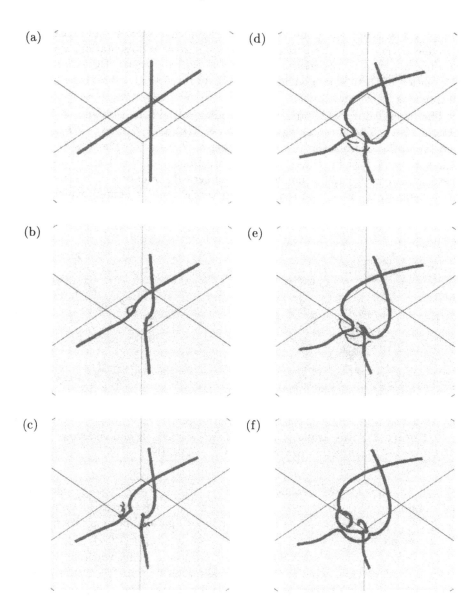

Figure 7. Reconnection of initially orthogonal vortex tubes. The axes of low-pressure vortices are drawn with lines. Thick ones denote the axes that have developed from the beginning, whereas thin ones denote newly born axes. (a) $t = 0$, (b) $t = 1$, (c) $t = 2$, (d) $t = 3$, (e) $t = 5$, (f) $t = 10$.

complex flow. Since the computational load for such a skeleton representation is reduced very much compared with the iso-surface representation, we can produce a movie of the motion of vortices relatively easily. By looking at the motion of vortices in a whole field, we have discovered several interesting phenomena as shown below.

The present flow field starts with a random initial condition and with no external force. The initial energy spectrum is $E(k) \propto (k/k_0)^2 \exp[-2(k/k_0)^2]$, and the initial micro-scale Reynolds number is $R_\lambda(0) = 97$. The flow is periodic with period 2π in three orthogonal directions, and the Fourier spectral method with modes 128^3 is employed for numerical simulation of the Navier–Stokes equation. More than ten thousand vortices are generated, which are moving about by self- and mutual-induction velocity.

(a)   (b)

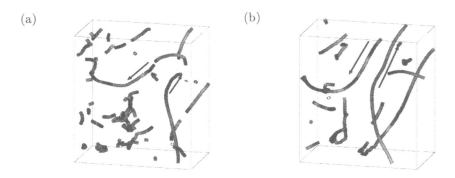

Figure 8. Anti-parallel approach of two vortices. A double arrow denotes the direction of vorticity. (a) $t = 281$, (b) $t = 363$.

It is quite often observed that two nearby vortices are arranged in parallel with opposite sense of rotation. By tracing these vortices back in time, we find that they have approached each other in anti-parallel fashion (see figure 8). A possible mechanism for this was discussed by Siggia (1985).

The second example is a bunching phenomenon of many vortices. Sometimes several vortices come together to make a cluster. In figure 9, we see four vortex axes are gathering at the central region of the box. The directions of vorticity of nearest-neighbor tubes are opposite. They seem to break down after approaching. However, the mechanism of bunching is unknown. It may be due to self-induction velocity of those four vortices or to a large-scale straining motion caused by all the other vortices. More detailed analysis of the flow is necessary to clarify it.

Many vortex axes which are advected about sometimes reconnect with each other. There are two types of reconnection processes observed: one is of tip-type and the other is of T-type. An example of reconnection of two vortices

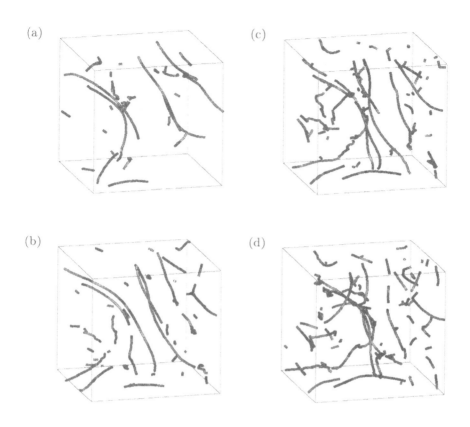

Figure 9. Bunching of vortices. Four vortices at the central region approach each other, and later they break down. (a) $t = 30$, (b) $t = 68$, (c) $t = 94$, (d) $t = 110$.

at the tips is depicted in figure 10. This is reminiscent of the relinking of vortices by pressure waves discussed by Verzicco *et al.* (1995). This process contributes to the increase of mean length of the vortex axes in the field.

Another type of reconnection, T-type, is shown in figure 11. There are two long vortex axes. The one which is running from the top of the (X, Y)-plane to the right in figure 11a has reconnected at the middle with another one which is going to the right in figure 11b. This type of reconnection is similar to the bridging of orthogonal vortex tubes discussed in the preceding section.

(a)   (c)

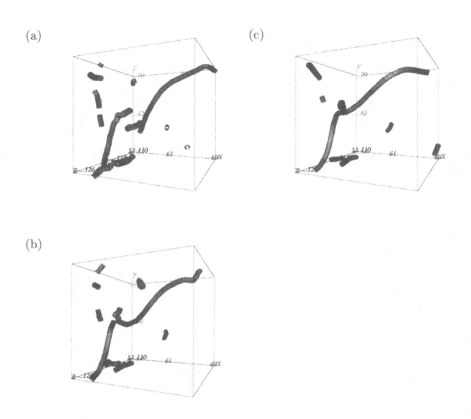

(b)

Figure 10. Tip-reconnection. (a) $t = 265$, (b) $t = 279$, (c) $t = 290$.

5   Concluding Remarks

The low-pressure vortex we have discussed is one of unambiguously defined useful structures. The skeleton representation of the spatial distribution of their axes is simple and helpful for grasping the global flow structure. In fact, several interesting phenomena have been observed such as the anti-parallel approach of two vortices, a bunching of many vortices, the tip-type and T-type reconnections. Applications to various kinds of flows with strong shear and/or solid boundaries will be the interesting next targets of this approach. An automatic scheme for the chasing of a vortex axis is developed and applied to a flow field of the interaction of two vortices.

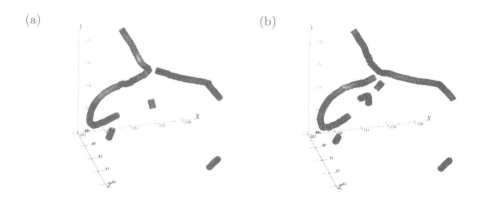

Figure 11. T-type reconnection. (a) $t = 602$, (b) $t = 611$.

References

Belin, F., Maurer, J., Tabeling, P., Willaime, H. (1996) 'Observation of intense filaments in fully developed turbulence', *J. Phys. II France* **6**, 573–583.

Bernal, L.P., Roshko, A. (1986) 'Streamwise vortex structure in plane mixing layers', *J. Fluid Mech.* **170**, 499–525.

Bonn, D., Couder, Y., van Dam, P.H.J., Douady, S. (1993) 'From small scales to large scales in three-dimensional turbulence: the effect of diluted polymers', *Phys. Rev. E* **47**, R28–R31.

Boratav, O.N., Pelz, R.B., Zabusky, N.J. (1992) 'Reconnection in orthogonally interacting vortex tubes: direct numerical simulations and quantifications', *Phys. Fluids A* **4**, 581–605.

Brachet, M.E., Meneguzzi, M., Politano, H., Sulem, P.L. (1988) 'The dynamics of freely decaying two-dimensional turbulence', *J. Fluid Mech.* **194**, 333–349.

Cantwell, B.J., Coles, D., Dimotakis, P. (1978) 'Structure and entrainment in the plane of symmetry of a turbulent spot', *J. Fluid Mech.* **87**, 641–672.

Chong, M.S., Perry, A.E., Cantwell, B.J. (1990) 'A general classification of three-dimensional flow fields', *Phys. Fluids A* **2**, 765–777.

Dallmann, U. (1983) 'Topological structures of three-dimensional flow separations', *DFVLR, Rep. No. 221-82-A07*, Göttingen, West Germany.

Jeong, J., Hussain, F. (1995) 'On the identification of a vortex', *J. Fluid Mech.* **285**, 69–94.

Jiménez, J., Wray, A.A., Saffman, P.G., Rogallo, R.S. (1993) 'The structure of intense vorticity in isotropic turbulence', *J. Fluid Mech.* **255**, 65–90.

Kida, S., Miura, H. (1998a) 'Identification and analysis of vortical structures', *Euro. J. Mech. B, Fluids* **17**, 471–488.

Kida, S., Miura, H. (1998b) 'Swirl condition in low-pressure vortices', *J. Phys. Soc. Japan* **67**, 2166–2169.

Kida, S. Takaoka, M. (1988) 'Reconnection of vortex tubes', *Fluid Dyn. Res.* **3**, 257–261.

Kim, H.T., Kline, S.J., Reynolds, W.C. (1971) 'The production of turbulence near a smooth wall in a turbulent boundary layer', *J. Fluid Mech.* **50**, 133–160.

Lesieur, M. (1997) *Turbulence in Fluids*, Kluwer.

McWilliams, J.C. (1984) 'The emergence of isolated coherent vortices in turbulent flow', *J. Fluid Mech.* **146**, 21–43.

Melander, M.V., Zabusky, N.J. (1988) 'Interaction and "apparent" reconnection of 3D vortex tubes via direct numerical simulations', *Fluid Dyn. Res.* **3**, 247–250.

Miura, H., Kida, S. (1997) 'Identification of tubular vortices in turbulence', *J. Phys. Soc. Japan* **66**, 1331–1334.

Miyake, Y., Ushiro, R., Morikawa, T. (1995) 'On regeneration of quasi-streamwise vortices of near-wall turbulence', *Trans. Japan Soc. Mech. Engin.* **61**, 1272–1278.

She, Z.S., Jackson, E., Orszag, S.A. (1990) 'Intermittent vortex structures in homogeneous isotropic turbulence', *Phys. Nature* **344**, 226–228.

Siggia, E.D. (1985) 'Collapse and amplification of a vortex filament', *Phys. Fluids* **28**, 794–805.

Tanaka, M., Kida, S. (1993) 'Characterization of vortex tubes and sheets', *Phys. Fluids A* **5**, 2079–2082.

Taneda, S. (1988) *Fluid Dynamics Learned from Pictures*, Asakura (in Japanese).

Verzicco, R., Jiménez, J., Orlandi, P. (1995) 'On steady columnar vortices under local compression', *J. Fluid Mech.* **299**, 367–388.

Weiss, J. (1991) 'The dynamics of enstrophy transfer in two-dimensional hydrodynamics', *Physica D.* **48**, 273–294.

Wu, J., Shwridan, J., Welsh, M.C., Hourigan, K. (1996) 'Three-dimensional vortex structures in a cylinder wake', *J. Fluid Mech.* **312**, 201–222.